COLLOQUIA MATHEMATICA
SOCIETATIS JÁNOS BOLYAI, 4.

Colloquium on

COMBINATORIAL THEORY
AND ITS APPLICATIONS II.

Edited by

P. Erdős, A. Rényi and Vera T. Sós

DISTRIBUTED BY
NORTH-HOLLAND PUBLISHING COMPANY
AMSTERDAM-LONDON

Distributed by
North-Holland Publishing Company
Amsterdam–London

ISBN 7204 2037 7

Responsible for publication Á. Császár,
Secretary general of the János Bolyai
Mathematical Society.

69-2080.-Felsőoktatási Jegyzetellátó Vállalat, Budapest

Printed in Hungary

Some remarks on Ramsey's and Turán's theorem

by

P. Erdős and Vera T. Sós
Budapest, Hungary

1. In this paper we are going to discuss some special cases of a general problem which might be considered as being on the one hand a generalisation of the problem raised and solved by the well-known theorem of Turán, on the other hand as the well known problem of the Ramsey-numbers.

Before going to explain this in details, we give the notations we shall use:

$G(n)$ is a graph with n vertices

$G(n;e)$ is a graph with n vertices and e edges

$e(G)$ denotes the number of edges of G

$\bar{G}$ is the complementary graph of G

$K(\nu)$ is the complete graph with ν vertices

$H(n;k,\ell)$ is the class of $G(n)$ graphs, where $G(n)$ contains no $K(k)$ and $\bar{G}(n)$ contains no $K(\ell)$

$H(n;k)$ is the class of $G(n)$ graphs, where $G(n)$ contains no $K(k)$

$$f(n;k,\ell) \stackrel{\text{def}}{=} \begin{cases} \max\limits_{G \in H(n;k,\ell)} e(G) & \text{if} \quad H(n;k,\ell) \neq \emptyset \\ 0 & \text{if} \quad H(n;k,\ell) = \emptyset \end{cases}$$

$$f(n;k) \stackrel{\text{def}}{=} \max_{G \in H(n;k)} e(G)$$

$G(x_1,\dots,x_k)$ denotes the subgraph of G spanned by the vertices $x_1,\dots,x_k$.

The well-known, special form of Ramsey's theorem [5] asserts that for any k,ℓ there exists a $N(k,\ell)$ such that if $n > N(k,\ell)$ then $H(n;k,\ell) = \emptyset$.

The well-known theorem of Turán [6] gives the exact value of $f(n;k)$ namely that

$$f(n;k) = \frac{1}{2}\,\frac{k-2}{k-1}\,(n^2 - r^2) + \binom{r}{2} \quad \text{where} \quad n \equiv r \bmod (k-1) \qquad 0 \leq r < k-1 .$$

The only "extreme graph" in $H(n;k)$ with $e = f(n;k)$ is the complete $k-1$ chromatic graph in each class having $\left[\frac{n}{k-1}\right]$ resp. $\left[\frac{n}{k-1}\right]+1$ vertices. It is worthy of note that for this graph $\bar{G}(n)$ contains a rather "large" complete graph (with $\left[\frac{n}{k-1}\right]$ vertices).

Now the general problem is to determine $f(n;k,\ell)$.

In the special - extremal-case when $\ell = n+1$ (i.e. if there is no condition on the complementary graph), $f(n;k,\ell) = f(n;k)$ is determined by Turán's theorem.

In the other special case, when k and ℓ are fixed and n is large enough, $f(n;k,\ell) = 0$ by Ramsey's theorem. The exact determination of $f(n,k,\ell)$ is probably hopeless, since this would imply the determination of the Ramsey-numbers. But one might expect - having in mind the remark in connection with Turán's theorem, - that $f(n;k,\ell)$ is essentially smaller, than $f(n;k)$ when ℓ is supposed to be much smaller than $\left[\frac{n}{k-1}\right]$.

It is easy to show that for every $c<1$

(1) $$f(n; k, c\frac{n}{k-1}) < g(c)\frac{1}{2}\cdot\frac{k-2}{k-1}n^2$$

with $g(c)<1$, but we cannot determine the exact value of $g(c)$. We do not prove (1) in this paper, but hope to return to it, and to other related questions, at another occasion.

2. In this paper we first investigate the case when k is fixed and $\ell = o(n)$.

Trivally $f(n;3,\ell) \le \frac{n\ell}{2}$* since if G contains no triangle and has a vertex of valency v, the v vertices joined to this vertex must be independent. Therefore $f(n;3,\ell) = o(n^2)$ if $\ell = o(n)$.

For the general case we prove

THEOREM 1. If $\ell = o(n)$ then

(4) $$f(n; 2r+1, \ell) = \frac{1}{2}(1-\frac{1}{r})n^2(1+o(1)).$$

REMARK:

We cannot settle the case $k=4$. Perhaps

(3) $$f(n;4,\ell) = o(n^2)$$

if $\ell = o(n)$. We only get crude upper bounds for $f(n;4,\ell)$

If (3) holds, we can deduce for each fixed r and $\ell = o(n)$

(4) $$f(n; 2r+2, \ell) = \frac{1}{2}(1-\frac{1}{r})n^2(1+o(1)).$$

Now we prove Theorem 1. First we prove it if $r=2$, i.e. we prove that if $\ell = o(n)$ then

---

*In some cases $f(n;3,\ell) = \frac{n\ell}{2}$. See [1], [2].

$$f(n;5,\ell) = (1+o(1))\frac{n^2}{4}. \tag{5}$$

First we show that for sufficiently large n

$$f(n;5,cn^{\frac{1}{2}}\log^2 n) > \frac{n^2}{4}. \tag{6}$$

It is well known [3] that there is a $G(m)$ which contains no triangle and for which $\bar{G}(m)$ contains no $K([cm^{1/2}\log^2 m])$.

Let $G_1([\frac{n}{2}])$ and $G_2([\frac{n+1}{2}])$ be two such graphs which do not have a common vertex. Join every vertex of G_1 to all the vertices of G_2. The resulting graph clearly proves (6).

To complete the proof of (5) we have to show that if $n > n_0(\varepsilon)$ and $G(n;[\frac{n^2}{4}(1+\varepsilon)])$ does not contain a $K(5)$ then $\bar{G}$ contains a $K([c_\varepsilon n])$ where c_ε depends only on ε.

First we show the following

LEMMA. Let $0<\alpha<\frac{1}{2}$ and $G(n;[\alpha n^2(1+\varepsilon)])$ be any graph. Then there is a subgraph $G(m)$, $m>c_{\varepsilon,\alpha}n$ each vertex of which has in $G(m)$ valency greater than $2\alpha m(1+\frac{\varepsilon}{4})$.

Let us assume that our Lemma is false. Then we can write the vertices in a sequence $x_1,\dots,x_n$ so that for every $k<(1-c)n$ the valency of x_k in $G(x_k,\dots,x_n)$ is less than $2\alpha(n-k)(1+\frac{\varepsilon}{2})$. But then

$$e(G(n)) = \left[\alpha n^2(1+\frac{\varepsilon}{4})\right] < 2\alpha(1+\frac{\varepsilon}{2})\sum_{k=0}^{n-1}(n-k) + \binom{[cn]}{2} <$$

$$< \alpha n^2(1+\frac{\varepsilon}{2}) + \frac{c^2n^2}{4}$$

which is an evident contradiction if $c<\sqrt{\alpha\cdot\varepsilon}$.

Now we use the Lemma with $\alpha=\frac{1}{4}$. Let $G(m)$, $m>cn$

be a subgraph of our $G(n;[\frac{n^2}{4}(1+\varepsilon)]$ each vertex of which has valency $> \frac{m}{2}(1+\frac{\varepsilon}{4})$.

Let $G(x_1, x_2, x_3)$ be a triangle of our $G(m)$ (clearly every edge of $G(m)$ is contained in a triangle). Let $y_1, \dots, y_{m-3}$ be the other vertices of our $G(m)$. Each vertex of $G(m)$ has valency at least $\frac{m}{2}(1+\frac{\varepsilon}{2})$, hence more than $\frac{3}{2}m$ edges of type (x_i, y_j) $1 \le i \le 3$, $1 \le j \le m-3$ are in our $G(m)$.

Thus more than $\frac{m}{6}$ y_i's are joined to the same two x_i's say x_1 and x_2. If these y_i's are independent we have found an $K([cn])$ in $\bar{G}(m)$.

If y_r and y_s are joined, then $G(x_1, x_2, y_r, y_s)$ is a $K(4)$ in our $G(m)$. Henceforth we can thus assume that $G(m)$ contains a $K(4)$.

Let $G(z_1, z_2, z_3, z_4)$ be a $K(4)$ of our $G(m)$ and $\omega_1, \dots, \omega_{m-4}$ are the other vertices of it. At least $2m(1+\frac{\varepsilon}{2}) + o(1)$ edges of the form (z_i, ω_j) belong to $G(m)$ $(1 \le i \le 4,\ 1 \le j \le m-4)$. Thus by a simple computation there are at least $\frac{\varepsilon m}{100}$ vertices ω_j which are joined to the same three z_i's. These ω_j's must be independent since otherwise $G(m)$ contains a $K(5)$ and this completes the proof of (5).

Now we prove (2) for general r. First we show

$$f(n; 2r+1, e) > \frac{1}{2}\left(1 - \frac{1}{r}\right)n^2 \tag{7}$$

The proof follows the proof of (6).

Let G_i; $1 \le i \le r$ be graphs of $\left[\frac{n}{r}\right]$ vertices (with disjoint set of vertices) which contains no triangle, and where $\bar{G}_i$ contains no $K([cn^{1/2}\log^2 n])$.

Join every vertex of G_i to every vertex of G_j for every $1 \le i < j \le r$. The resulting graph proves (7).

To complete the proof of (2), assume that it holds for $2r-1$ and we prove it for $2r+1$. Thus we have to prove that every

$$G(n; [\frac{1}{2}(1-\frac{1}{r}+\varepsilon)n^2])$$

either contains a $K(2r+1)$ or $\bar{G}$ contains a $K([cn])$ where c depends only on ε and r. The proof will be very similar to that of (5). First of all, from our Lemma we obtain that we can assume that our $G(n)$ contains a subgraph $G(m)$ with $m > c_{\varepsilon,r} n$ each vertex of which has the valency $> m(1-\frac{1}{r}+\frac{\varepsilon}{2})$ Clearly for this $G(m)$

$$e(G(m)) > \frac{1}{2}(1-\frac{1}{r}+\frac{\varepsilon}{2})m^2.$$

Hence by our induction hypothesis we can assume that our $G(m)$ contains a $K(2r-1)$ whose vertices are $x_1, \ldots, x_{2r-1}$.

Denote by $y_1, \ldots, y_{m-2r+1}$ the other vertices of $G(m)$. At least

$$(2r-1)(1-\frac{1}{r}+\frac{\varepsilon}{2})m + O(1) > (2r-3)m + \frac{m}{r}$$

edges of type (x_i, y_j), $1 \le i \le 2r-1$, $1 \le j \le m-2r+1$ belong to $G(m)$.

Thus as in the proof of (5) we obtain that there are at least $c_1 m$ $(c_1 = c_1(r))$ vertices of $G(m)$ which are joined to the same $2r-2$ x_i's, since all these vertices cannot be independent, two of them must be joined, thus our $G(m)$ contains a $K(2r)$.

Let now $z_1, \ldots, z_{2r}$ be the vertices of this $K(2r)$ and let $\omega_1, \ldots, \omega_{m+2r}$ be the other vertices of $G(m)$. At least

$$2r(1-\frac{1}{r}-\frac{\varepsilon}{2})m + O(1) = (2r-2)m + \varepsilon rm + O(1)$$

of the edges (z_i, ω_j), $1 \le i \le 2r$, $1 \le j \le m-2r$ belongs to our $G(m)$. Hence by the same argument as used in the proof of (5) at least $c_{\varepsilon,r} m$ vertices w_i are joined to the same $2r-1$ z_i's. If two of these z_i's are joined, $G(m)$ contains a $K(2r+1)$, if no two of them are joined, $G(m)$ contains a $K([c_{\varepsilon,r} m])$ and since $m > c_1 n$ the proof of Theorem 1 is complete.

3. <u>We remark that (6) is nearly best possible</u>. In fact we prove

(8) $$f(n;5,[cn^{1/2}]) < \frac{1}{8}(1+\varepsilon)n^2$$

for every c and ε if $n > n_0(\varepsilon, c)$.

Let $G(n;[\frac{1}{8}(1+\varepsilon)n^2])$ be any graph for which $\bar{G}$ does not contain a $K([cn^{1/2}])$. We will show that it must contain a $K(5)$. First of all, observe that by our Lemma it must contain a subgraph $G(m)$, $m > c_\varepsilon n$ each vertex of which has valency $> \frac{1}{4}(1+\frac{\varepsilon}{2})m$ and therefore

(9) $$e(G(m)) > \frac{1}{8}(1+\frac{\varepsilon}{2})m^2.$$

Secondly observe that

(10) $$f(n;4,cn^{1/2}) = o(n^2).$$

Namely if (10) would be false, there would exist a $G(n;[\delta n^2])$ which contains no $K(4)$ and $\bar{G}$ contains no $K([cn^{1/2}])$. G clearly contains a vertex of valency $[2\delta n]$ i.e. G has a vertex x which is joined to $y_1, \ldots, y_s$, $s \geq [2\delta n]$.

By a result of Graver and Jackel [4] $G(y_1, \ldots, y_s)$ must either contain a triangle or $\bar{G}(y_1, \ldots, y_s)$ contains a $K([c, n^{1/2}])$. Both assumptions clearly lead to a contradiction. Thus (10) is proved.

(9) and (10) clearly imply that $G(m)$ contains a $K(4)$ with vertices (x_1, x_2, x_3, x_4). Since each of the x_i's $(1 \leq i \leq 4)$ have valency $> \frac{1}{4}(1+\frac{\varepsilon}{2})m$, there clearly are $c\varepsilon m > c_1 \varepsilon n$ vertices $y_1, \ldots, y_\ell$ $(\ell > c_1 \varepsilon m)$ which are joined to the same two x_i's say to x_1 and x_2. $G(y_1, \ldots, y_\ell)$ cannot contain a $K([c\sqrt{n}])$ thus by [4] $G(y_1, \ldots, y_\ell)$ contains a triangle, say $G(y_1, y_2, y_3)$ but then $G(x_1, x_2, y_1, y_2, y_3)$ is a $K(5)$ of our $G(n)$, which completes the proof of (8).

Perhaps

$$f(n;5,[cn^{1/2}]) = o(n^2)$$

is true, but we could not prove it.

4. As to the case $k = 2r$, we prove that assuming $f(n;4,\ell) = o(n^2)$ for $\ell = o(n)$ we have for every fixed r

(11) $$f(n;2r+2,\ell) = \frac{1}{2}\left(1-\frac{1}{r}\right)n^2(1+o(1))$$

For the sake of simplicity we only prove (11) for $r = 2$.

The proof of the general case is the same, only slightly more complicated.

$f(n;6,\ell) > \frac{n^2}{4}$ is trivial, (it follows from $f(n;5,\ell) > \frac{n^2}{4}$). Thus to prove (11) for $r = 2$ we only have to show that for every $\varepsilon > 0$ there is a $c_\varepsilon > 0$ so that for every $G(n;[\frac{n^2}{4}(1+\varepsilon)])$ which contains no $K(6)$ $\bar{G}$ contains a $K([c_\varepsilon n])$ (we of course assume $f(n;4,\ell) = o(n^2)$).

From Lemma it follows that our $G(n)$ has a subgraph $G(m)$ with $m > c_1 n$ so that every vertex of $G(m)$ has in $G(m)$ valency greater than $\frac{1}{2}(1+\frac{\varepsilon}{2})m$. Let x be any vertex of $G(m)$, denote by $S(x)$ the set of vertices of $G(m)$ joined to x.

We evidently have

(12) $$|S(x) \cap S(y)| > \frac{\varepsilon m}{2}.$$

Put

$$M = \max |S(x) \cap S(y)|$$

where the maximum is taken over every two vertices x and y of $G(m)$ which are joined. By (12) we have $M > \frac{\varepsilon m}{2}$.

Assume that for x_1 and x_2 we have $|S(x_1) \cap S(x_2)| = M$ and let $y_1, \dots, y_M$ be the vertices of $G(m)$ joined to both x_1 and x_2. Our assumption $f(M;4,\ell) = o(M^2)$ clearly implies

$$e(G(z_1, \dots, z_M)) = o(M^2). \tag{13}$$

To see (13), observe that $G(z_1, \dots, z_M)$ cannot contain a $K(4)$ thus if (13) would not hold, then $\bar{G}(z_1, \dots, z_M)$ would contain a $K([c_\varepsilon m])$, which is impossible.

From (13) it immediately follows that for all but $o(m) = o(M)$ vertices the valency (in $G(z_1, \dots, z_M)$) is $o(M)$. Hence there is a subgraph $G(z_1, \dots, z_N)$ of $G(z_1, \dots, z_M)$ with $N = (1+o(1))M$ each vertex of which (in $G(z_1, \dots, z_N)$) has valency $o(N)$. Since $N > \frac{\varepsilon m}{4}$ we can assume that the vertices $z_1, \dots, z_N$ are not all independent, without loss of generality we can assume that z_1 and z_2 are joined.

Now we prove

$$|S(z_1) \cap S(z_2)| > M$$

and this contradiction will prove our assertion.

Let $y_1, \dots, y_s$ be the vertices of our $G(m)$ different from $z_1, \dots, z_N$. Clearly both z_1 and z_2 are joined to at least $\frac{1}{2}(1+\frac{\varepsilon}{2})m + o(m)$ of the y_i's. Thus we evidently have

$$|S(z_1) \cap S(z_2)| > m(1+\frac{\varepsilon}{2}) - s + o(m) = M + \frac{\varepsilon}{2} m + o(m).$$

This contradiction completes the proof of (11).

Incidentally it is easy to see that if $f(n;4,\ell) \neq o(n^2)$ then $f(n;6,\ell) > \frac{n^2}{4}(1+\varepsilon)$ for infinitely many n and $\ell = o(n)$.

To see this let G_1 and G_2 both have n vertices, every vertex of G_1 is joined to every vertex of G_2, G_1 contains no triangles, G_2 no $K(4)$, G_2 has more than εn^2 edges and both $\bar{G}_1$ and $\bar{G}_2$ do not contain a $K(\ell)$.

REFERENCES

[1] B. ANDRÁSFAI, Graphentheoretische Extremalprobleme, Acta Math. Acad. Sci. Hungar. 15 (1964) 413-438.

[2] P. ERDŐS, On the construction of certain graphs, J. of Combinatorial Theory 1 (1966) 149-153.

[3] P. ERDŐS, Graph theory and probability, II. Canad. J. of Math. 13 (1961) 346-352.

[4] J.E. GRAVER and J. JACKEL, Some graph theoretic results associated with Ramsey's theorem, J. Combinatorial Theory 4 (1968) 125-175.

[5] F.F. RAMSEY, On a problem of formal logic, Collected papers, 82-111, see also P. Erdős and G. Szekeres, A combinatorial problem in geometry, Compositio Math. 2 (1935) 463-470.

[6] P. TURÁN, Eine Extremalaufgabe aus der Graphentheorie (in Hungarian), Mat. és Fiz. Lapok 48 (1941), 436-452.

Nonaveraging sets II

by

P. Erdős and E. G. Straus

Budapest, Hungary Los Angeles, USA

1. INTRODUCTION

We wish to consider sets of integers $A = \{a_1, \cdots, a_n\}$ so that $0 \leq a_1 < a_2 < \cdots < a_n \leq x$ and no a_i is the arithmetic mean of any subset of A consisting of two or more elements. In a previous paper [3] one of us has initiated the study of the maximal number of elements in nonaveraging sets and sets which satisfy related conditions.

Using the notation of [3] we define $f(x)$ as the maximal number of elements in a nonaveraging set; $h(x)$ as the number of elements of a maximal set of integers in the interval $[0, x]$ such that no two distinct subsets have the same arithmetic mean; and $h^*(x)$ as the number of elements of a maximal set of integers in $[0, x]$ such that no two subsets with a relatively prime number of elements have the same arithmetic mean. In [3] we proved ($\log_r x = \log x / \log r$):

$$\log_2 f(x) > \sqrt{2 \log_2 x} + \frac{1}{2} + o(1/\sqrt{\log x}) \tag{1.1}$$

(1.2) $$(1+o(1))\log x/\log\log x < h(x) < \log_2 x + O(\log\log x)$$

(1.3) $$\log_2 h^*(x) \geq \sqrt{\log_2 x} - 1 + O(1/\sqrt{\log x})$$

In the present note we prove in § 2 that (1.2) can be replaced by

(1.4) $$-1+\log_4 x \leq h(x) < \log_2 x + O(\log\log x)$$

Next, in § 3, we prove that even if we ease the restriction on our sets so that only subsets with different numbers of elements must have different averages then the maximal number, $h^{**}(x)$, of elements satisfies

(1.5) $$h^{**}(x) < c(\log x)^2 \qquad \text{for some constant } c.$$

Finally in § 4 we get an upper bound for

(1.6) $$f(x) < cx^{3/4}$$

2. SETS FOR WHICH DIFFERENT SUBSETS HAVE DIFFERENT AVERAGES

The new lower bound (1.4) for $h(x)$ is obtained inductively as follows. Assume that $h(x)=k$ and that $\{a_1,\dots,a_k\}$ is a set with the desired property. A $(k+1)$-st element a_{k+1} must satisfy $<4^{k+1}$ inequalities

(2.1) $$t(a_{i_1}+\dots+a_{i_s}) \neq s(a_{j_1}+\dots+a_{j_t})$$

for each pair of subsets $\{a_{i_1},\dots,a_{i_s}\}$ and $\{a_{j_1},\dots,a_{j_s}\}$ of $\{a_1,\dots,a_{k+1}\}$ (where at least one of the elements $a_{i_1},\dots,a_{i_s},a_{j_1},\dots,a_{j_s}$ must be a_{k+1}).

Hence it is possible to choose a_{k+1} in the interval $[0, 4^{k+1}]$ so that $h(4^{k+1}) \geq k+1$ and in general $h(x) \geq [\log_4 x] > \log_4 x - 1$.

The conjecture has been communicated to us that the sequence $\{3^k\}$ has the property discussed in this section. On the other hand it is easy to see that the sequence $\{2^k\}$ does not. Nevertheless it seems likely to us that correct asymptotic value is

$$h(x) = (1 + o(1)) \log_2 x. \tag{2.2}$$

3. SETS FOR WHICH SUBSETS OF DIFFERENT CARDINALITY HAVE DIFFERENT AVERAGES

Let $h^{**}(x) = n$ and let $\{a_1, \dots, a_n\}$ be a set with the desired property. We form all possible subsets $\{i_1, \dots, i_k\} \subset \{1, \dots, n\}$ and consider the $\binom{n}{k}$ sums $a_{i_1} + \dots + a_{i_k}$. Since all these sums are in the interval $\{0, kx\}$ there are at least

$$\binom{n}{k} / kx = N \tag{3.1}$$

such sums which have a common value t.

Now there cannot exist three subsets $\{a_{i_1}, \dots, a_{i_k}\}$, $\{a_{j_1}, \dots, a_{j_k}\}, \{a_{\ell_1}, \dots, a_{\ell_k}\}$ whose sums are t and whose index sets have pairwise equal intersections. For otherwise, deleting the common part from all three sets, the average of the first set will equal the average of the union of the second and third set.

However, according to a theorem of P. Erdős and R. Rado [1, Theorem III] it follows that whenever

$$N \geq k! \, 3^{k+1} \tag{3.2}$$

there do exist three sets with pairwise equal intersections. Combining (3.1) and (3.2) we get

$$(3.3) \qquad \binom{n}{k} < k.k!\,3^{k+1}x$$

and hence, if we choose $k = \frac{1}{2}\log x$ we have

$$\frac{(n-k)^k}{k!} < \binom{n}{k} < \left((1+o(1))\frac{3k}{c}\right)^k x$$

or

$$n-k < (1+o(1))\frac{3k^2}{c^2}x^{1/k} < \left(\frac{3}{4}+o(1)\right)(\log x)^2$$

so that

$$(3.4) \qquad h^{**}(x) < \left(\frac{3}{4}+o(1)\right)(\log x)^2 .$$

It was conjectured in [1] that Theorem III could be improved so that the lower bound in (3.2) could be replaced by c^k for a suitable c. In that case it would follow that $h^{**}(x) < c_1 \log x$ for suitable c_1 and this would, according to § 2, give the correct order of magnitude.

4. UPPER BOUNDS FOR THE NUMBER OF ELEMENTS IN A NONAVERAGING SET

As was pointed out in [3] we can find upper bounds for $f(x)$ by finding upper bounds for the maximum number, $F(x)$ of elements in two sets of integers A, B in $[0, x]$ so that A and B have the same number of elements and the sums of elements of nonempty subsets of A and B are distinct. It was conjectured there that $F(x) \le \sqrt{2x}$ and it was observed that $f(x) \le 2F(x)+1$.

In this section we want to obtain upper bounds for $F(x)$. Assume $n = F(x) \geq cx^{3/4} + 1$. Then there are at least $\frac{1}{2} c^2 x^{3/2}$ sums $a_i + a_j$, $a_i, a_j \in A$ with $1 \leq i < j \leq n$ all lying in the interval $(0, 2x)$. Thus there exists an integer with more than $\frac{1}{4} c^2 x^{1/2}$ representations of the form $a_i + a_j$ and similarly there exists an integer with more than $\frac{1}{4} c^2 x^{1/2}$ representations of the form $a_i + a_j$.

Now let M be the maximal integer which has at least $\frac{1}{8} c^2 x^{1/2}$ representations either as $a_i + a_j$ or as $b_i + b_j$. Without loss of generality we may assume that it has representations of the form $b_i + b_j$. There can be no more than $\frac{1}{2} cx^{3/4}$ elements of A which exceed M since otherwise more than $\frac{1}{4} c^2 x^{3/2}$ sums $a_i + a_j$ would exceed M and therefore some integer greater than M would have more than $\frac{1}{8} c^2 x^{1/2}$ representations as $a_i + a_j$ contrary to hypothesis. Thus there are more than $\frac{1}{2} cx^{3/4}$ elements of A below M.

According to a theorem of Szemerédi [4] which will appear in Acta Arithmetica (for a slightly weaker result see Ryavec [2]) it follows that if

$$\frac{1}{2} cx^{3/4} > c_1 M^{1/2} \tag{4.1}$$

then there exists a sum of distinct elements of A (all less than M) which is divisible by M. Here c_1 is an absolute constant.

Now if

$$a_{i_1} + \cdots + a_{i_m} = LM < mM < c_1 M^{3/2} \tag{4.2}$$

then $L < c_1 M^{1/2} < 2c_1 x^{1/2}$ (since $M < 2x$). Thus, if

$$\frac{1}{8} c^2 x^{1/2} \geq 2c_1 x^{1/2} > L \tag{4.3}$$

then LM can be represented as a sum of different elements of A contrary to hypothesis when (4.2) is satisfied. Thus we get

$$c \leq \max\{2c_1, 4c_1^{1/2}\}$$

and $F(x) < cx^{3/4} + 1, \quad f(x) < 2cx^{3/4} + 3$.

Assume that the following result holds: Let $a_1, \dots, a_k$, $k > c_1 M^{1/2}$ be k distinct residues $\bmod M$. Then there are ℓ distinct a's, satisfying

(4.4) $$a_{i_1} + \dots + a_{i_\ell} \equiv 0 \pmod M, \quad \ell < c_2 M/k$$

This result combined with the above proof gives immediately $F(x) < c_3 x^{2/3}$. Szemerédi just informed us that his proof gives (4.4) without any difficulty.

As we conjectured in [3] it is probable that

$$f(x) = \exp(c\sqrt{\log x}) = o(x^{\varepsilon}) .$$

The method of estimating $F(x)$ can of course yield nothing better than $F(x) = O(x^{1/2})$ and hence $f(x) = O(x^{1/2})$.

REFERENCES

[1] P. ERDŐS and R. RADO, Intersection theorems for systems of sets, J. London Math. Soc. 35 (1960), 85-90.

[2] C. RYAVEC, The addition of residue classes modulo n, Pacific J. of Math. 26 (1968), 367-373.

[3] E.G. STRAUS, Nonaveraging sets, Proc. Symposia in Pure Math. A.M.S. (1967)

[4] E. SZEMERÉDI, On a conjecture of Erdős and Heillronn, Acta Arithmetica

On the rook polynomials of Ferrers relations

by

D. Foata and M. P. Schützenberger

Strasbourg, France Paris, France

1. INTRODUCTION

A quasi-permutation on a set I is a relation on I that is contained in a bijective relation. Formally $Q \subset I \times I$ is a quasi-permutation if and only if

$$(i,j),\ (i',j') \in Q \Longrightarrow (i,j) = (i',j') \quad \text{or } i \neq i' \text{ and } j \neq j'$$

For Q finite the weight $\lambda(Q)$ of Q is the number of elements $(i,j) \in Q$ and for any relation $R \subset I \times I$ we shall denote $Q_k(R)$ the set of the quasi-permutations of weight k contained in R. Then assuming R finite, the rook polynomial $\rho(R)$ of R is the generating function

$$1 + \sum_{0<k} t^k \operatorname{card} Q_k(R).$$

We refer the reader to the last two chapters of Riordan's book on combinatorial analysis [4] for the general theory of rook polynomials and their applications. Riordan gives several theorems stating conditions for two relations to have the same rook polynomial or, as we shall say, to be rook-equivalent. Our theorem 3 belongs to this type. It generalizes the obvious fact that any relation is rook-equivalent with its transpose.

The case when R is a Ferrers relation plays a role in the applications since it relates to the Laguerre polynomials, the Eulerian polynomials, the Shanks polynomials, the Poussin polynomials and more generally to the so-called "Newcomb's problem for arbitrary specification". Our main result (theorem 11) states that a cross-section ("minimal set of representatives") of the Ferrers relations with respect to the rook-equivalence is provided by those relations which are strictly decreasing, i.e. which correspond to partitions into unequal parts.

In the last section we effectively compute the rook-equivalent decreasing Ferrers relations for those relations which are a total preorder deprived for an arbitrary subset of its equivalence classes (theorem 19). Since this family is closed under complementation, one might apply Riordan's theory of "complementary boards" to deduce non-trivial identities by using the rather explicit expressions for the rook polynomial that are given in our property 5.

2. A GENERAL PROPERTY OF ROOK EQUIVALENCE

We use the standard notation $[n]$ for the ordered set $\{1,2,\dots,n\}$ $([0]=\emptyset)$. Thus we can say in short that α is a (m,n)-injection if and only if it is a map sending each pair $(i,j)\in[m]\times[m]$ onto the pair $(\alpha_1(i),\alpha_2(j))\in[n]\times[n]$ where α_1 and α_2 are two injections of $[m]$ into $[n]$. We write then $\alpha=\alpha_1\times\alpha_2$, $A_i=\alpha_i([m])$ $(i=1,2)$.

Definition 1.

The (m,n)-injection $\alpha:[m]\times[m]\to[n]\times[n]$ is compatible with the relation $R\subset[n]\times[n]$ if and only if there exist subsets $\bar{A}_i\subset[n]\setminus A_i$ $(i=1,2)$ such that $R\cap(A_1\times[n])=R\cap(A_1\times A_2)\cup(A_1\times\bar{A}_2)$ and symmetrically $R\cap([n]\times A_2)=R\cap(A_1\times A_2)\cup(\bar{A}_1\times A_2)$.

Definition 2.

With the same notations, the α-<u>transpose</u> of R is the relation

$$R' = R \setminus (A_1 \times A_2) \cup \alpha(\tilde{S}) \subset [n] \times [n]$$

where $\tilde{S}$ is the ordinary transpose
$\tilde{S} = \{(i,j) \in [m] \times [m] : (j,i) \in S\} \subset [m] \times [m]$ of the inverse image

$$S = \alpha^{-1}(R \cap (A_1 \times A_2)) \subset [m] \times [m] .$$

THEOREM 3.

If the injection α is compatible with the relation R, then R and its α-transpose R' are rook-equivalent.

Proof.

Note that reciprocally R is the α-transpose of R'. Accordingly, if suffices to construct an injective map of $\underset{\sim}{Q}(R)$ into $\underset{\sim}{Q}(R')$.

Consider any given quasi-permutation Q contained in R. The inverse image $P = \alpha^{-1}(Q \cap (A_1 \times A_2))$ is a quasi-permutation contained in $S = \alpha^{-1}(R \cap (A_1 \times A_2)) \subset [m] \times [m]$.

Let B_1 and B_2 be the least subsets of $[m]$ that satisfy $P \subset B_1 \times B_2$. Since P is a quasi-permutation, we have $\lambda(P) = \underset{\sim}{\mathrm{Card}}\, B_1 = \underset{\sim}{\mathrm{Card}}\, B_2$.

We define a (m,m)-injection $\sigma = \sigma_1 \times \sigma_2$ by the following two conditions where $i, i' = 1,2$ and $i' \neq i$.

(1) The restriction of σ_i to $[m] \setminus B_i$ is the unique order-preserving bijection of this set onto $[m] \setminus B_{i'}$;

(2) For each $(k,k') \in P$ we set $\sigma_1(k) = k'$ and $\sigma_2(k') = k$.

Clearly $\sigma(P)$ is a quasi-permutation contained in the transpose $\tilde{S}$ of S .

We now extend σ to a (n,n)-injection $\tau = \tau_1 \times \tau_2$ by letting

$\tau_i(j) = j$ for any $j \in [n] \setminus A_i$;

$\tau_i(j) = \alpha_i(\sigma_i(\alpha_i^{-1}(j)))$ for any $j \in A_i$.

Because of the compatibility condition $R \setminus (A_1 \times A_2)$ is invariant under τ. Thus $Q' = \tau(Q)$ is a quasi-permutation contained in R'. Finally $Q \longrightarrow Q'$ is injective because in view of our canonical choice of σ, this injection, hence Q itself, are determined by Q' without ambiguity.

3. APPLICATION TO FERRERS RELATIONS

Let $\mathbb{P} = \mathbb{N} \setminus \{0\}$ denote the set of all positive integers. A Ferrers relation is a relation $F(\varphi) \subset \mathbb{P} \times \mathbb{P}$ that is defined by a non-increasing map $\varphi : \mathbb{P} \longrightarrow \mathbb{N}$ such that the sum $\sum_i \varphi(i)$ is finite and the condition

$$F(\varphi) = \{(i,j) \in \mathbb{P} \times \mathbb{P} : \; j \leq \varphi(i)\} .$$

Its transpose $\widetilde{F}(\varphi)$ is another Ferrers relation $F(\tilde{\varphi})$ where $\tilde{\varphi} : \mathbb{P} \longrightarrow \mathbb{N}$ is defined by

$$\begin{aligned} \tilde{\varphi}(i) &= \operatorname{Max}\{ j \in \mathbb{P} : \; \varphi(j) \geq i \} \qquad \text{if} \quad i \leq \varphi(i) \\ &= 0 \qquad \text{otherwise} \end{aligned}$$

We shall say shortly that φ is <u>decreasing</u> if and only if it is so on the non-trivial part of its domain, i.e. if and only if $\varphi(i) > 0$ implies $\varphi(i) > \varphi(i+1)$. Then $F(\varphi)$ will be called a <u>decreasing</u> Ferrers relation.

In the next two lemmas we consider a fixed decreasing map φ and for convenience we set

$$p = \tilde{\varphi}(1) \quad (= \operatorname{Max}\{ i \in \mathbb{P} : \; \varphi(i) > 0 \})$$

$$x_i = \varphi(i) \qquad (i \in [p]);$$

$$r_0 = 1 ;$$

$$r_k = \sum_{(k)} \prod_{1 \leq j \leq k} (x_{i_j} - k + j) \qquad (k \in [p])$$

$$r_k = 0 \quad \text{for} \quad k > 0$$

where $\sum_k$ indicates a summation over the set $[p]^{(k)}$ of the $\binom{p}{k}$ strictly increasing sequences $(i_1 < i_2 < \ldots < i_k)$ of length k with elements in $[p]$.

Lemma 4. The rook polynomial of $F(\varphi)$ is

$$\varrho(F(\varphi)) = \sum_{0 \le k \le p} t^k r_k .$$

Proof. For any relation $R \subset [p] \times \mathbb{P}$ the set $\underline{Q}_k(R)$ of the quasi-permutations of weight k contained in R is empty for $k > p$; and for $k \le p$ it is the disjoint union over all sequences (of length k) $I \subset [p]^{(k)}$ of the sets $\underline{Q}_k((R \cap I) \times \mathbb{P})$.

Further, when $R = F(\varphi)$, each restriction $F(\varphi) \cap (I \times \mathbb{P})$ is rook-equivalent with a decreasing Ferrers relation $F(\varphi')$ where

$$\varphi'(j) = \varphi(i_j) \quad (I = (i_1, i_2, \ldots, i_k)) \quad \text{and} \quad \tilde{\varphi}'(1) = k.$$

Thus it suffices to consider the special case of $k = p$. Then $r_p = (x_1 - p + 1)(x_2 - p + 2) \ldots (x_{p-1} - 1) x_p$ and the equality of this quantity with $\text{Card } \underline{Q}_p(F(\varphi))$ is trivial since this last number is the number of injections $\eta : [p] \to \mathbb{P}$ such that $\eta(i) \le \varphi(i)$ identically.

Q.E.D.

PROPERTY 5. For each $k \in [p]$ one has the identity

$$r_k = \sum_{0 \le j \le k} a_j \, S(p-j, p-k) \qquad (0 \le k \le p)$$

where the $S(i,j)$'s are Stirling numbers of the second kind[*] and where a_j $(0 \le j \le p)$ are the symmetric functions $a_0 = 1$ and

$$a_j = \sum_{(j)} y_{i_1} y_{i_2} \cdots y_{i_j} \quad (j \in [p]) \quad \text{in the variables } y_i = x_i - p + i \; (i \in [p]).$$

Proof. Let $p' = p - 1$; $x'_i = x + 1$ $(i \in [p'])$ and denote by

---

[*] We make the usual convention that $S(0,0) = 1$ and $S(i,j) = 0$ if exactly one of i and j is zero.

primed letters the quantities defined with respect to p' and the x'_i's in the same manner as the corresponding quantities (r, y or a) were defined with respect to p and the x_i's.

Thus, by definition

$$r_k = (x_1 - k + 1) r'_{k-1} + r'_k = (y_1 + p - k) r'_{k-1} + r'_k$$

and

$$y_{i+1} = x_{i+1} - p + i + 1 = x'_i - p' + i = y'_i \qquad (i \in [p'])$$

Using induction on p, the first relation gives

$$r_k = (y_1 + p - k) \sum_j a'_j S'(p'-j, p'-k+1) + \sum_j a'_j S'(p'-j, p'-k) =$$

$$= \sum_j y_1 a'_j S(p'-j, p'-k+1) + \sum_j a'_j ((p-k) S(p'-j, p'-k+1) + S(p'-j, p'-k))$$

that is

$$r_k = \sum_j (y_1 a'_{j-1} + a'_j) S(p-j, p-k)$$

because of the classical identity

$$S(p-j, p-k) = (p-k) S(p-j-1, p-k) + S(p-j-1, p-k-1)$$

and the result follows since the second relation implies the identity $a_j = y_1 a'_{j-1} + a'_j$.

Q.E.D.

Corollary 6. The Ferrers relations $F(\varphi)$ and $F(\varphi')$ defined by two decreasing maps φ and φ' are rook-equivalent only if $\varphi = \varphi'$.

Proof. Let $\tilde{\varphi}(1) = p \geq \tilde{\varphi}'(1) = p'$. If $p \neq p'$, t^p has a positive coefficient in $\rho(F(\varphi))$ and a zero coefficient in $\rho(F(\varphi'))$. Thus we can assume $p = p'$.

Because of the strictly decreasing character of the sequence

$\underset{\sim}{x} = (x_1, x_2, \dots, x_p)$, the sequence $\underset{\sim}{y} = (y_1, y_2, \dots, y_p)$ is non-increasing. Its members are positive because $y_p = x_p > 0$.

Since the correspondence $\underset{\sim}{x} \longrightarrow \underset{\sim}{y}$ is injective for each given p, it follows that $\underset{\sim}{x} \longrightarrow \{y_1, y_2, \dots, y_p\}$ is also injective. Now the set $\{y_1, y_2, \dots, y_p\}$ is unequivocally determined by the symmetric functions a_k and the result follows from lemma 5.

Q.E.D.

Observation 7. Because of the well-known orthogonality relations between the Stirling numbers of the first kind $s(i,j)$ and those of the second kind $S(i,j)$, the formula of lemma 5 is equivalent with

$$a_k = \sum_j r_j \, s(p-j, p-k)$$

As a side remark we may note that the formula of Property 6 does not depend upon the fact that the x_i's form a decreasing sequence of positive integers. By taking all the y_i's equal to 0 (resp. to 1) and by using a straightforward computation this formula gives directly the known identities

$$S(p, p-k) = \sum_{(k)} i_1 (i_2 - 1) \dots (i_k - k + 1)$$

$$S(p+1, p+1-k) = \sum_{0 \le j \le k} \binom{p}{j} S(p-j, p-k).$$

We now return to our main argument. To simplify notations, each Ferrers relation $F(\varphi)$ is considered as a relation in $[n] \times [n]$ where $n = \sum_i \varphi(i)$.

Definition 8.

Let $\varphi : \mathbb{P} \longrightarrow \mathbb{N}$ be non-increasing. The element $(k,k') \in \mathbb{P} \times \mathbb{P}$ is <u>admissible</u> if and only if the following inequalities hold

(1) $$0 \le \varphi(k) - k' \le \tilde{\varphi}(k') - k \le \varphi(k-1) - k'$$

where by convention $\varphi(k-1) - k' = +\infty$ for $k = 1$.

Definition 8 bis.

If (k,k') is admissible, the (k,k')-transform of φ is the map $\varphi' = \Theta(k,k';\varphi)$ defined by $\varphi'(i) = k'-k+\tilde{\varphi}(k'-k+i)$ if $k \le i \le \tilde{\varphi}(k')$ and $\varphi'(i) = \varphi(i)$ otherwise.

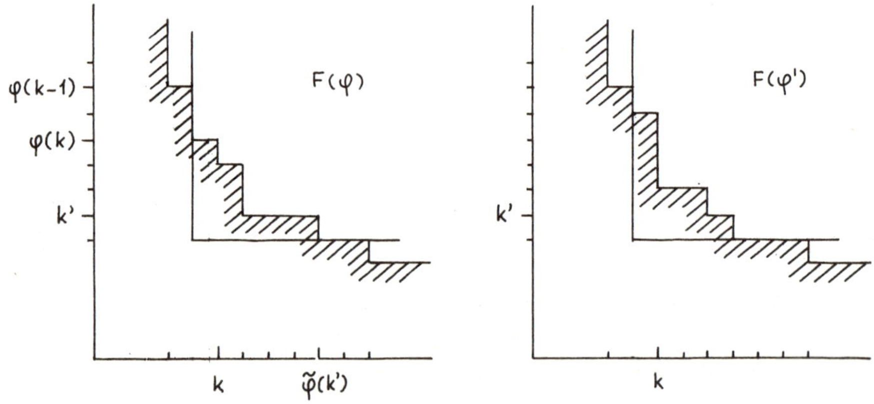

Fig. 1.

As can be seen from figure 1, where (k,k') is admissible, the relation $F(\varphi')$ is obtained from $F(\varphi)$ by transposing the set $F(\varphi) \cap B$ where $B = \{(i,j) \in \mathbb{P} \times \mathbb{P} : i \ge k, \ j \ge k'\}$ and leave $F(\varphi) \setminus B$ invariant. As shown in lemma 5 below, the admissibility conditions will insure that the relation $F(\varphi')$ obtained from $F(\varphi)$ is still a Ferrers relation having the same rook polynomial as $F(\varphi)$.

Lemma 9.

Let φ, (k,k') and $\varphi' = \Theta(k,k';\varphi)$ as above and assume $\sum \varphi(i) \le n$. Then φ' is a non-increasing map such that the Ferrers relations $F(\varphi)$ and $F(\varphi')$ are rook-equivalent.

Proof.

Using inequalities (1) and the fact that φ is non-increasing, we have $k' \le \varphi(i) \le \varphi(k)$ for each i such that $k \le i \le \tilde{\varphi}(k')$. Accordingly the set $F(\varphi) \cap B = \{(i,j) \in F(\varphi) : i \ge k, \ j \ge k'\}$ is entirely contained in $[k, \tilde{\varphi}(k')] \times [k', \varphi(k)]$.

Let $m = \tilde{\varphi}(k') - k + 1$ and consider the (m,n)-injection α such that $\alpha(i,j) = (k-1+i, k'-1+j)$ identically.

Keeping the same notations as in section 2 we have

$$A_1 = \alpha_1([m]) = [k, \tilde{\varphi}(k')] \quad \text{and} \quad A_2 = \alpha_2([m]) = [k', k'-1+m] .$$

Again using inequalities (1) we get $k'-1+m = \tilde{\varphi}(k') + k' - k \geq \varphi(k) \geq k'$. Hence $F(\varphi) \cap B$ is a subset of $A_1 \times A_2$. Therefore $\alpha^{-1}(F(\varphi) \cap (A_1 \times A_2))$ is the Ferrers relation $F(\psi) \subset [m] \times [m]$ where $\psi(i) = \varphi(k-1+i) - k'+1$ for each $i \in [m]$.

For the same reason $F(\varphi) \setminus (A_1 \times A_2)$ is the Ferrers relation $F(\bar{\varphi})$ where $\bar{\varphi}(i) = k'-1$ or $= \varphi(i)$ depending upon $k \leq i \leq k-1+m = \tilde{\varphi}(k')$ or not. It follows that $F(\bar{\varphi}) \cap (A_1 \times [n]) = A_1 \times [k'-1]$.

Because of inequalities (1) we have

$F(\bar{\varphi}) \cap ([n] \times A_2) = [k-1] \times A_2$ since either $[k-1] = \phi$ or $\varphi(k-1) \geq \tilde{\varphi}(k') + k' - k = k'-1+m$.

Thus the compatibility condition is satisfied and $F(\varphi)$ is rook-equivalent with its α-transpose $F' = F(\bar{\varphi}) \cup \alpha(F(\tilde{\psi}))$.

Now since $\bar{\varphi}(i) = k'-1$ for $k \leq i \leq k-1+m$ we have

$$F' = \{(i,j) : j \leq \bar{\varphi}(i) + \psi'(i)\} \quad \text{where} \quad \psi'(i) = \tilde{\psi}(i-k+1) \text{ if } k \leq i \leq k-1+m;$$

$$= \quad 0 \text{ otherwise.}$$

Direct computation shows that $\bar{\varphi} + \psi'$ is in fact the (k,k')-transform φ' of φ.

Further, inequalities (1) imply $\varphi'(k-1) \geq \varphi'(k)$ if $k > 1$. Since both $\bar{\varphi}$ and ψ' (for $i \geq k$) are non-increasing, this establishes that φ' is also non-increasing, and that, accordingly, $F' = F(\varphi')$ is a Ferrers relation.

Q.E.D.

Remark 10.

Consider any non-increasing map φ. It follows from definition 4 that the pair $(1, \varphi(1))$ is always admissible (with respect to φ) since inequalities (1) reduce to

$$0 \leq \varphi(1) - \varphi(1) \leq \tilde{\varphi}(\varphi(1)) - 1.$$

Let $m = \tilde{\varphi}(\varphi(1))$. The $(1, \varphi(1))$-transform φ' of φ satisfies

$$\varphi'(1) = \varphi(1) + m - 1$$

$$\varphi'(i) = \varphi(i) - 1 \qquad \text{for } 1 \leq i \leq m;$$

$$\varphi'(i) = \varphi(i) \qquad \text{otherwise}.$$

In particular $\varphi = \varphi'$ if and only if $m = 1$, that is, if and only if $\varphi(1) > \varphi(2)$ since $m = \tilde{\varphi}(\varphi(1))$ is characterized by $\varphi(1) = \varphi(2) = \dots$ $\dots = \varphi(m) > \varphi(m+1)$.

We shall say that $\bar{\varphi}$ is a transform of the non-increasing map φ if and only if $\bar{\varphi}$ can be obtained by a succession of (k, k')-transformations (in which, of course, the admissibility conditions are satisfied).

We now come to our main theorem.

THEOREM 11.

Each Ferrers relation is rook-equivalent with exactly one decreasing Ferrers relation.

Proof. In view of Corollary 6, it suffices to show that any Ferrers relation has a transform which is decreasing.

Consider a Ferrers relation $F(\varphi)$ which is not decreasing and let j be the least value such that $\varphi(j) = \varphi(j+1)$. Setting $k' = \varphi(j)$ and $j' = \tilde{\varphi}(k') = \mathrm{Max}\{i \in [n] : \varphi(i) = \varphi(j)\}$ we have $j < j'$, hence

$$0 = \varphi(j) - k' < j' - j = \tilde{\varphi}(k') - j.$$

proving that $k = \mathrm{Min}\{i \in [n] : \varphi(i) - k' < \tilde{\varphi}(k') - i\}$ is positive.

Because of the minimal character of k, the pair (k, k') is admissible. Thus φ admits a (k,k') -transform $\varphi' = \Theta(k, k'; \varphi)$. Further φ' precedes φ in the sense that the sequence $(\varphi'(1), \varphi'(2), \ldots, \varphi'(n))$ precedes the sequence $(\varphi(1), \varphi(2), \ldots, \varphi(n))$ in lexicographic order because by construction

$$\varphi'(i) = \varphi(i) \quad \text{for} \quad i < k \quad \text{and}$$

$$\varphi'(i) = \tilde{\varphi}(k') + k' - k > \varphi(i) .$$

Using induction on this order, it concludes the proof.

Q.E.D.

The subsequent results require some further definitions.

For $a > 0$ and $b \geq 0$ we shall denote $\eta_{a,b}$ the non-increasing map such that $\eta_{a,b}(i) = b$ or 0 depending upon $i \leq a$ or not. Let φ be a non-increasing map. The transpose of $\tilde{\varphi} + \eta_{a,b}$ will be called the (a,b)-translate of φ and denoted $\Delta(a, b; \varphi)$.

From the geometrical point of view the graph of $\Delta(a,b;\varphi)$ is obtained from $F(\varphi)$ by first considering $F(\varphi)$ as a subset of $\mathbb{P} \times [a]$ and then making a b-length translation of $F(\varphi)$ to the right.

In particular if $a \geq \varphi(1)$, the $(a-b)$-translate of φ is the non-increasing map φ' such that

$$\varphi'(i) = a \qquad \text{for} \quad i \leq b \quad \text{and}$$

$$\varphi'(i) = \varphi(i-b) \quad \text{for} \quad i > b .$$

From the proof of Theorem 11 it follows that the unique decreasing map $\hat{\varphi}$ in which the Ferrers relations $F(\varphi)$ and $F(\hat{\varphi})$ are rook-equivalent is a transform of φ; we shall then say that $\hat{\varphi}$ is the decreasing transform of φ.

The following lemma is a special case of a theorem in Riordan ([4] p. 181, theorem 3).

Lemma 12. Let $\bar{\varphi}$ (resp. $\bar{\varphi}'$) be a transform of φ (resp. φ') Then

(i) $\bar{\varphi} + \eta_{a,b}$ is a transform of $\varphi + \eta_{a,b}$ if we have $a \geq \tilde{\varphi}(1)$;

(ii) $\Delta(a,b;\bar{\varphi})$ is a transform of $\Delta(a,b;\varphi)$ if the inequality $a \geq \varphi(1)$ holds;

(iii) $\Delta(a,b;\bar{\varphi}) + \bar{\varphi}'$ is a transform of $\Delta(a,b;\varphi) + \varphi'$ if the inequalities $a \geq \varphi(1)$ and $b \geq \tilde{\varphi}'(1)$ both hold.

Proof.

For proving (i) and (ii), is suffices to consider the case when $\bar{\varphi} = \Theta(k,k';\varphi)$. Direct verification shows that

(i) if $a \geq \tilde{\varphi}(1)$, the pair $(k, k'+b)$ is admissible for $\varphi_1 = \varphi + \eta_{a,b}$ and then we have $\bar{\varphi}_1 = \Theta(k, k'+b; \varphi_1)$;

(ii) if $a \geq \varphi(1)$, the pair $(k+b, k')$ is admissible for $\varphi_2 = \Delta(a,b;\varphi)$ and we get $\bar{\varphi}_2 = \Theta(k+b, k'; \varphi_2)$.

Now let $\psi = \Delta(a,b;\varphi) + \varphi'$ and $\bar{\psi} = \Delta(a,b;\bar{\varphi}) + \bar{\varphi}'$. We already know from part (ii) that $\bar{\varphi}_1 = (a,b;\bar{\varphi})$ is a transform of $\varphi_1 = \Delta(a,b;\varphi)$. Note also that $\varphi_1(i) = a$ for $i \leq b$ and $\tilde{\varphi}_1(i) = 0$ for $i > a$; this implies

$$\tilde{\psi}(i) = \tilde{\varphi}'(i-a) \quad \text{for } i > a.$$

To prove (iii) it suffices to consider the case when $\bar{\varphi}' = \Theta(k,k';\varphi')$. Since $b \geq \tilde{\varphi}'(1)$, we necessarily have $k \leq b$ and then $\varphi_1(k) = a$. It then follows that

$$\begin{aligned} \psi(k) - (k'+a) &= \varphi_1(k) + \varphi'(k) - (k'+a) \\ &= \varphi'(k) - k'. \end{aligned}$$

We also have

$$\tilde{\psi}(k'+a) - k = \tilde{\varphi}'(k') - k \qquad \text{and}$$

$$\psi(k-1) - (k'+a) = \varphi'(k-1) - k'.$$

These last three equations show that the pair $(k, k'+a)$ is admissible for ψ and then we have $\bar{\psi} = \Theta(k, k'+a; \psi)$.

Q.E.D.

We consider now a very special case needed in the study of Newcomb's problem.

Definition 13.

Let $\underset{\sim}{d} = (d_1, d_2, \dots, d_r)$ be a sequence of $r > 0$ positive integers and set $\bar{d}_s = d_1 + d_2 + \dots + d_s \quad (1 \le s \le r)$.

The <u>special map of type</u> $\underset{\sim}{d}$, denoted by $\underset{\sim}{\mathrm{Spec}}\ \underset{\sim}{d}$, is the mapping φ from $\mathbb{P}$ into $\mathbb{N}$ that is defined by the following conditions:

$$\varphi(i) = 0 \qquad \text{for} \qquad i > \bar{d}_r$$

$$\varphi(\bar{d}_r) = 1 \qquad \text{and}$$

$$\varphi(i) = \varphi(i+1) + 2 \quad \text{or} \quad = \varphi(i+1) + 1 \quad \text{depending upon}$$

$$i \in \{\bar{d}_1, \bar{d}_2, \dots, \bar{d}_{r-1}\} \qquad \text{or not for} \qquad 1 \le i < \bar{d}_r.$$

The definition implies that any special map is <u>decreasing</u>. One computes easily that the value at 1 of $\underset{\sim}{\mathrm{Spec}}\ \underset{\sim}{d} = \varphi$ is $\bar{d}_r + r - 1$ and that $\tilde{\varphi}(1) = d_r$.

For $\underset{\sim}{d} = (d_1)$ the Ferrers relation $F(\underset{\sim}{\mathrm{Spec}}\,(d))$ is a "triangular board" in Riordan's terminology (see [4] p. 213). Then its rook polynomial is given by Riordan's formula ([4] p. 214) which involves the Stirling numbers $S(i,j)$ as it could be shown directly from our property 5, since it corresponds to the case when $y_i = 1$ identically.

We now give two results which show that "translation" preserves the "special" character of a Ferrers relation.

Lemma 14. Let $\psi = \underset{\sim}{\text{Spec}}(d_1, d_2, \dots, d_r)$ and $q \le r$. The special map $\bar{\psi} = \underset{\sim}{\text{Spec}}(d_1+1, d_2+1, \dots, d_q+1, d_{q+1}, \dots, d_r)$ is the decreasing transform of $\Delta(d, q; \psi)$ where $d = 1 + \psi(1)$.

Proof.

Let $\varphi = \Delta(d, q; \psi)$; we have $\varphi(i) = d$ for $1 \le i \le q$ and $\varphi(i) = \psi(i-q)$ for $q < i \le \bar{d}_r + q$. On the other hand $\bar{\psi}(i) = \psi(i-q)$ for $\bar{d}_q + q \le i \le \bar{d}_r + q$.

Consequently, if $q = 1$, it is readily verified that $\Delta(d, 1; \psi) = \bar{\psi}$ and the lemma is proved in this case.

If $q > 1$, we construct the $(1, \varphi(1))$-transform φ' of φ. We have $\varphi'(1) = \varphi(1) + q - 1 \ (= d')$, $\varphi'(i) = \varphi(1) - 1 = \psi(1)$ for $1 < i \le q$ and $\varphi'(i) = \varphi(i)$ for $i > q$.

We now distinguish two cases:

(i) suppose $d_1 > 1$, and let $\psi' = \underset{\sim}{\text{Spec}}(d_1 - 1, d_2, \dots, d_r)$; one can verify that φ' is obtained by the following two transformations. We first let $\psi'' = \Delta(\psi(1), q; \psi')$; then we have $\varphi' = \Delta(d', 1; \psi'')$. On the other hand let $\bar{\psi}' = \underset{\sim}{\text{Spec}}(d_1 - 1 + 1, d_2 + 1, \dots, d_q + 1, d_{q+1}, \dots, d_r)$; as we have $d' = 1 + \psi'(1) + q = 1 + \bar{\psi}'(1)$, we get $\bar{\psi} = \Delta(d', 1; \bar{\psi}')$.

By induction on $d_1 + d_2 + \dots + d_r$ the special map $\bar{\psi}'$ is a transform of ψ'' and by lemma 12 $\bar{\psi} = \Delta(d', 1; \bar{\psi}')$ is a transform of $\varphi' = \Delta(d', 1; \psi'')$. This concludes the proof in this case.

(ii) suppose $d_1 = 1$ and construct the $(2, \varphi'(2))$-transform φ'' of φ'. Using the same device as above, let $\psi' = \underset{\sim}{\text{Spec}}(d_2, d_3, \dots, d_r)$; we successively form $\psi'' = \Delta(1 + \psi'(1), q-1; \psi')$ and $\psi''' = \Delta(d'-1, 1; \psi'')$ where $d' = \psi(1) + q$. Then we obtain $\varphi'' = \Delta(d', 1; \psi'')$. On the other hand, let $\bar\psi' = \underset{\sim}{\mathrm{Spec}}\,(d_2+1, d_3+1, \ldots, d_q+1, d_{q+1}, \ldots, d_r)$; we obtain $\bar\psi = \Delta(d', 1; \chi)$ where $\chi = \Delta(d'-1, 1; \bar\psi')$.

Again by induction and using lemma 12 as above, we verify that $\bar\psi$ is a transform of φ'', hence of φ.

Q.E.D.

Lemma 15.

Let be ψ as above, $q \geq r \geq 0$, $d = \psi(1) + q - r + 1$ and $\varphi = \Delta(d, q; \psi)$. Then the decreasing transform of φ is $\bar\psi = \underset{\sim}{\mathrm{Spec}}\,(d'_1, d'_2, \ldots, d'_q)$ where $d'_i = 1$ for $i \leq q-r$ and $d'_i = 1 + d_{i-q+r}$ for $i > q-r$.

Proof.

Lemmas 14 and 15 coincide for $q = r$. Suppose $q > r$, and take the $(1, \varphi(1))$-transform φ' of φ. As in the proof of lemma 14, we verify that φ' is obtained from ψ by the two successive transformations. First let $\psi'' = \Delta(\psi(1)+q-r, q-1; \psi)$; then $\varphi' = \Delta(\bar\psi(1), 1; \psi'')$. By induction on q a transform of $\psi'' = \Delta(\psi(1) + (q-1) - r+1, q-1; \psi)$ is given by $\bar\psi'' = \underset{\sim}{\mathrm{Spec}}\,(d''_1, d''_2, \ldots, d''_{q-1})$ where $d''_i = 1$ for $i \leq q-1-r$ and $d''_i = 1 + d_{i-q+1+r}$ for $i > q-1-r$. Therefore it follows from lemma 12 that $\bar\psi = \Delta(\bar\psi(1), 1; \bar\psi'')$ is a transform of $\varphi' = \Delta(\bar\psi(1), 1; \psi'')$ (hence of φ).

Q.E.D.

It is to be noted that for $r = 0$, i.e. for $\varphi = \eta_{q,q}$, one has simply $\bar\psi = \sum(1, 1, \ldots, 1)$. This is a special case of Riordan's formula (34) ([4] p. 211).

4. APPLICATION TO NEWCOMB'S PROBLEM

Let $\gamma : [n] \longrightarrow [p]$ be an order-preserving surjection and J a subset of $[p]$. Then the relation $\bar R(\gamma, J) \subset [n] \times [n]$ is defined by the conditions $(i,j) \in \bar R(\gamma, J)$ if and only if $\gamma(i) < \gamma(j)$ or $\gamma(i) = \gamma(j) \in J$. Thus for $J = [p]$ the relation $\bar R(\gamma, J)$ is simply the total <u>preorder</u>

induced by γ, and $\bar{R}(\gamma, J)$ is the complement of a total preorder when $J = \emptyset$. Both extreme cases occur naturally in Newcomb's problem that involves the determination of the rook polynomial of $\bar{R}(\gamma,[p])$ or $\bar{R}(\gamma,\emptyset)$. In the generalization of this problem studied by one of us ([1] & [2]) it appears just as natural to consider the rook polynomial of $\bar{R}(\gamma, J)$ for any subset J of [p]. This is the purpose of this last section.

Keeping the same notations we let $\pi : [p] \longrightarrow \mathbb{P}$ be the map defined by

$$\pi(j) = \underset{\sim\sim}{\mathrm{card}}\, \gamma^{-1}(j) \qquad (j \in [p])$$

and prove the following property (Cf. Riordan [4], Ex. 4, p. 185).

Property 16. For any permutation σ of $[p]$, the relations $\bar{R}(\gamma, J)$ and $\bar{R}(\sigma\gamma, \sigma J)$ are rook-equivalent.

Proof.

It suffices to verify the property in the special case where σ is the transposition exchanging two consecutive values q and $q+1$ of $[p]$.

Set
$$m = \pi(q) + \pi(q+1);$$
$$a' = \mathrm{Min}\,\{i \in [n] : \gamma(i) = q\};$$
$$a = \mathrm{Max}\,\{i \in [n] : \gamma(i) = q+1\};$$

and define the (m,n)-injection α by $\alpha(i,j) = (a+1-i, a'+j-1)$.

The verification of the compatibility condition is trivial and one sees that $\bar{R}(\sigma\gamma, \sigma J)$ is the α-transpose of $\bar{R}(\gamma, J)$.

Q.E.D.

Observe now that the relation

$$R(\gamma, J) = \{(i,j) \in [n] \times [n] : (i, n+1-j) \in \bar{R}(\gamma, J)\}$$

is obviously a Ferrers relation which is rook-equivalent to $\bar{R}(\gamma, J)$. We let $\varphi = \phi(\gamma, J)$ denote the non-increasing map such that $R(\gamma, J)$ is the Ferrers relation $F(\varphi)$. According to property 16 we can assume that J is the interval $[p']$ with $0 \leq p' \leq p$ and that the two restrictions of π respectively to $[p']$ and $[p] \setminus [p']$ are <u>non-increasing</u>.

Throughout this section φ will designate the non-increasing map $\varphi = \phi(\gamma, J)$ with $\gamma : [n] \longrightarrow [p]$ and $J = [p']$ $(0 \leq p' \leq p)$.

We first consider the case when $J = [p]$, that is when $\bar{R}(\gamma, [p])$ is the <u>total preorder</u> induced by γ. With our conventions the map $\pi : [p] \longrightarrow \mathbb{P}$ is then non-increasing and we can define the non-increasing map $\tilde{\pi}$ its transpose, as defined in the beginning of section 3.

Lemma 17.

The decreasing transform of the defining map $\varphi = \phi(\gamma, [p])$ of the Ferrers relation $R(\gamma, [p])$ is the special map

$$\psi = \underset{\sim}{\mathrm{Spec}}(\tilde{\pi}(q), \tilde{\pi}(q-1), \ldots, \tilde{\pi}(1))$$

where $q = \pi(1)$.

Proof.

For $p = 1$, we have on the one hand

$$\pi(1) = n = q, \quad \tilde{\pi}(1) = \tilde{\pi}(2) = \ldots = \tilde{\pi}(n) = 1, \quad \tilde{\pi}(n+1) = 0.$$

On the other hand, $\varphi = \eta_{n,n}$ since $R(\gamma, [p]) = [n] \times [n]$. Thus the result is covered by Lemma 15 and we can use induction on $p \geq 2$.

Define $\pi' : [p-1] \longrightarrow \mathbb{P}$ by letting $\pi'(i) = \pi(i+1)$ for each $i \in [p-1]$. Thus let $r = \pi'(1)$ $(= \pi(2))$; we have

(1) $$r \leq q.$$

Moreover, let

$$d'_1 = \tilde{\pi}'(r), \quad d'_2 = \tilde{\pi}'(r-1), \ldots, d'_r = \tilde{\pi}'(1)$$

and $\psi' = \underset{\sim}{\mathrm{Spec}}(\underset{\sim}{d'})$; then, by the induction hypothesis, ψ' is the decreasing transform of $\varphi' = \phi(\gamma', [p-1])$ where γ' is the unique order-preserving surjection of $[n']$ $(n' = n - \pi(1))$ onto $[p-1]$ that satisfies

$\pi'(j) = \underset{\sim}{\mathrm{card}}\ \gamma'^{-1}(j)$ $(j \in [p-1])$ identically.

As we know, the value of ψ' at 1 is

$$\psi'(1) = d'_1 + d'_2 + \dots + d'_r + r - 1 = \sum \tilde{\pi}'(i) + r - 1 .$$

On the other hand

$$\sum \tilde{\pi}'(i) = \sum \pi'(i) = \sum \pi(i+1) = n - \pi(1) = n - q ;$$

hence

(2) $$n = \psi'(1) + q - r + 1 .$$

We also note that the sequence $(\tilde{\pi}(q), \tilde{\pi}(q-1), \dots, \tilde{\pi}(1))$ is the sequence obtained when putting $q-r$ elements equal to 1 in front of $\underset{\sim}{d}'$ and increasing by 1 each term of $\underset{\sim}{d}'$. In other words

$$\psi = \underset{\sim}{\mathrm{Spec}}\,(1, \dots, 1, 1+d'_1, \dots, 1+d'_r)$$

where the 1's are repeated $(q-r)$ times in the sequence.

Furthermore φ is seen to be the (n,q)-translate of φ', i.e.

(3) $$\varphi = \Delta(n, q; \varphi') .$$

As $n \geq \varphi'(1)$, the conditions of lemma 12 are fulfilled and we conclude that φ is a transform of $\bar{\varphi} = \Delta(n, q; \psi')$.

In view of (1), (2) and (3) we can apply lemma 15 by taking ψ' instead of ψ, $d = n$ and $\bar{\varphi}$ in place of φ and we deduce that ψ is a transform of $\bar{\varphi}$, hence of φ.

Q.E.D.

We now study the case of <u>complements of total preorders</u>. With the same notations as above, the set J is assumed to be empty. In the following lemma we characterize the decreasing Ferrers relation that is rook-equivalent to $R(\gamma,\emptyset)$. We can assume $p>1$, since otherwise $R(\gamma,\emptyset) = \bar{R}(\gamma,\emptyset) = \emptyset$. We set $b = \pi(1) - \pi(2)$ and $r = \pi(2)$. Since π is non-increasing, we can consider the transpose $\tilde{\pi}$ of π. We then have $r = \mathrm{Max}\,\{i : \tilde{\pi}(i) > 1\}$.

Lemma 18.

The decreasing transform of the defining map $\varphi = \phi(\gamma,\emptyset)$ of $R(\gamma,\emptyset)$ is $\psi + \eta_{m,b}$ where $m = n - \pi(1)$ and where ψ is the special map $\text{Spec}\;(\tilde{\pi}(1)-1,\ \tilde{\pi}(2)-1,\dots,\ \tilde{\pi}(r)-1)$.

Proof.

Instead of $\phi(\gamma,\emptyset)$ and $R(\gamma,\emptyset)$ we will also write $\phi(\pi)$ and $R(\pi)$. Since the lemma is trivial for $n \le 2$, we can use induction on $n \ge 3$.

We distinguish two cases.

Case 1: $b = \pi(1) - \pi(2) > 0$.

Let $n' = n-b$ and define $\pi' : [p] \longrightarrow \mathbb{P}$ by letting $\pi'(1) = \pi(1) - b = \pi(2)$; $\pi'(i) = \pi(i)$ otherwise. By construction we have $b' = \pi'(1) - \pi'(2) = 0$ and then

$$\begin{aligned} \tilde{\pi}'(i) &= \tilde{\pi}(i) \quad \text{for} \quad 1 \le i \le r; \\ &= 0 \quad \text{for} \quad i > r. \end{aligned}$$

Thus since $b' = 0$, the induction hypothesis implies that the special map ψ defined in the theorem is a transform of the defining map $\varphi' = \phi(\pi')$ of the Ferrers relation $R(\pi')$. Now from the definition of $R(\pi)$ and $R(\pi')$ we have that

$$\varphi = \varphi' + \eta_{m,b}$$

where $m = n - \pi(1)$. As $m = n - \pi(1) = n-b-r = \sum\{\tilde{\pi}(i) - 1 : 1 \le i \le r\} = \tilde{\psi}(1)$, the conditions of lemma 12 are satisfied and $\psi + \eta_{m,b}$ is indeed a transform of φ.

Case 2: $b = \pi(1) - \pi(2) = 0$.

By hypothesis we have $\pi(1) = \pi(2) = \ldots = \pi(s) = r$ where $s \geq 2$ and either $s = p$ or $s < p$ and $\pi(s) > \pi(s+1)$. Define maps π' and π'' of p into $\mathbb{P}$ by letting $\pi'(s) = \pi(s) - 1$; $\pi'(i) = \pi(i)$ for $i \neq s$;

$\pi''(1) = \pi(1) - 1$; $\pi''(i) = \pi(i)$ for $i \neq 1$.

Thus $\sum \pi'(i) = \sum \pi''(i) = n-1$. The map π' is non-increasing and $\tilde{\pi}'(r) = r-1$, $\tilde{\pi}'(i) = \tilde{\pi}(i)$ for $i < r$ $(= \pi(1) = \pi'(1))$. Hence by the induction hypothesis φ' $\phi(\pi')$ has for transform

$\psi' = \underset{\sim}{\text{Spec}}\,(\tilde{\pi}(1) - 1, \tilde{\pi}(2) - 1, \ldots, \tilde{\pi}(r-1) - 1, \tilde{\pi}(r) - 2)$

where eventually $\tilde{\pi}(r) - 2 = 0$.

Now by property 16, $R(\pi')$ and $R(\pi'')$ that is $F(\varphi')$ and $F(\varphi'')$ (where $\varphi'' = \phi(\pi'')$ are rook-equivalent. Further $\varphi = \phi(\pi)$ is equal to $\varphi'' + \eta_{m,1}$. Thus observing that $\tilde{\psi}'(1) = m-1$, we can again conclude from lemma 12 that the special map ψ defined in the lemma is a transform of φ since it is equal to $\psi' + \eta_{m,1}$.

Q.E.D.

We now come to the general case when J is not necessarily equal to $[p]$ or empty.

Theorem 19.

Let $\varphi = \phi(\gamma, J)$ be the defining map of the Ferrers relation $R(\gamma, J)$ where $\gamma : [n] \longrightarrow [p]$ is an order-preserving surjection and $J = |p'|$ with $0 \leq p' \leq p$.

Moreover let

$$\begin{aligned} \pi'(i) &= \underset{\sim}{\text{card}}\ \gamma^{-1}(i) && \text{for } i \in [p'] \\ &= 0 && \text{for } i > p', \\ \pi''(i) &= \underset{\sim}{\text{card}}\ \gamma^{-1}(p'+i) && \text{for } i \in [p-p'] \\ &= 0 && \text{for } i > p-p'. \end{aligned}$$

Then the decreasing transform of φ is

$$\psi + \eta_{m,b}$$

where $m = n - \pi''(1)$, $b = \pi''(1) - \pi''(2)$ and

$$\psi = \underset{\sim}{\mathrm{Spec}}\,(\tilde{\pi}'(q), \dots, \tilde{\pi}'(1), \tilde{\pi}''(1)-1, \dots, \tilde{\pi}''(r)-1)$$

with $q = \pi'(1)$ and $r = \pi''(2)$.

Proof.

The case $J = [p]$ or $J = \emptyset$ has been considered in lemmas 17 and 18. We shall then assume $1 \le p' < p$, and will also use the following notations:

$$n' = \pi'(1) + \dots + \pi'(p'), \quad p'' = p - p', \quad n'' = n - n',$$

$$\gamma'(i) = \gamma(i) \quad \text{for} \quad i \in [n'] \quad \text{and} \quad \gamma''(i) = \gamma(n'+i) \quad \text{for} \quad i \in [n''].$$

Let us define the two maps φ' and φ'' as follows:

$$\begin{aligned} \varphi'(i) &= \varphi(i) - n'' && \text{for} \quad i \in [n'] \\ &= 0 \quad \text{for} \quad i > n' \\ \varphi''(i) &= \varphi(i+n') && \text{for} \quad i \in \mathbb{P}. \end{aligned}$$

By construction we have

$$\varphi = \Delta(n'', n'; \varphi'') + \varphi'.$$

Moreover we clearly have $\varphi' = \phi(\gamma', [p']$ and $\varphi'' = \phi(\gamma'', \emptyset)$ if $p'' > 1$ and $\varphi'' = 0$ if $p'' = 1$. According to lemma 17 the decreasing transform of φ' is then the special map

$$\bar{\varphi}' = \underset{\sim}{\mathrm{Spec}}\,(\tilde{\pi}'(q), \tilde{\pi}'(q-1), \dots, \tilde{\pi}'(1))$$

where $q = \pi'(1)$. In the same manner lemma 18 asserts that if $p'' > 1$, the decreasing transform of φ'' is $\bar{\varphi}'' = \psi'' + \eta_{m'',b}$ where $m'' = n'' - \pi''(1)$ and $\psi'' = \underset{\sim}{\mathrm{Spec}}\,(\tilde{\pi}''(1)-1, \tilde{\pi}''(2)-1, \dots, \tilde{\pi}''(r)-1)$ with $r = \pi''(2)$.
If $p'' = 1$, we let $\bar{\varphi}'' = \varphi'' = 0$.

Now since $n'' > \varphi''(1)$ and $n' = \tilde{\varphi}'(1)$, the conditions of lemma 12 are fulfilled and we can conclude that

$$\bar{\varphi} = \Delta(n'', n'; \bar{\varphi}'') + \bar{\varphi}'$$

is a transform of $\varphi = \Delta(n'', n'; \varphi'') + \varphi'$. As one has $n' = \pi(1) + \dots + \pi(p') = \tilde{\pi}'(q) + \dots + \tilde{\pi}'(1)$, the value of $\bar{\varphi}'$, at n' is equal to 1. On the other hand the value of $\bar{\varphi}''$ at 1 is equal to 0 if $p'' = 1$ and if $p'' > 1$, we have

$$\bar{\varphi}''(1) = b + \sum_{1 \le i \le r} (\tilde{\pi}''(i) - 1) + r - 1$$
$$= n'' - 1 .$$

Accordingly we have $\bar{\varphi}(n') = n'' + 1$ and $\bar{\varphi}(n'+1) = n'' - 1$ if $p'' > 1$ and $\bar{\varphi}(n'+1) = 0$ if $p'' = 1$. As $\bar{\varphi}'$ and ψ'' are special, and this shows, in the case when $p'' > 1$, that $\bar{\varphi}$ is equal to $\eta_{m,b} + \psi$ where $m = n' + m'' = n - \pi''(1)$, $b = \pi''(1) - \pi''(2)$ and $\psi = \underset{\sim}{\mathrm{Spec}}\,(\tilde{\pi}'(q), \dots, \tilde{\pi}'(1), \tilde{\pi}''(1) - 1, \dots, \tilde{\pi}''(r) - 1)$ with $r = \pi''(2)$. If $p'' = 1$, one has $\bar{\varphi} = \eta_{n', n''} + \psi'$ where $\psi' = \underset{\sim}{\mathrm{Spec}}\,(\tilde{\pi}'(q), \dots, \tilde{\pi}'(1))$. In fact, in this last case we can also write $\bar{\varphi} = \eta_{m,b} + \psi$ since then $r = \pi''(2) = 0$ and ψ is reduced to $\psi = \underset{\sim}{\mathrm{Spec}}\,(\tilde{\pi}'(q), \dots, \tilde{\pi}'(1))$.

Q.E.D.

Remark 20.

The sequence $(\tilde{\pi}'(q), \dots, \tilde{\pi}'(1), \tilde{\pi}''(1) - 1, \dots, \tilde{\pi}''(r) - 1)$ just defined is unimodal, namely it is the juxtaposition of a non-decreasing sequence (of length q) and a non-increasing sequence (of length r). Thus the decreasing transform of a map $\varphi = \phi(\gamma, J)$ is apart from the function $\eta_{m,b}$ a special map $\psi = \underset{\sim}{\mathrm{Spec}}(d_1, \dots, d_s)$ whose defining sequence $(d_1, \dots, d_s)$ is unimodal. Conversely, let $\underset{\sim}{d} = (d_1, \dots, d_s)$ be an unimodal sequence of positive integers and b be a non-negative integer. Then form the decreasing map $\psi = \eta_{m,b} + \underset{\sim}{\mathrm{Spec}}\,(\underset{\sim}{d})$. It is readily verified that ψ is the decreasing transform of (at least) one non-increasing map φ of the form $\phi(\gamma, [p'])$.

A Ferrers relation $R(\gamma, [p'])$ in fact depends on four parameters n, p, γ and p'. Let us write $R(n,p,\gamma,p')$ instead of $R(\gamma,p')$. Thus it is easily proved that $F(\psi)$ is rook-equivalent to exactly one Ferrers relation $R(n,p,\gamma,p')$ such that $p \geq 2p'$.

Example 21.

Let $n = 14$, $p = 6$, $p' = 3$; $\pi(1) = 4$, $\pi(2) = 3$, $\pi(3) = 2$, $\pi(4) = 2$, $\pi(5) = 2$, $\pi(6) = 1$. The sequence of positive values of $\varphi = \phi(\gamma, [3])$ where γ is determined by the map π just defined, is $(14, 14, 14, 14, 10, 10, 10, 7, 7, 3, 3, 1, 1)$. The decreasing transform of φ is $\bar{\varphi} = \underset{\sim}{\mathrm{Spec}}(1,2,3,3,2,1) + \eta_{12,0}$ and the sequence of positive values of $\bar{\varphi}$ is then $(17, 15, 14, 12, 11, 10, 8, 7, 6, 4, 3, 1)$.

REFERENCES

[1] P.CARTIER & D.FOATA: Problèmes combinatoires de commutation et réarrangements, Lecture Notes in Math. n°85, Springer-Verlag, Berlin (1969).

[2] D.FOATA: Etude algèbrique de certains problèmes d'analyse combinatoire et du calcul des probabilités, Publ. Inst. Statist. Univ. Paris 14 (1965), 81-241.

[3] F.POUSSIN: Sur une propriété arithmétique de certains polynômes associés aux nombres d'Euler, C.R.Acad. Sc. Paris 266 (1968), 392-393.

[4] J.RIORDAN: An Introduction to Combinatorial Analysis, Wiley, New York (1958).

An upper bound on the chromatic number of a graph

by

J. H. Folkman*

Santa Monica, USA

1. INTRODUCTION

By a graph we will mean a finite undirected graph with no edge joining a vertex to itself and at most one edge joining any pair of distinct vertices. Such a graph G may be regarded as consisting of a finite set $V(G)$ (the vertices of G) together with a collection $E(G)$ of two element subsets of $V(G)$ (the edges of G).

If G is a graph, a set $S \subseteq V(G)$ is <u>independent</u> in G if $\{x,y\} \notin E(G)$ for every $x, y \in S$. If r is a positive integer, an r-<u>coloring</u> of the graph G is a function C from $V(G)$ to the first r positive integers such that if $\{x,y\} \in E(G)$ then $C(x) \neq C(y)$. We denote by $\chi(G)$ the smallest positive integer r such that G has an r-coloring.

Our objective here is to prove the following theorem which was conjectured by Erdős and Hajnal [1].

---

* Due to the tragic and untimely death of the author he was unable to deliver his lecture.

THEOREM. Let G be a graph and let k be a nonnegative integer. Suppose that for each set $S \subseteq V(G)$ there is a set $S' \subseteq S$ such that S' is independent in G and $|S'| \geq \frac{1}{2}(|S|-k)$. Then $\chi(G) \leq k+2$.

2. PRELIMINARY LEMMAS

Let G be a graph. Let $k(G)$ be the smallest nonnegative integer k such that for any set $S \subseteq V(G)$ there is a set $S' \subseteq S$ such that S' is independent in G and $|S'| \geq \frac{1}{2}(|S|-k)$. (There clearly is such an integer k. We may, for example, take $k = |V(G)|$ and then take S' to be empty for every $S \subseteq V(G)$).

Let G be a graph and let $\mathcal{P}$ be a collection of subsets of $V(G)$ such that if $S, T \in \mathcal{P}$ and $S \neq T$ then $S \cap T = \emptyset$. Define $G/\mathcal{P}$, the weak quotient of G by $\mathcal{P}$, to be the graph H_1 given by $V(H_1) = \mathcal{P}$ and $E(H_1) = \{\{S,T\} \mid S,T \in \mathcal{P},\ S \neq T$ and for some $x \in S$ and some $y \in T$, $\{x,y\} \in E(G)\}$. Define $G/\!/\mathcal{P}$, the strong quotient of G by $\mathcal{P}$, to be the graph H_2 given by $V(H_2) = \mathcal{P}$ and $E(H_2) = \{\{S,T\} \mid S,T \in \mathcal{P},\ S \neq T$ and for every $x \in S$ and every $y \in T$, $\{x,y\} \in E(G)\}$.

Recall that if G is a graph, a set $S \subseteq V(G)$ is a clique in G if $\{x,y\} \in E(G)$ for every $x, y \in S$ with $x \neq y$.

LEMMA 2.1. Let G be a graph and let r be a nonnegative integer. Let $B \subseteq V(G)$ be a clique in G with $|B| = r+1$. Let a_1, a_2 be distinct elements of $V(G)-B$ such that $\{a_i, x\} \in E(G)$ for every $x \in B$ and for $i = 1$ or 2. Let $A = \{a_1, a_2\}$ and let $\mathcal{P} = \{A\} \cup \{\{x\} \mid x \in V(G)-B-A\}$. Let $H = G/\mathcal{P}$. Then $k(H) \leq k(G) - r$.

PROOF. By the definition of $k(G)$ there is a set $S \subseteq \{a_1\} \cup B$ such that S is independent in G and $|S| \geq \frac{1}{2}(|\{a_1\} \cup B| - k(G)) = \frac{1}{2}(r+2-k(G))$. Now $\{a_1\} \cup B$ is a clique so $|S| \leq 1$. Hence, $2 \geq r+2-k(G)$ so $k(G) - r \geq 0$. It now suffices to show that if $S \subseteq V(H)$ there is a set $S' \subseteq S$ with S' independent in H and $|S'| \geq \frac{1}{2}(|S| - (k(G)-r))$.

Case 1. $A \notin S$.

Let $T = \{a_1\} \cup B \cup \{x \in V(G) - B - A \mid \{x\} \in S\}$. Then $|T| = r+2+|S|$. There is set $T' \subseteq T$ with T' independent in G and $|T'| \geq \frac{1}{2}(|T| - k(G)) = \frac{1}{2}(r+2+|S| - k(G)) = 1 + \frac{1}{2}(|S| - (k(G)-r))$. Let $S' = \{\{x\} \in S \mid x \in T'\}$. Then S' is a subset of S that is independent in H. Since T' is independent in G and $\{a_1\} \cup B$ is a clique in G, $|T' \cap (\{a_1\} \cup B)| \leq 1$. Therefore, $|S'| = |T' - (\{a_1\} \cup B)| = |T'| - |T' \cap (\{a_1\} \cup B)| \geq \frac{1}{2}(|S| - (k(G)-r))$.

Case 2. $A \in S$.

Let $T = A \cup B \cup \{x \in V(G) - A - B \mid \{x\} \in S\}$. Then $|T| = |S| - 1 + |A \cup B| = |S| + r + 2$. There is a set $T' \subseteq T$ such that T' is independent in G and

$$|T'| \geq \frac{1}{2}(|T| - k(G)) = \frac{1}{2}(|S| + r + 2 - k(G)) = 1 + \frac{1}{2}(|S| - (k(G) - r)).$$

Let $S' = \{\{x\} \in S \mid x \in T'\}$. Then S' is a subset of S and S' is independent in H. Furthermore,

$$|S'| = |T' - (A \cup B)| = |T'| - |T' \cap (A \cup B)| \geq$$

$$\geq \frac{1}{2}(|S| - (k(G) - r)) + 1 - |T' \cap (A \cup B)|.$$

If $|T' \cap (A \cup B)| \leq 1$ then $|S'| \geq \frac{1}{2}(|S| - (k(G) - r))$.

Now suppose that $|T' \cap (A \cup B)| > 1$. Then $T' \cap (A \cup B) = A$ since T' is independent and all two-element subsets of $A \cup B$ other than A belong to $E(G)$. Since $A \subseteq T'$ and T' is independent in G, $S' \cup \{A\}$ is independent in H. Furthermore,

$$|S' \cup \{A\}| = |S'| + 1 \geq \frac{1}{2}(|S| - (k(G) - r)) + 1 - |T' \cap (A \cup B)| + 1 =$$

$$= \frac{1}{2}(|S| - (k(G) - r)).$$

LEMMA 2.2. Let G be a graph and let r be an integer with

$r \geq 2$. Let $a_1, a_2, \ldots, a_r, b_1, b_2, \ldots, b_r$ be $2r$ distinct elements of $V(G)$. Suppose that

$$\{a_i, b_i\} \in E(G) \quad \text{for } 1 \leq i \leq r,$$

$$\{b_i, a_{i+1}\} \in E(G) \quad \text{for } 1 \leq i \leq r-1, \quad \text{and}$$

$$\{b_r, a_1\} \in E(G).$$

Let $A = \{a_1, a_2, \ldots, a_r\}$ and $B = \{b_1, b_2, \ldots, b_r\}$.

Let $S \subseteq V(G)$ be independent in G and suppose that $|S \cap (A \cup B)| \geq r$. Then $S \cap (A \cup B) = A$ or $S \cap (A \cup B) = B$.

PROOF. We have

$$r \leq |S \cap (A \cup B)| = \sum_{i=1}^{r} |S \cap \{a_i, b_i\}|.$$

Now S is independent in G and $\{a_i, b_i\} \in E(G)$ for $1 \leq i \leq r$ so $|S \cap \{a_i, b_i\}| \leq 1$ for $1 \leq i \leq r$. Hence, $r = |S \cap (A \cup B)|$ and $|S \cap \{a_i, b_i\}| = 1$ for $1 \leq i \leq r$.

Suppose that $S \cap A = \phi$. Then $|S \cap B| = |S \cap (A \cup B)| = r$. But $|B| = r$ so $B = S \cap B = S \cap (A \cup B)$. Similarly, if $S \cap B = \phi$ than $S \cap (A \cup B) = A$. Now suppose that $S \cap A \neq \phi$ and $S \cap B \neq \phi$. Let i be the smallest integer such that $1 \leq i \leq r$ and $a_i \in S$. Let j be the largest integer such that $1 \leq j \leq r$ and $b_j \in S$. If $i > 1$ then $1 \leq i-1 < i \leq r$ so $a_{i-1} \notin S$ by the choice of i. Now $|S \cap \{a_{i-1}, b_{i-1}\}| = 1$ so $b_{i-1} \in S$. But $\{b_{i-1}, a_i\} \in E(G)$ and S is independent in G so a_i and b_{i-1} cannot both be in S. Hence, $i = 1$ so $a_1 \in S$. If $j < r$ then $1 \leq j < j+1 \leq r$ so $b_{j+1} \notin S$ by the choice of j. Now $|\{a_{j+1}, b_{j+1}\} \cap S| = 1$ so $a_{j+1} \in S$. But $\{b_j, a_{j+1} \in E(G)$ and S is independent in G so b_j and a_{j+1} cannot both be in S. Hence, $j = r$ so $b_r \in S$. Now we have $a_1, b_r \in S$ and $\{a_1, b_r\} \in E(G)$. This contradicts the fact that S is independent in G.

LEMMA 2.3. Let $G, r, a_1, a_2, \ldots, a_r, b_1, b_2, \ldots, b_r$, A and B be as in Lemma 2.2. Let $\mathcal{P} = \{A, B\} \cup \{\{x\} \mid x \in V(G) - A - B\}$ and let $H = G/\mathcal{P}$. Then $k(H) \leq k(G)$.

PROOF. Let $k = k(G)$ and let $S \subseteq V(H)$. It suffices to show that there is a set $S' \subseteq S$ such that S' is independent in H and $|S'| \geq \frac{1}{2}(|S| - k)$.

Case 1. $A, B \notin S$.

Let $T = \{x \in V(G) - A - B \mid \{x\} \in S\}$. Then $|T| = |S|$. There is a set $T' \subseteq T$ such that T' is independent in G and $|T'| \geq \frac{1}{2}(|T| - k) = \frac{1}{2}(|S| - k)$. Let $S' = \{\{x\} \mid x \in T'\}$. Then $S' \subseteq S$, S' is independent in H and $|S'| = |T'| \geq \frac{1}{2}(|S| - k)$.

Case 2. $A \in S$, $B \notin S$.

Let $T = A \cup \{b_1, b_2, \ldots, b_{r-1}\} \cup \{x \in V(G) - A - B \mid \{x\} \in S\}$. Then $|T| = r + r - 1 + |S| - 1 = |S| + 2(r-1)$. There is a set $T' \subseteq T$ such that T' is independent in G and

$$|T'| \geq \frac{1}{2}(|T| - k) = \frac{1}{2}(|S| + 2(r-1) - k) = \frac{1}{2}(|S| - k) + r - 1.$$

Let $S' = \{\{x\} \mid x \in T' - A - B\}$. Then $S' \subseteq S$ and S' is independent in H. Furthermore

$$|S'| = |T' - (A \cup B)| = |T'| - |T' \cap (A \cup B)| \geq \frac{1}{2}(|S| - k) + r - 1 - |T' \cap (A \cup B)|.$$

If $|T' \cap (A \cup B)| \leq r - 1$ then $|S'| \geq \frac{1}{2}(|S| - k)$. Now suppose that $|T' \cap (A \cup B)| \geq r$. By Lemma 2.2, $T' \cap (A \cup B) = A$ or $T' \cap (A \cup B) = B$. Now $T' \subseteq T$ and $b_r \in B$ but $b_r \notin T$ so we must have $T' \cap (A \cup B) = A$. Since $A \subseteq T'$ and T' is independent in G, $S' \cup \{A\}$ is independent in H. Furthermore, $S' \cup \{A\} \subseteq S$ and

$$|S' \cup \{A\}| = |S'| + 1 \geq \frac{1}{2}(|S| - k) + r - 1 - |T' \cap (A \cup B)| + 1 =$$

$$= \frac{1}{2}(|S| - k) + r - |A| = \frac{1}{2}(|S| - k).$$

Case 3. $A \notin S$, $B \in S$.

Let $T = \{a_1, a_2, \dots, a_{r-1}\} \cup B \cup \{x \in V(G) - A - B \mid \{x\} \in S\}$ and argue as in Case 2.

Case 4. $A, B \in S$.

Let $T = A \cup B \cup \{x \in V(G) - A - B \mid \{x\} \in S\}$. Then $|T| = 2r + |S| - 2 = |S| + 2(r-1)$. There is a set $T' \subseteq T$ such that T' is independent in G and

$$|T'| \geq \frac{1}{2}(|T| - k) = \frac{1}{2}(|S| + 2(r-1) - k) = \frac{1}{2}(|S| - k) + r - 1.$$

Let $S' = \{\{x\} \mid x \in T' - A - B\}$. Then $S' \subseteq S$, S' is independent in H and $|S'| = |T' - A - B| = |T'| - |T' \cap (A \cup B)| \geq \frac{1}{2}(|S| - k) + r - 1 - |T' \cap (A \cup B)|$. If $|T' \cap (A \cup B)| \leq r-1$ then $|S'| \geq \frac{1}{2}(|S| - k)$. Now suppose $|T' \cap (A \cup B)| \geq r$. By Lemma 2.2, $T' \cap (A \cup B) = A$ or $T' \cap (A \cup B) = B$. If $T' \cap (A \cup B) = A$ then $S' \cup \{A\}$ is independent in H, $S' \cup \{A\} \subseteq S$ and

$$|S' \cup \{A\}| = |S'| + 1 \geq \frac{1}{2}(|S| - k) + r - 1 - |T' \cap (A \cup B)| + 1 =$$

$$= \frac{1}{2}(|S| - k) + r - |A| =$$

$$= \frac{1}{2}(|S| - k).$$

Similarly, if $T' \cap (A \cup B) = B$ then $S' \cup \{B\}$ is independent in H, $S' \cup \{B\} \subseteq S$ and $|S' \cup \{B\}| \geq \frac{1}{2}(|S| - k)$.

LEMMA 2.4. Let G be a graph and let m be a positive integer. Let $U \subseteq V(G)$ with $|U| = 2m-1$. Suppose that for each $x \in U$ there is a set $U' \subseteq U - \{x\}$ such that U' is independent in G and $|U'| = m$. Then there is a set $W \subseteq U$ such that W is independent in G and $|W| \geq m+1$.

PROOF. Let C be the collection of all sets $U' \subseteq U$ such that U' is independent in G and $|U'| = m$. By hypothesis, $C \neq \emptyset$. In fact, for each $x \in U$ there is a $U' \in C$ such that $x \notin U'$. Let $\mathfrak{F}$ be the family of all collections $D \subseteq C$ such that $|U - \cup D| < |\cap D|$. Let $U' \in C$ and let $D = \{U'\}$. Then $\cup D = U' = \cap D$ so

$$|\mathcal{U}-\cup D| = |\mathcal{U}-\mathcal{U}'| = |\mathcal{U}|-|\mathcal{U}'| = 2m-1-m = m-1 < m = |\mathcal{U}'| = |\cap D|.$$

Hence, $D \in \mathfrak{F}$ so $\mathfrak{F}$ is nonempty. For each $D \in \mathfrak{F}$, $|D| \le |C| \le \binom{2m-1}{m}$. Hence, $\mathfrak{F}$ contains a maximal member D^*. Since $D^* \in \mathfrak{F}$, $0 \le |\mathcal{U}-\cup D^*| < |\cap D^*|$ so $\cap D^* \ne \phi$. Let $x \in \cap D^*$. By hypothesis there is a set $\mathcal{U}'_0 \in C$ such that $x \notin \mathcal{U}'_0$. Now $x \in \cap D^*$ so $x \in \mathcal{U}'$ for every $\mathcal{U}' \in D^*$. Hence, $\mathcal{U}'_0 \notin D^*$ so D^* is a proper subset of $D = D^* \cup \{\mathcal{U}'_0\}$. Since D^* is a maximal member of $\mathfrak{F}$, D cannot be a member of $\mathfrak{F}$.

Now $D \subseteq C$ so D must fail to be a member of $\mathfrak{F}$ because

$$|\mathcal{U}'_0 \cap (\cap D^*)| = |\cap D| \le |\mathcal{U} - \cup D| = |\mathcal{U} - (\mathcal{U}'_0 \cup (\cup D^*))|.$$

Let $W = \cap D^* \cup (\mathcal{U}'_0 \cap (\cup D^*))$. Since $|\mathcal{U}'_0 \cap (\cap D^*)| \le |\mathcal{U} - (\mathcal{U}'_0 \cup (\cup D^*))|$, we have $-|\mathcal{U}'_0 \cap (\cap D^*)| \ge -|\mathcal{U} - (\mathcal{U}'_0 \cup (\cup D^*))|$. Since $D^* \in \mathfrak{F}$ we have $|\cap D^*| > |\mathcal{U} - \cup D^*|$. Hence,

$$\begin{aligned}
|W| &= |\cap D^* \cup (\mathcal{U}'_0 \cap (\cup D^*))| = \\
&= |\cap D^* - \mathcal{U}'_0| + |\mathcal{U}'_0 \cap (\cup D^*)| = \\
&= |\cap D^*| - |\mathcal{U}'_0 \cap (\cap D^*)| + |\mathcal{U}'_0 \cap (\cup D^*)| > \\
&> |\mathcal{U} - \cup D^*| - |\mathcal{U} - (\mathcal{U}'_0 \cup (\cup D^*))| + |\mathcal{U}'_0 \cap (\cup D^*)| = \\
&= |\mathcal{U}| - |\cup D^*| - (|\mathcal{U}| - |\mathcal{U}'_0 \cup (\cup D^*)|) + |\mathcal{U}'_0 \cap (\cup D^*)| = \\
&= |\mathcal{U}'_0 \cup (\cup D^*)| + |\mathcal{U}'_0 \cap (\cup D^*)| - |\cup D^*|.
\end{aligned}$$

Now $|S \cup T| + |S \cap T| = |S| + |T|$ for any pair of sets S, T. Taking $S = \mathcal{U}'_0$ and $T = \cup D^*$ in the preceding inequality we have

$$|W| > |S \cup T| + |S \cap T| - |T| = |S| = m.$$

Hence, $|W| \ge m+1$.

Suppose that W is not independent in G. Then there are elements $x, y \in W$ such that $\{x, y\} \in E(G)$. Now $\mathcal{U}'_0 \cap (\cup D^*) \subseteq \mathcal{U}'_0$ and $\mathcal{U}'_0$ is independent in G so x and y cannot both belong to $\mathcal{U}'_0$. Hence, either $x \in \cap D^*$ or $y \in \cap D^*$. Interchanging x and y if necessary, we may assume that $x \in \cap D^*$. Let $\mathcal{U}' \in D^*$. Then $x \in \mathcal{U}'$. Now $\mathcal{U}'$ is independent in G so $y \notin \mathcal{U}'$. Hence, $y \notin \cup D^*$. But $y \in W = (\cap D^*) \cup (\mathcal{U}'_0 \cap (\cup D^*)) \subseteq \cup D^*$. This is a contradiction. Hence, W is the required independent subset of $\mathcal{U}$.

LEMMA 2.5. Let G be a graph and let r be a nonnegative integer. Let $B \subseteq V(G)$ be a clique in G with $|B| = r+1$. Suppose there is a set $A \subseteq V(G) - B$ such that $|A| \geq k(G) - r + 2$ and $\{x, y\} \in E(G)$ for every $x \in A$ and every $y \in B$. Let $\mathcal{P} = \{B\} \cup \{\{x\} \mid x \in V(G) - B\}$. Let $H = G /\!/ \mathcal{P}$. Then $k(H) \leq k(G) - r$.

PROOF. By the definition of $k(G)$ there is a set $B' \subseteq B$ such that B' is independent in G and $|B'| \geq \frac{1}{2}(|B| - k(G))$. Now B is a clique so $1 \geq |B'| \geq \frac{1}{2}(|B| - k(G)) = \frac{1}{2}(r+1-k(G))$. Therefore, $k(G) \geq r-1$. Now $|A| \geq k(G) - r + 2 \geq r - 1 - r + 2 = 1$ so $A \neq \phi$.

Let $a \in A$. Then $\{a\} \cup B$ is a clique and $|\{a\} \cup B| = r+2$. Reasoning as above we have

$$1 \geq \frac{1}{2}(|\{a\} \cup B| - k(G)) = \frac{1}{2}(r+2-kG))$$

so $k(G) \geq r$. Hence, $k(G) - r \geq 0$ so it suffices to show that for each set $S \subseteq V(H)$ there is a set $S' \subseteq S$ such that S' is independent in H and $|S'| \geq \frac{1}{2}(|S| - (k(G) - r))$.

CASE 1. $B \in S$.

Let $T = B \cup \{x \in V(G) - B \mid \{x\} \in S\}$. Then $|T| = |B| + |S| - 1 = |S| + r$. There is a set $T' \subseteq T$ such that T' is independent in G and $|T'| \geq \frac{1}{2}(|T| - k(G)) = \frac{1}{2}(|S| - (k(G) - r))$. Let $S' = \{\{x\} \mid x \in T' - B\}$. Then $S' \subseteq S$, S' is independent and

$$|S'| = |T' - B| = |T'| - |T' \cap B| \geq \frac{1}{2}(|S| - (k(G) - r)) - |T' \cap B|.$$

If $T' \cap B = \phi$ then $|S'| \geq \frac{1}{2}(|S| - (k(G) - r))$. Now suppose that $T' \cap B \neq \phi$. Let $b \in T' \cap B$. Since T' is independent in G, $\{x, b\} \notin E(G)$ for every

$x \in T' - B$. Hence, since $H = G//\mathcal{P}$, $\{\{x\}, B\} \in E(H)$ for every $\{x\} \in S'$. Therefore, $S' \cup \{B\}$ is independent in H. Now T' is independent in G and B is a clique in G so $|T' \cap B| \le 1$. Therefore,

$$|S' \cup \{B\}| = |S'| + 1 \ge \frac{1}{2}(|S| - (k(G) - r)) - |T' \cap B| + 1 \ge \frac{1}{2}(|S| - (k(G) - r)).$$

Case 2. $B \notin S$ and there is an $a \in A$ such that $\{a\} \notin S$.

Let $T = B \cup \{a\} \cup \{x \in V(G) \mid \{x\} \in S\}$. Then $|T| = |B \cup \{a\}| + |S| = |S| + r + 2$. There is a set $T' \subseteq T$ such that T' is independent in G and $|T'| \ge \frac{1}{2}(|T| - k(G)) = \frac{1}{2}(|S| - (k(G) - r)) + 1$. Now T' is independent in G and $B \cup \{a\}$ is a clique in G so $|T' \cap (B \cup \{a\})| < 1$. Therefore,

$$|T' - (B \cup \{a\})| \ge |T'| - 1 \ge \frac{1}{2}(|S| - (k(G) - r)).$$

Let $S' = \{\{x\} \mid x \in T' - (B \cup \{a\})\}$. Then $S' \subseteq S$, S' is independent in H and $|S'| = |T' - (B \cup \{a\})| \ge \frac{1}{2}(|S| - (k(G) - r))$.

Case 3. $B \notin S$, for every $a \in A$, $\{a\} \in S$ and $|S| - (k(G) - r)$ is even. Let $2m = |S| - (k(G) - r)$. Let $T = B \cup \{x \in V(G) - B \mid \{x\} \in S\}$. Then $|T| = |B| + |S| = |S| + r + 1$. There is a set $T' \subseteq T$ such that T' is independent in G and

$$|T'| \ge \frac{1}{2}(|T| - k(G)) = \frac{1}{2}(|S| - (k(G) - r) + 1) = \frac{1}{2}(2m + 1) = m + \frac{1}{2}.$$

Now $|T'|$ is an integer so $|T'| \ge m + 1 = \frac{1}{2}(|S| - (k(G) - r)) + 1$. Since T' is independent in G and B is a clique in G, $|T' \cap B| \le 1$. Therefore, $|T' - B| \ge |T'| - 1 \ge \frac{1}{2}(|S| - (k(G) - r))$. Let $S' = \{\{x\} \mid x \in T' - B\}$. Then $S' \subseteq S$, S' is independent in H and $|S'| = |T' - B| \ge \frac{1}{2}(|S| - (k(G) - r))$.

Case 4. $B \notin S$, for every $a \in A$, $\{a\} \in S$ and $|S| - (k(G) - r)$ is odd.

Let $2m + 1 = |S| - (k(G) - r)$. If $m < 0$ then $|\emptyset| = 0 \ge \frac{1}{2}(2m + 1) = \frac{1}{2}(|S| - (k(G) - r))$, $\emptyset \subseteq S$ and $\emptyset$ is independent in H.

If $m = 0$ then $|S| = k(G) - r + 1$. Now $k(G) - r \geq 0$ so $|S| \geq 1$ so $S \neq \phi$. Let $x \in S$ and let $S' = \{x\}$. Then $S' \subseteq S$, S' is independent in H and $|S'| = 1 \geq \frac{1}{2} = \frac{1}{2}(|S| - (k(G) - r))$.

Now suppose that $m > 0$. Let $T = \{x \mid \{x\} \in S\}$. By the hypotheses of this case $A \subseteq T$. Hence,

$$|T - A| = |T| - |A| = |S| - |A| \leq |S| - (k(G) - r) - 2 = 2m - 1 .$$

On the other hand, $|T| = |S| = k(G) - r + 2m + 1 \geq 2m + 1 > 2m - 1$. Therefore, there is a set U such that $T - A \subseteq U \subseteq T$ and $|U| = 2m - 1$. Suppose that for every set $T' \subseteq T$, if T' is independent in G then $|T'| \leq m$. Let $x \in U$. We have

$$|(T - \{x\}) \cup B| = |T| - 1 + |B| = k(G) - r + 2m + 1 - 1 + r + 1 = k(G) + 2m + 1 .$$

There is a set $T' \subseteq (T - \{x\}) \cup B$ such that T' is independent in G and $|T'| \geq \frac{1}{2}(|T - \{x\}) \cup B| - k(G)) = \frac{1}{2}(2m + 1) = m + \frac{1}{2}$. Since $|T'|$ is an integer, $|T'| \geq m + 1$. Now $|T' \cap B| \leq 1$ because T' is independent and B is a clique. By our assumption, $|T' \cap (T - \{x\})| \leq m$ because $T' \cap (T - \{x\})$ is an independent subset of T. Hence, we have $m + 1 \leq |T'| = |T' \cap B| + |T' \cap (T - \{x\}| \leq 1 + m$ so $|T' \cap B| = 1$ and $|T' \cap (T - \{x\})| = m$. Let $b \in T' \cap B$. For each $a \in A$, $\{a, b\} \in E(G)$. Since T' is independent in G and $b \in T'$, $T' \cap A = \phi$. Therefore, $T' \cap (T - \{x\}) \subseteq T - A \subseteq U$. If we let $U' = T' \cap (T - \{x\})$ then U' is independent in G, $|U'| = m$ and $U' \subseteq U - \{x\}$. By Lemma 2.4, there is a set $W \subseteq U$ such that W is independent in G and $|W| \geq m + 1$. Now $W \subseteq U \subseteq T$ so this contradicts our assumption that every independent subset of T contains at most m elements. Hence, there is a set $T' \subseteq T$ such that T' is independent in G and $|T'| \geq m + 1$. Let $S' = \{\{x\} \mid x \in T\}$. Then $S' \subseteq S$, S' is independent in H and $|S'| = |T'| \geq m + 1 \geq \frac{1}{2}(2m + 1) = \frac{1}{2}(|S| - (k(G) - r))$.

LEMMA 2.6. Let G be a graph and let r be a nonnegative integer.

Let $B \subseteq V(G)$ be a clique in G with $|B| = r+2$. Let $\mathcal{P} = \{\{x\} \mid x \in V(G) - B\}$ and let $H = G/\mathcal{P}$. Then $k(H) \leq k(G) - r$.

PROOF. By the definition of $k(G)$ there is a set $B' \subseteq B$ such that B' is independent in G and $|B'| \geq \frac{1}{2}(|B| - k(G)) = \frac{1}{2}(r - k(G)) + 1$. Since B is a clique and B' is independent, $|B'| \leq 1$. Therefore, $1 \geq |B'| \geq \frac{1}{2}(r - k(G)) + 1$ so $k(G) - r \geq 0$. Hence, it suffices to show that for any set $S \subseteq V(H)$ there is a set $S' \subseteq S$ such that S' is independent in H and $|S'| \geq \frac{1}{2}(|S| - (k(G) - r))$.

Let $T = B \cup \{x \in V(G) - B \mid \{x\} \in S\}$. Then $|T| = |B| + |S| = |S| + r + 2$, There is a set $T' \subseteq T$ such that T' is independent in G and $|T'| \geq \frac{1}{2}(|T| - k(G)) = \frac{1}{2}(|S| - (k(G) - r)) + 1$. Now B is a clique in G and T' is independent in G so $|T' \cap B| \leq 1$. Hence, $|T' - B| = |T'| - |T' \cap B| \geq \frac{1}{2}(|S| - (k(G) - r))$. Let $S' = \{\{x\} \mid x \in T' - B\}$. Then $S' \subseteq S$, S' is independent in H and $|S'| = |T' - B| \geq \frac{1}{2}(|S| - (k(G) - r))$.

Let B be a graph. By a <u>circuit in</u> G we mean a sequence $(x_1, x_2, \ldots, x_r)$ of elements of $V(G)$ such that $r \geq 3$, $x_i \neq x_j$ for $1 \leq i < j \leq r$, $\{x_r, x_1\} \in E(G)$ and $\{x_i, x_{i+1}\} \in E(G)$ for $1 \leq i \leq r-1$. We call the integer r the <u>length</u> of the circuit $(x_1, x_2, \ldots, x_r)$.

LEMMA 2.7. Let G be a graph and let r be a positive integer. Let $(x_1, x_2, \ldots, x_{2r+1})$ be a circuit in G and let $C = \{x_1, x_2, \ldots, x_{2r+1}\}$. If $S \subseteq V(G)$ is independent in G then $|S \cap C| \leq r$.

PROOF. We have $2|S \cap C| = |S \cap \{x_{2r+1}, x_1\}| + \sum_{i=1}^{2r} |S \cap \{x_i, x_{i+1}\}|$. Since S is independent in G, $\{x_{2r+1}, x_1\} \in E(G)$ and $\{x_i, x_{i+1}\} \in E(G)$ for $1 \leq i \leq 2r$, each of the $2r+1$ terms on the r.h.s. of the above equation is at most one. Hence $2|S \cap C| \leq 2r+1$ so $|S \cap C| \leq r + \frac{1}{2}$. Now $|S \cap C|$ is an integer so $|S \cap C| \leq r$.

LEMMA 2.8. Let G be a graph and let r be a positive integer. Let $(x_1, x_2, \ldots, x_{2r+1})$ be a circuit in G. Let $C = \{x_1, x_2, \ldots, x_{2r+1}\}$, $\mathcal{P} = \{\{x\} \mid x \in V(G) - C\}$ and $H = G/\mathcal{P}$. Then $k(H) \leq k(G) - 1$.

PROOF. There is a set $C' \subseteq C$ such that C' is independent in G and $|C'| \geq \frac{1}{2}(|C| - k(G)) = \frac{1}{2}((2r+1) - k(G))$. By Lemma 2.7, $r \geq |C' \cap C| = |C'| \geq \frac{1}{2}(2r+1) - k(G))$ so $k(G) > 0$. Hence, $k(G) - 1 \geq 0$ so it suffices to show that if $S \subseteq V(H)$ then there is a set $S' \subseteq S$ such that S' is independent in H and $|S'| \geq \frac{1}{2}(|S| - (k(G) - 1))$.

Let $T = C \cup \{x \in V(G) - C \mid \{x\} \in S\}$. Then $|T| = |C| + |S| = |S| + 2r+1$. There is a set $T' \subseteq T$ such that T' is independent in G and $|T'| \geq \frac{1}{2}(|T| - k(G)) = \frac{1}{2}(|S| - (k(G) - 1)) + r$. By Lemma 2.7, $|T' \cap C| \leq r$. Hence,

$$|T' \dot{-} C| = |T'| - |T' \cap C| \geq \frac{1}{2}(|S| - (k(G) - 1)).$$

Let $S' = \{\{x\} \mid x \in T' - C\}$. Then $S' \subseteq S$, S' is independent in H and $|S'| = |T' - C| \geq \frac{1}{2}(|S| - (k(G) - 1))$.

3. PROOF OF THE THEOREM

In the notation of Section 2, the theorem asserts that $\chi(G) \leq k(G) + 2$ for every graph G. Suppose there is a graph G such that $\chi(G) \geq k(G) + 3$. Let G be such a graph and choose G with $|V(G)|$ as small as possible.

LEMMA 3.1. Let $\{b_1, b_2\} \in E(G)$. Let

$$A = \{a \in V(G) - \{b_1, b_2\} \mid \{a_1, b_1\}, \{a_2, b_2\} \in E(G)\}.$$

Let $\mathfrak{Z} = \{\{x\} \mid x \in V(G) - \{b_1, b_2\}\}$ and let $G_1 = G/\mathfrak{Z}$. Let $k = k(G)$. If C is a $(k+1)$-coloring of G_1 then for each integer i with $1 \leq i \leq k+1$ there is an $a \in A$ such that $C(\{a\}) = i$.

PROOF. Suppose there is a $(k+1)$-coloring C of G_1 and an integer i with $1 \leq i \leq k+1$ such that $C(\{a\}) \neq i$ for every $a \in A$. Without loss of generality we may assume that $i = k+1$. Hence, C is a $(k+1)$-coloring of G_1 and $C(\{a\}) \leq k$ for every $a \in A$.

Define a function $\hat{C}$ from $V(G)$ to the first $k+2$ positive integers as follows:

$$\hat{C}(b_1) = k+1 .$$

$$\hat{C}(b_2) = k+2 \quad \text{and}$$

for $x \in V(G) - \{b_1, b_2\}$,

$$\hat{C}(x) = \begin{cases} k+2 & \text{if } C(\{x\}) = k+1 \text{ and } \{x, b_1\} \in E(G) \\ C(\{x\}) & \text{otherwise.} \end{cases}$$

Now $\chi(G) \geq k(G)+3 = k+3$ so $\hat{C}$ cannot be a $(k+2)$-coloring of G. Hence, there is a set $\{x,y\} \in E(G)$ such that $\hat{C}(x) = \hat{C}(y)$.

Suppose that $x, y \in V(G) - \{b_1, b_2\}$. Then $\{x\}, \{y\} \in 2 = V(G_1)$ and $\{\{x\}, \{y\}\} \in E(G_1)$. Since C is a $(k+1)$-coloring of G_1, $C(\{x\}) \neq C(\{y\})$. Hence, either $\hat{C}(x) \neq C(\{x\})$ or $\hat{C}(y) \neq C(\{y\})$. We may as well assume that $\hat{C}(x) \neq C(\{x\})$. Then we must have $C(\{x\}) = k+1$ and $\hat{C}(x) = k+2$. But then $\hat{C}(y) = \hat{C}(x) = k+2$. Since $y \in V(G) - \{b_1, b_2\}$, $\hat{C}(y) = k+2$ implies that $C(\{y\}) = k+1 = C(\{x\})$. This is a contradiction. Hence, $\{x, y\}$ is not a subset of $V(G) - \{b_1, b_2\}$.

Now suppose that $b_2 \in \{x, y\}$ say $b_2 = x$. Then $\{b_2, y\} \in E(G)$ and $\hat{C}(y) = \hat{C}(b_2) = k+2$. Now $y \neq b_2$ because $\{b_2, y\} \in E(G)$. Also, $y \neq b_1$ because $\hat{C}(b_1) = k+1 \neq k+2 = \hat{C}(y)$. Hence, $y \in V(G) - \{b_1, b_2\}$. Now $\hat{C}(y) = k+2$ so we must have $C(\{y\}) = k+1$ and $\{y, b_1\} \in E(G)$. Hence, $y \in A$ so $k+1 = C(\{y\}) \leq k$. This is a contradiction.

We have now shown that $b_2 \notin \{x, y\}$ and $\{x, y\}$ is not a subset of $V(G) - \{b_1, b_2\}$. The only remaining possibility is that $\{x, y\} = \{b_1, z\}$ where $z \in V(G) - \{b_1, b_2\}$. Since $z \in V(G) - \{b_1, b_2\}$, either $\hat{C}(z) = k+2$ or $\hat{C}(z) = C(\{z\})$. Now $\hat{C}(z) = \hat{C}(b_1) = k+1 \neq k+2$ so $\hat{C}(z) = C(\{z\})$. Hence, we have $\{z, b_1\} \in E(G)$ and $C(\{z\}) = \hat{C}(z) = \hat{C}(b_1) = k+1$. From the definition of $\hat{C}$, this implies that $\hat{C}(z) = k+2 \neq k+1 = \hat{C}(b_1)$. This final contradiction shows that our assumption that the lemma did not hold was false.

LEMMA 3.2. Let b_1, b_2 and A be as in Lemma 3.1. If $a_1 a_2 \in A$

and $a_1 \neq a_2$ then $\{a_1, a_2\} \in E(G)$.

PROOF. Suppose there are distinct elements $a_1, a_2 \in A$ such that $\{a_1, a_2\} \notin E(G)$. Let $r = 1$ and let $B = \{b_1, b_2\}$. Then B is a clique in G and $|B| = 2 = r+1$. Since $a_1, a_2 \in A$, $\{a_1, b\}, \{a_2, b\} \in E(G)$ for every $b \in B$. Let

$$\mathcal{P} = \{\{a_1, a_2\}\} \cup \{\{x\} \mid x \in V(G) - B - \{a_1, a_2\}\} \quad \text{and let } H = G/\mathcal{P}$$

By Lemma 2.1, $k(H) \leq k(G) - r = k(G) - 1$. Now

$|V(H)| = 1 + |V(G)| - 4 < |V(G)|$. Hence, $\chi(H) \leq k(H) + 2 \leq k(G) + 1$.

Let $k = k(G)$. Then there is a $(k+1)$-coloring C of H.

Let $\mathcal{Z}$ and G_1 be as in Lemma 3.1. Define a function $\hat{C}$ from $\mathcal{Z} = V(G_1)$ to the first $k+1$ positive integers by $\hat{C}(\{a_1\}) = \hat{C}(\{a_2\}) = $
$= C(\{a_1, a_2\})$ and

$$\hat{C}(\{x\}) = C(\{x\}) \quad \text{for } x \in V(G) - \{b_1, b_2, a_1, a_2\}.$$

Since $\{a_1, a_2\} \notin E(G)$ and C is a $(k+1)$-coloring of H, it is easily verified that $\hat{C}$ is a $(k+1)$-coloring of G_1. By Lemma 3.1, for each integer i with $1 \leq i \leq k+1$ there is an $a \in A$ such that $\hat{C}(\{a\}) = i$. Hence, $|A| \geq k+1 = k(G) - r + 2$.

Now let $\mathcal{R} = \{B\} \cup \{\{x\} \mid x \in V(G) - B\}$ and let $L = G/\!/\mathcal{R}$. By Lemma 2.5, $k(L) \leq k(G) - r = k(G) - 1$. We have $|V(L)| = |\mathcal{R}| =$ $1 + |V(G)| - 2 < |V(G)|$. Hence $\chi(L) \leq k(L) + 2 \leq k(G) + 1 = k+1$. Let C be a $(k+1)$-coloring of L. Now $V(G_1) = \mathcal{Z} \subseteq \mathcal{R} = V(L)$. Furthermore, $E(G_1) \subseteq E(L)$ so the restriction of C to $V(G_1)$ is a $(k+1)$-coloring of G_1. We have $1 \leq C(B) \leq k+1$ and $C(B)$ is an integer. By Lemma 3.1 there is an $a \in A$ such that $C(\{a\}) = C(B)$. But $\{\{a\}, B\} \in E(L)$ because $a \in A$ so this contradicts the fact that C is a $(k+1)$-coloring of L.

LEMMA 3.3. Let r be an integer with $r \geq 2$

Let $a_1, a_2, \ldots, a_r, b_1, b_2, \ldots, b_r$ be $2r$ distinct elements of $V(G)$ such that

$\{a_i, b_i\} \in E(G)$ for $1 \leq i \leq r$,

$\{b_i, a_{i+1}\} \in E(G)$ for $1 \leq i \leq r-1$, and

$\{b_r, a_1\} \in E(G)$.

Then there are integers i and j with $1 \leq i < j \leq r$ such that either $\{a_i, a_j\} \in E(G)$ or $\{b_i, b_j\} \in E(G)$.

PROOF. Let $A = \{a_1, a_2, \ldots, a_r\}$ and $B = \{b_1, b_2, \ldots, b_r\}$. Let $\mathcal{P} = \{A, B\} \cup \{\{x\} \mid x \in V(G) - A - B\}$ and let $H = G/\mathcal{P}$. By Lemma 2.3, $k(H) \leq k(G)$. Now $|V(H)| = |\mathcal{P}| = 2 + |V(G)| - 2r < |V(G)|$ so $\chi(H) \leq k(H) + 2 \leq k(G) + 2$. Let $k = k(G)$ and let C be a $(k+2)$-coloring of H.

For each $x \in V(G)$ there is a unique set $S \in \mathcal{P}$ such that $x \in S$. For $x \in V(G)$ let $\hat{C}(x) = C(S)$ where $S \in \mathcal{P} = V(H)$ is the unique set such that $x \in S$. Now $\hat{C}$ is a function from $V(G)$ to the first $k+2$ positive integers. Since $\chi(G) \geq k(G) + 3 = k+3 > k+2$, $\hat{C}$ cannot be a $(k+2)$-coloring of G. Hence, there are elements $x, y \in V(G)$ such that $\{x, y\} \in E(G)$ and $\hat{C}(x) = \hat{C}(y)$. Let $S, T \in \mathcal{P}$ be the sets such that $x \in S$ and $y \in T$. Then $C(S) = \hat{C}(x) = \hat{C}(y) = C(T)$. Since C is a $(k+2)$-coloring of H, $\{S, T\} \notin E(H)$. Now $E(H) = \{\{S, T\} \mid S, T \in \mathcal{P}$ there are elements $x \in S$ and $y \in T$ such that $\{x, y\} \in E(G)$ and $S \neq T\}$. We have $S, T \in \mathcal{P}$ and there are elements $x \in S$ and $y \in T$ such that $\{x, y\} \in E(G)$. Hence, since $\{S, T\} \notin E(H)$, we must have $S = T$. Therefore, $\{x, y\} \subseteq S = T$. Since $\{x, y\} \in E(G)$, $2 = |\{x, y\}| \leq |S| = |T|$. Now A and B are the only sets in $\mathcal{P}$ containing more than one element so either $S = T = A$ or $S = T = B$. Hence, either A or B contains the two element set $\{x, y\}$ which is a member of $E(G)$.

LEMMA 3.4. Let $B \subseteq V(G)$ be a maximal clique in G. If $x \in V(G) - B$ then there is at most one element $b \in B$ such that $\{x, b\} \in E(G)$.

PROOF. Let $x \in V(G) - B$ and suppose there are elements $b_1, b_2 \in B$

such that $b_1 \neq b_2$ and $\{x, b_1\}, \{x, b_2\} \in E(G)$. Let $A = \{a \in V(G) - \{b_1, b_2\} \mid \{a, b_1\}, \{a, b_2\} \in E(G)\}$. Now $\{b_1, b_2\} \in E(G)$ because b_1 and b_2 are distinct elements of the clique B. By Lemma 3.2, $\{a_1, a_2\} \in E(G)$ for any $a_1, a_2 \in A$ with $a_1 \neq a_2$. We also have $\{b_1, b_2\} \in E(G)$ and $\{a, b_1\}, \{a, b_2\} \in E(G)$ for every $a \in A$ by the definition of A so $A \cup \{b_1, b_2\}$ is a clique in G. Clearly, B is a subset of $A \cup \{b_1, b_2\}$. Furthermore, $x \notin B$ but $x \in A \subseteq A \cup \{b_1, b_2\}$ so B is a proper subset $A \cup \{b_1, b_2\}$. This contradicts the maximality of B.

LEMMA 3.5. Let $B \subseteq V(G)$ be a maximal clique in G. Let $u, v \in B$ and $x, y \in V(G) - B$ and suppose that $u \neq v$, $x \neq y$ and $\{x, u\}, \{y, v\} \in E(G)$. Then $\{x, y\} \notin E(G)$.

PROOF. By Lemma 3.4, $\{x, v\}, \{y, u\} \notin E(G)$. Suppose $\{x, y\} \in E(G)$. Let $a_1 = x$, $a_2 = v$, $b_1 = u$ and $b_2 = y$. Then a_1, a_2, b_1, b_2 are four distinct elements of $V(G)$. By hypothesis, $\{a_1, b_1\} = \{x, u\} \in E(G)$ and $\{a_2, b_2\} = \{v, y\} \in E(G)$. Since u and v are distinct elements of the clique B, $\{b_1, a_2\} = \{u, v\} \in E(G)$. By assumption, $\{b_2, a_1\} = \{y, x\} \in E(G)$. However, $\{a_1, a_2\} = \{x, v\} \notin E(G)$ and $\{b_1, b_2\} = \{u, y\} \notin E(G)$. This contradicts Lemma 3.3. Hence, $\{x, y\} \notin E(G)$.

LEMMA 3.6. Every clique in G contains at most two elements.

PROOF. Suppose there is a clique in G containing three or more elements. Every clique is a subset of a maximal clique so it follows that there is a positive integer r and a set $B \subseteq V(G)$ such that B is a maximal clique in G and $|B| = r+2$. Let $\mathcal{P} = \{\{x\} \mid x \in V(G) - B\}$ and let $H = G/\mathcal{P}$. By Lemma 2.6, $k(H) \leq k(G) - r$. Now $|V(H)| = |\mathcal{P}| = |V(G)| - r - 2 < |V(G)|$ so $\chi(H) \leq k(H) + 2 \leq k(G) - r + 2$.
Let $k = k(G)$ and let C be a $(k-r+2)$-coloring of H.

Let $B = \{b_1, b_2, \ldots, b_{r+2}\}$. Define a function $\hat{C}$ from $V(G)$ to the first $k+2$ positive integers as follows:

Let $\hat{C}(b_i) = k - r + i$ for $1 \leq i \leq r+2$. If $x \in V(G) - B$ let $\hat{C}(x) = k+2$

if $C(\{x\}) = k-r+1$ and $\{x,b_1\} \in E(G)$ or if $C(\{x\}) = k-r+2$ and $\{x,b_2\} \in E(G)$ and let $\hat{C}(x) = C(\{x\})$ otherwise.

Now $\chi(G) \geq k(G)+3 = k+3 > k+2$ so $\hat{C}$ cannot be a $(k+2)$-coloring of G. Hence, there is a set $\{x,y\} \in E(G)$ such that $\hat{C}(x) = \hat{C}(y)$. Suppose $x,y \in V(G)-B$. Then $\{\{x\},\{y\}\} \in E(H)$ so $C(\{x\}) \neq C(\{y\})$ and $C(\{x\}), C(\{y\}) \leq k-r+2 < k+2$ since C is a $(k-r+2)$-coloring of H. Hence, the integers $C(\{x\})$, $C(\{y\})$ and $k+2$ are distinct. Now $\hat{C}(x) = C(\{x\})$ or $k+2$ and $\hat{C}(y) = C(\{y\})$ or $k+2$ so we must have $\hat{C}(x) = \hat{C}(y) = k+2$. From the definition of $\hat{C}$, each of the integers $C(\{x\})$ and $C(\{y\})$ is either $k-r+1$ or $k-r+2$. Since $C(\{x\}) \neq C(\{y\})$ we may, by interchanging x and y if necessary, assume that $C(\{x\}) = k-r+1$ and $C(\{y\}) = k-r+2$. Since $\hat{C}(x) = \hat{C}(y) = k+2$ we must have $\{x,b_1\}, \{y,b_2\} \in E(G)$. Now $b_1 \neq b_2$ and $x \neq y$. By Lemma 3.5, $\{x,y\} \notin E(G)$. This is a contradiction.

Now suppose that $x,y \in B$. Then there are integers i and j with $1 \leq i,j \leq r+2$ such that $x = b_i$ and $y = b_j$. Hence, $k-r+i = \hat{C}(b_i) = \hat{C}(x) = \hat{C}(y) = \hat{C}(b_j) = k-r+j$ so $x = b_i = b_j = y$. This contradicts the fact that $\{x,y\} \in E(G)$.

We have now shown that $\{x,y\}$ is not a subset of $V(G)-B$ or of B. The only remaining possiblity is that $\{x,y\} = \{b_i,z\}$ where $z \in V(G)-B$ and i is an integer with $1 \leq i \leq r+2$. If $i = r+2$ then $i \neq 1,2$ so $b_i \neq b_1$ or b_2. Now $z \in V(G)-B$ and $\{z,b_i\} \in E(G)$. By Lemma 3.4, $\{z,b_1\},\{z,b_2\} \notin E(G)$. Hence, $\hat{C}(z) = C(\{z\}) \leq k-r+2 < k+2 = k-r+r+2 = \hat{C}(b_{r+2}) = \hat{C}(b_i)$. This is a contradiction so $i<r+2$. Hence, $\hat{C}(z) = \hat{C}(b_i) = k-r+i < k-r+r+2 = k+2$ so $\hat{C}(z) = C(\{z\})$. Now $k-r+i = \hat{C}(b_i) = \hat{C}(z) = C(\{z\}) \leq k-r+2$ so $i \leq 2$. If $i=2$ we have $C(\{z\}) = \hat{C}(z) = \hat{C}(b_2) = k-r+2$ and $\{z,b_2\} \in E(G)$ so $\hat{C}(z) = k+2 > k-r+2 = k-r+i = \hat{C}(b_i)$. This is a contradiction so we must have $i=1$. But then $C(\{z\}) = \hat{C}(z) = \hat{C}(b_1) = k-r+1$ and $\{z,b_1\} \in E(G)$ so again we have the contradiction $\hat{C}(z) = k+2 > k-r+1 = k-r+i = \hat{C}(b_i)$. This completes the proof of the lemma.

Since $\chi(G) \geq k(G)+3 \geq 3 > 2$, the graph G is not bipartite.

Hence, G contains a circuit of odd length so, in particular, G contains a circuit. Let r be the smallest integer such that G contains a circuit of length r and let $(x_1, x_2, \ldots, x_r)$ be a circuit of length r in G. Let $C = \{x_1, x_2, \ldots, x_r\}$.

LEMMA 3.7. If $1 \le i < j \le r$ and $\{x_i, x_j\} \in E(G)$ then either $i = 1$ and $j = r$ or $j = i+1$. The integer r is odd and $r \ge 5$.

PROOF. Let $1 \le i < j \le r$ and suppose that $\{x_i, x_j\} \in E(G)$. The sequence $(x_1, x_2, \ldots, x_i, x_j, x_{j+1}, \ldots, x_r)$ contains $r-j+i+1$ terms. If $r-j+i+1 \le 2$ then $j \ge i+r-1$. Since $i \ge 1$ and $j \le r$ this implies that $i = 1$ and $j = r$. On the other hand, if $r-j+i+1 \ge 3$ then $(x_1, x_2, \ldots, x_i, x_j, x_{j+1}, \ldots, x_r)$ is a circuit in G of length $r-j+i+1$. By the minimality of r, $r-j+i+1 \ge r$ so $j \le i+1$. We also have $i < j$ so $j = i+1$.

By the definition of a circuit, $r \ge 3$. If it were true that $r = 3$, then $\{x_1, x_2, x_3\}$ would be a clique in G containing more than two elements. This would contradict Lemma 3.6 so $r \ge 4$. Suppose that r is even. Then $r = 2s$ where s is an integer with $s \ge 2$. For $1 \le i \le s$ let $a_i = x_{2i-1}$ and let $b_i = x_{2i}$. By Lemma 3.3 there are integers i and j with $1 \le i < j \le s$ such that either

$$\{x_{2i-1}, x_{2j-1}\} = \{a_i, a_j\} \in E(G) \quad \text{or} \quad \{x_{2i}, x_{2j}\} = \{b_i, b_j\} \in E(G).$$

Now $r = 2s \not\equiv 1 \pmod 2$ and $i \not\equiv i+1 \pmod 2$ for any integer i. Hence, by the first part of this lemma, if $1 \le k, \ell \le r$ and $\{x_k, x_\ell\} \in E(G)$, then $k \not\equiv \ell \pmod 2$. But $2i-1 \equiv 2j-1 \pmod 2$ and $2i \equiv 2j \pmod 2$ so we have a contradiction.

Therefore, r is odd. Since $r \ge 4$, we must have $r \ge 5$.

LEMMA 3.8. Let $u \in V(G) - C$. There is at most one element $x \in C$ such that $\{u, x\} \in E(G)$.

PROOF. Suppose not. By cyclically relabeling the elements of C we may assume that $\{u, x_1\}, \{u, x_i\} \in E(G)$ where i is an integer with $2 \le i \le r$. The sequences $(u, x_1, x_2, \ldots, x_i)$ and $(x_1, u, x_i, x_{i+1}, \ldots, x_r)$

are both circuits in G. They have lengths $i+1$ and $r-i+3$, respectively. By the minimality of r, $i+1 \geq r$ and $r-i+3 \geq r$. Hence, $3 \geq i \geq r-1$ so $4 \geq r$. This contradicts Lemma 3.7.

LEMMA 3.9. Let $u,v \in V(G)-C$ with $u \neq v$. Suppose that there are integers i and j such that $1 \leq i < j \leq r$ and $\{u,x_i\},\{v,x_j\} \in E(G)$. Then $\{u,v\} \notin E(G)$.

PROOF. Suppose $\{u,v\} \in E(G)$. Then the sequences $\{u,x_i,x_{i+1},\dots,x_j,v\}$ and $\{x_1,x_2,\dots,x_i,u,v,x_j,x_{j+1},\dots,x_r\}$ are both circuits in G. They have lengths $\ell_1 = j-i+3$ and $\ell_2 = r+i-j+3$, respectively. By the minimality of r, $\ell_1,\ell_2 \geq r$. Therefore $r+6 = j-i+3+r+i-j+3 = \ell_1+\ell_2 \geq 2r$ so $6 \geq r$. By Lemma 3.7, $r = 5$. Hence, $\ell_1,\ell_2 \geq r = 5$ and $\ell_1+\ell_2 = r+6 = 11$ so either $\ell_1 = 6$ or $\ell_2 = 6$. Therefore, G contains a circuit $(y_1,y_2,y_3,y_4,y_5,y_6)$ of length 6. For $1 \leq i \leq 3$, let $a_i = y_{2i-1}$ and $b_i = y_{2i}$. Note that if $1 \leq i < j \leq 3$ then there is an integer k such that $1 \leq k \leq 3$ and $\{a_i,b_k\},\{a_j,b_k\} \in E(G)$ and there is an integer ℓ such that $1 \leq \ell \leq 3$ and $\{b_i,a_\ell\},\{b_j,a_\ell\} \in E(G)$. Hence, by Lemma 3.6, $\{a_i,a_j\},\{b_i,b_j\} \notin E(G)$. This contradicts Lemma 3.3 and completes the proof of Lemma 3.9.

Let $\mathcal{P} = \{\{x\} \mid x \in V(G)-C\}$ and let $H = G/\mathcal{P}$. By Lemmas 3.7 and 2.8, $k(H) \leq k(G)-1$. Let $k = k(G)$. We have $|V(H)| = |\mathcal{P}| = |V(G)|-r < |V(G)|$ so $\chi(H) \leq k(H)+2 \leq k(G)+1 = k+1$.
Let C be a $(k+1)$-coloring of H. Define a function $\hat{C}$ from $V(G)$ to the first $k+2$ positive integers as follows:

$$\hat{C}(x_r) = k+2,$$

$$\text{for} \quad 1 \leq i \leq r-1, \quad \hat{C}(x_i) = \begin{cases} k & \text{if } i \text{ is odd} \\ k+1 & \text{if } i \text{ is even.} \end{cases}$$

For $x \in V(G)-C$, let $\hat{C}(x) = k+2$ if there is an integer i such that $1 \leq i \leq r-1$ and $\{x,x_i\} \in E(G)$ and let $\hat{C}(x) = C(\{x\})$ otherwise.

We claim that $\hat{C}$ is a $(k+2)$-coloring of G. To prove this it suffices to show that if $\{x,y\} \in E(G)$ then $\hat{C}(x) \neq \hat{C}(y)$. We consider three cases:

Case 1. $x,y \in C$. If $\{x,y\} = \{x_i, x_r\}$ where $i = 1$ or $i = r-1$ then $\hat{C}(x_r) = k+2 > \hat{C}(x_i)$. By Lemma 3.7 the only other possibility is that $\{x,y\} = \{x_i, x_{i+1}\}$ where i is an integer with $1 \leq i \leq r-2$. If i is odd then $i+1$ is even so $\hat{C}(x_i) = k \neq k+1 = \hat{C}(x_{i+1})$. If i is even then $i+1$ is odd so $\hat{C}(x_i) = k+1 \neq k = \hat{C}(x_{i+1})$.

Case 2. $\{x,y\} = \{x_i, z\}$ where $z \in V(G)-C$ and i is an integer with $1 \leq i \leq r$. If $i \leq r-1$ then $\hat{C}(z) = k+2 > \hat{C}(x_i)$. If $i = r$ then, by Lemma 3.8, $\{z, x_j\} \notin E(G)$ for $1 \leq j \leq r-1$ so $\hat{C}(z) = C(\{z\}) \leq k+1 <$ $< k+2 = \hat{C}(x_i)$.

Case 3. $x,y \in V(G)-C$. Then $\{\{x\},\{y\}\} \in E(H)$ so $C(\{x\}) \neq C(\{y\})$. Hence, $\hat{C}(x) \neq \hat{C}(y)$ if $\hat{C}(x) = C(\{x\})$ and $\hat{C}(y) = C(\{y\})$. If $\hat{C}(x) \neq C(\{x\})$ then $\hat{C}(x) = k+2$ and $\{x, x_i\} \in E(G)$ for some integer i with $1 \leq i \leq r-1$. By Lemma 3.6, $\{y, x_i\} \notin E(G)$. By Lemma 3.9, $\{y, x_j\} \in E(G)$ for every integer j with $1 \leq j \leq r$ and $j \neq i$. Hence, $\hat{C}(y) = C(\{y\}) \leq k+1 < k+2 = \hat{C}(x)$. Similarly, $\hat{C}(y) \neq C(\{y\})$ implies that $\hat{C}(x) \neq \hat{C}(y)$.

We have now shown that $\chi(G) \leq k+2 = k(G)+2 < k(G)+3$, contradicting our original assumption that there was a graph G with $\chi(G) \geq k(G)+3$.

REFERENCES

[1] ERDŐS, P., and A. HAJNAL, Problem 3, p. 362, Theory of Graphs; Proceedings of the Colloquium held at Tihany, Hungary, September 1966, edited by P. Erdős and G. Katona, Akadémiai Kiadó, Budapest, 1968.

The allocation of power resources on networks

by

Z. Fónyad, Gy. Gyimesi and M. Hosszú

Miskolc, Hungary

NOTATIONS

i :	vertex, event
(i,j) :	edge, activity
t_{ij} :	activity duration time
$t_{ij} = \varepsilon \rightarrow 0$:	dummy activity
M_{ij} :	power
$\bigcirc, \square$:	design of power resources
$c_{ij} = M_{ij}/t_{ij}$:	loading
t° :	earliest start time
t^{1} :	latest finish time
T :	slack
i_j :	visual event (finish of (i,j))

Let us consider a total project. Suppose that the set of all actions can be decomposed into invidual activities. Figure 1 shows such a network.

Each vertex i is called an event $(i = 1, 2, \dots, n)$ and each (vector) edge (i, j) represents an activity. The quantities associated with edges in Fig. 1 represent activity duration times t_{ij}.

There is a 1 -1 correspondence between networks and matrices $[t_{ij}]$.

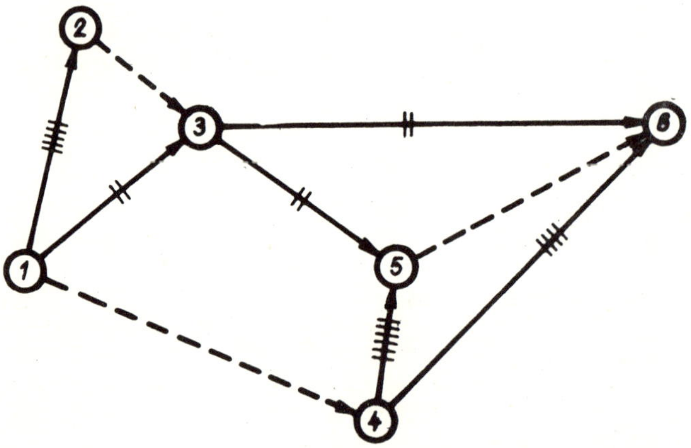

Fig. 1.

If value of t_{ij} is ε $(0 < \varepsilon << 1)$, the activity is called dummy activity. This does not need any effective time but determines the order of activities. The dummy activity is noted by a broken line on the network.

$[t_{ij}] =$

i \ j	1	2	3	4	5	6
1	0	3	2	ε	0	0
2	0	0	ε	0	0	0
3	0	0	0	0	2	2
4	0	0	0	0	6	4
5	0	0	0	0	0	ε
6	0	0	0	0	0	0

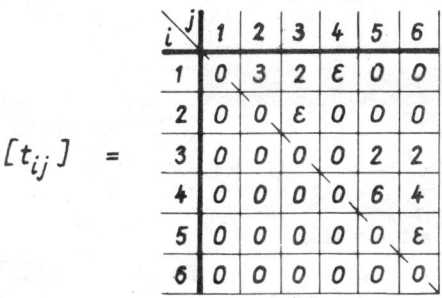

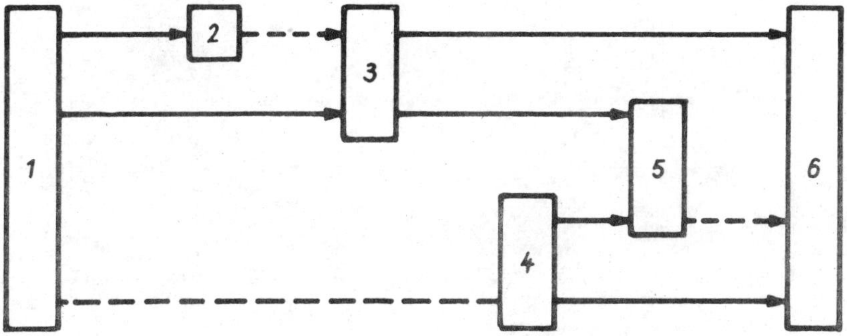

Fig. 2.

Note that by its nature a directed activity network is acyclic. By changing the indices we can attain that the edges tend from a vertex of lower index to a vertex of a greater one. Then the matrix t_{ij} is an upper triangular matrix.

The advantage of the above-mentioned network is that the order which is suitable for the applied technology can be seen well on the network. This network is useful for leaders and managers who operate all the activities. For manufacturers there is a more useful time - activity (Gant) diagram. This is a cartesian system of coordinates where the time is shown on the horizontal axis and the activites on a perpendicular one.

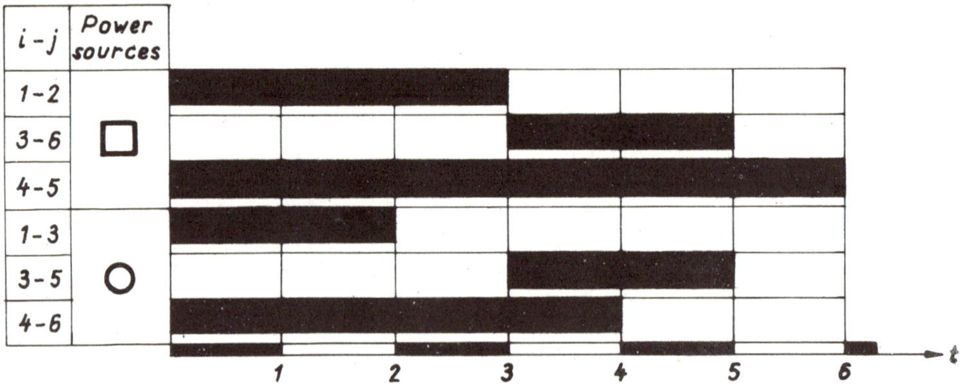

Fig. 3.

However, it is expedient for designers to use a time - event diagram.

Starting from the initial event (events) the time - event diagram can be constructed on the following principles:

1) Time is indicated on the horizontal axis and the single events on the perpendicular one.

2) The activites are represented by directed edges. Every edge starts from a point showing the initial time of the respective activity and the event after which it begins, while the endpoint of an edge shows the finish time of the respective activity and the resulting event for which this activity is necessary.

3) Starting from a point and using 2), we draw all edges which tend from this point to another, further we continue this method from a following point in the increasing order of indices.

4) Horizontal lines denote slacks. They occur when two or more activities are necessary for an event i and they are finished at different times. The latest time is the earliest time of the event i.

5) Drawing all the activities, the series of critical activities can be denoted. This contains activities which have no slacks. On the time - event diagram the critical path has no horizontal line.

The above time - event diagram can be easily drawn in the case of about 200 events and 1000 activities, too.

With the help of the time - event diagram the time - activity diagram can be easily made in the case of more non-concurrent power resources, too.

With the help of the diagram, the following changes can be carried out.

1) The shortening of slacks by the lengthening of activity durations.

2) The shortening of the total critical time.

3) Ensuring a smooth loading $\dot{c} = M/t$. A suitable allocation of power resources among activities.

4) The decomposition of the total network into subnetworks. The optimization of the loading, etc.

Fig. 5 shows the time - event diagram of our example. Fig. 6 shows a practical rearrangement of the power sources, which gives a more suitable loading. This can be seen by comparing the time - loading diagrams of Fig. 4. resp. 7.

The results of the optimization are:

1) The critical time is shortened.

2) The highest value of the loading decreased.

3) The loading became more suitable.

For a detailed mathematical treatement, it is useful to introduce the following notations:

Let t°_{ij} denote the earliest start time of the activity (i,j). It is at the same time the earliest finish time of the event i. Let t^{1}_{ij} be the latest finish time of the activity (i,j), which does not array the earliest beginning of other activities. The slack of the activity is

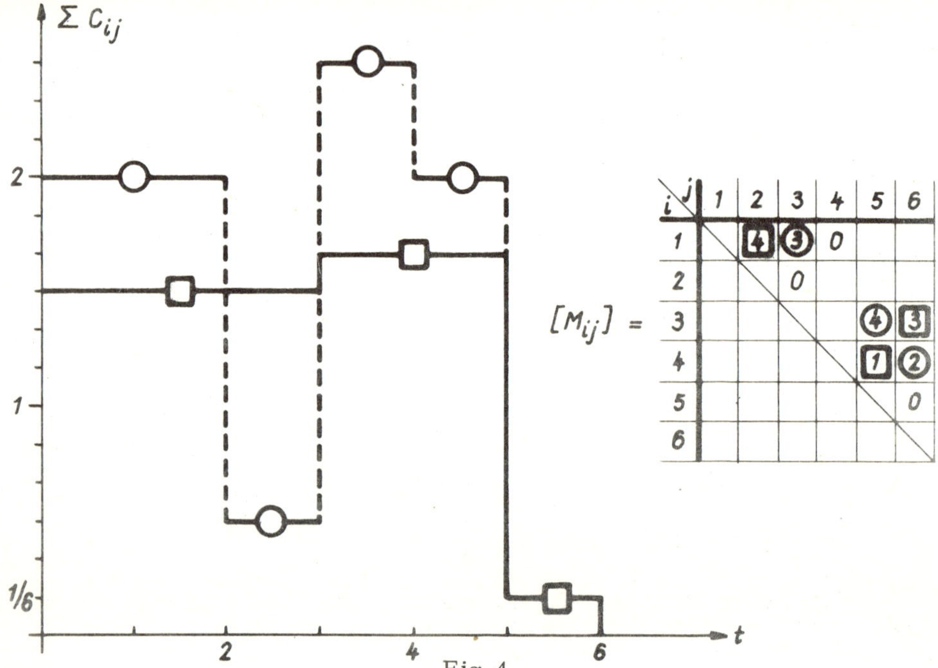
Σ c_{ij}
2
1
1/6
2
4
6
t
$[M_{ij}]$ =
j
i
1 2 3 4 5 6
1 4 3 0
2 0
3 4 3
4 1 2
5 0
6

Fig. 4.

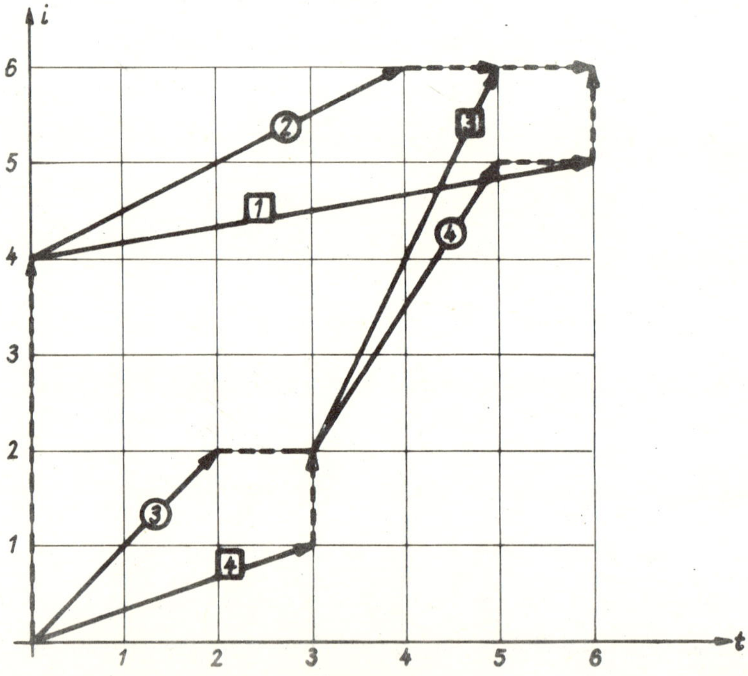
i
6
5
4
3
2
1
2
1
4
3
3
4
1 2 3 4 5 6
t

Fig. 5.

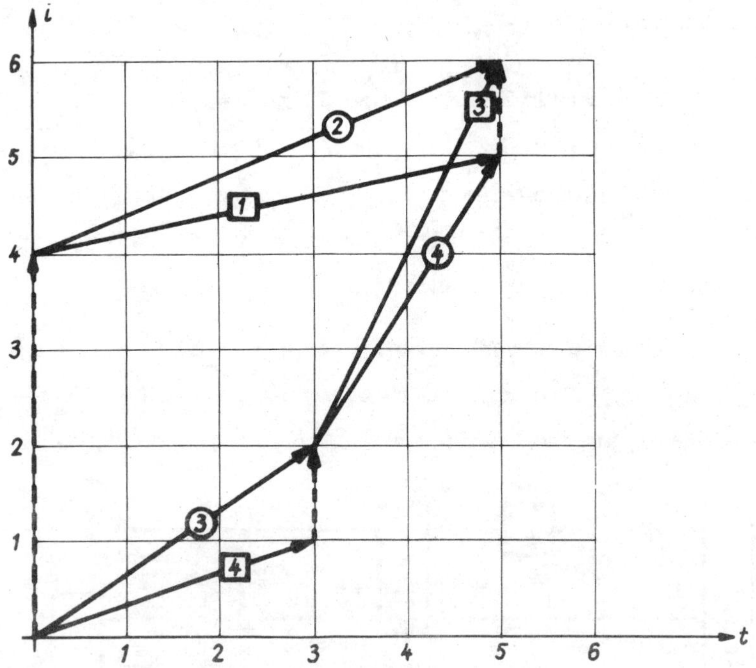

Fig. 6.

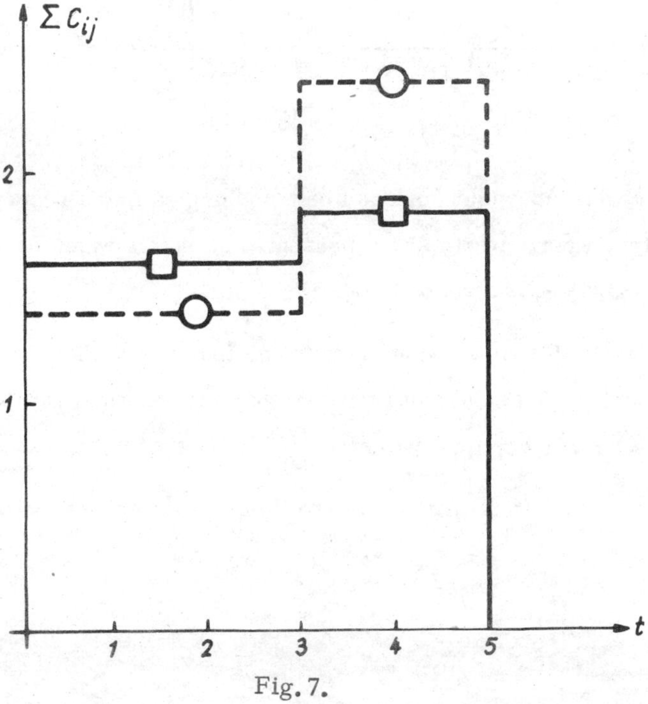

Fig. 7.

$$T_{ij} = t^1_{ij} - t_{ij} - t^o_{ij} \ .$$

Each activity (i,j) can be considered as the sum of an acitivity duration t_{ij} and a forced waiting time T_{ij} (Fig. 8). According to this, a vertex P^i_j can be set in the edge (i,j) before the vertex j. P^i_j shows the finish of the activity (i,j). In short it can be called visual event i_j, and we can speak about its earliest occurence $t^o_{i_j j}$.

Thus the network in Fig. 1 resp. 2. can be completed with vertices P^i_j and edges of T_{ij} value. The new edges obtained in this way can be distinguished by the indices (i, i_j) and (i_j, j) respectively (Fig. 8).

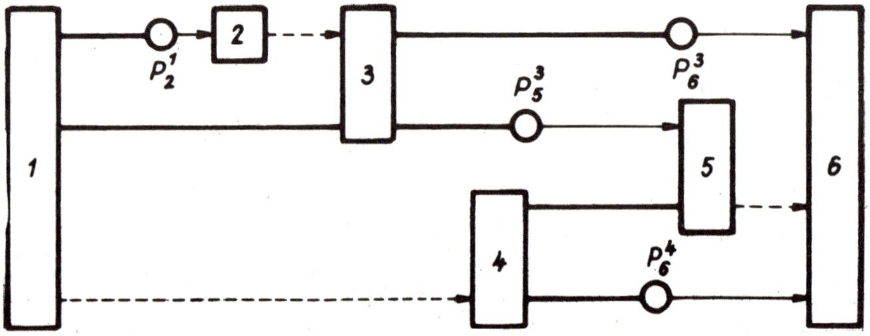

Fig. 8.

The above network can be transformed into a rectangular system of coordinates, where events are represented as points which have coordinates (t_{ij}, i) and (t_{ij}, j), respectively (Fig. 5).

On this time-event diagram we find a certain series of activites without slacks. This series of activities gives the critical path. In general, there may be more critical paths.

The critical paths give the minimal finishing time of the total set of events.

On homogeneous tournaments

by

E. Fried

Budapest, Hungary

Let $\mathcal{G}$ be a directed graph, i.e. a pair of non-empty sets $(\mathcal{P}(\mathcal{G}), \mathcal{E}(\mathcal{G}))$, where $\mathcal{E}(\mathcal{G})$ is a subset of the cartesian product $\mathcal{P}(\mathcal{G}) \times \mathcal{P}(\mathcal{G})$. The elements of $\mathcal{P}(\mathcal{G})$ and $\mathcal{E}(\mathcal{G})$ are called the points and the edges of the graph $\mathcal{G}$, respectively. A permutation of the set $\mathcal{P}(\mathcal{G})$ is called an automorphism of $\mathcal{G}$ if it maps $\mathcal{E}(\mathcal{G})$ onto itself. The automorphisms of $\mathcal{G}$ form, clearly, a subgroup, denoted by $\mathcal{A}(\mathcal{G})$, of the group of permutations of $\mathcal{P}(\mathcal{G})$. The element a^{σ} will denote the image of $a \in \mathcal{P}(\mathcal{G})$ under the automorphism $\sigma \in \mathcal{A}(\mathcal{G})$. The directed graphs $\mathcal{G}$ and $\mathcal{G}'$ are called similar to each other if there is a one-to-one mapping $\varphi : \mathcal{P}(\mathcal{G}) \rightarrow \mathcal{P}(\mathcal{G}')$ such that $(\varphi(a), \varphi(b)) \in \mathcal{E}(\mathcal{G}')$ if and only if $(a,b) \in \mathcal{E}(\mathcal{G})$. Let, further on, $\mathcal{G}$ be a tournament which means that, for any pair of elements a,b of $\mathcal{P}(\mathcal{G})$, exactly one of the relations

$$a = b \qquad (a,b) \in \mathcal{E}(\mathcal{G}) \qquad (b,a) \in \mathcal{E}(\mathcal{G})$$

holds.

Let $\mathcal{A}$ be a group of permutations of the set $\mathcal{P}$. If, to any pair of elements a,b of $\mathcal{P}$, a permutation $\sigma \in \mathcal{A}$ for which $a^{\sigma} = b$ is fulfilled, exists then one says that $\mathfrak{A}$ acts transitively on $\mathfrak{S}$. Let, for $a \in \mathfrak{S}$, denote by $\mathfrak{A}_a$ the subgroup of $\mathfrak{A}$ consisting of the elements which map $\underline{a}$ into itself. For a group $\mathfrak{A}$ which acts transitively on $\mathfrak{S}$, the condition that to any pair of elements a,b of $\mathfrak{S}$ there exists, exactly one permutation $\sigma \in \mathfrak{A}$, which maps $\underline{a}$ to $\underline{b}$, is clearly equivalent to the proposition that for any element $a \in \mathfrak{A}$ the unity is the only element of $\mathfrak{A}_a$. Of such a group we say that it acts 1-transitively on $\mathfrak{S}$.

The torunament $\mathfrak{G}$ will be called:

a) homogeneous, if $\mathfrak{A}(\mathfrak{G})$ acts transitively on $\mathfrak{E}(\mathfrak{G})$

b) 1-homogeneous, if $\mathfrak{A}(\mathfrak{G})$ has a subgroup $\mathfrak{B}(\mathfrak{G})$ acting 1-transitively on $\mathfrak{E}(\mathfrak{G})$

c) stiff (and homogeneous), if $\mathfrak{A}(\mathfrak{G})$ itself acts transitively on $\mathfrak{E}(\mathfrak{G})$.

Clearly, a stiff tournament is 1-homogeneous and a 1-homogeneous tournament is a homogeneous one.

Finally, let $|\mathfrak{S}|$ denote the number of elements of the finite set $\mathfrak{S}$.

PROPOSITION 1: If $|\mathfrak{S}(\mathfrak{G})| > 2$ and $\mathfrak{A}$ is a subgroup of $\mathfrak{A}(\mathfrak{G})$, acting transitively on $\mathfrak{E}(\mathfrak{G})$, then $\mathfrak{A}$ also acts transitively on $\mathfrak{S}(\mathfrak{G})$.

PROOF: Let a,b,c be distinct elements of $\mathfrak{S}(\mathfrak{G})$ and, for example, $(a,b) \in \mathfrak{E}(\mathfrak{G})$. If either (c,a) or (b,c) belongs to $\mathfrak{E}(\mathfrak{G})$ then there is either an automorphism σ in $\mathfrak{A}$, sending (c,a) to (a,b), or an automorphism ϱ in $\mathfrak{A}$, sending (a,b) to (b,c), since $\mathfrak{A}$ acts transitively on $\mathfrak{E}(\mathfrak{G})$. In these cases either $a^\sigma = b$ or $a^\varrho = b$ is valid, respectively. If neither (c,a) nor (b,c) belongs to $\mathfrak{E}(\mathfrak{G})$, then, because $\mathfrak{G}$ is a tournament, both (a,c) and (c,b) are elements of $\mathfrak{E}(\mathfrak{G})$. Hence, the automorphism $\tau \in \mathfrak{A}$, sending (c,b) to (a,b), maps (a,c) also, to an element of $\mathfrak{E}(\mathfrak{G})$, i.e. $(a^\tau, a) \in \mathfrak{E}(\mathfrak{G})$. Thus, the automorphism of $\mathfrak{A}$, sending (a^τ, a) to (c,b) maps $\underline{a}$ to $\underline{b}$ proving that $\mathfrak{A}$ acts transitivaly on $\mathfrak{S}(\mathfrak{G})$.

PROPOSITION 2: $\mathfrak{A}(\mathfrak{G})$ has no element of order 2.

PROOF: Let us suppos that, for $\sigma \in \mathcal{A}(\mathcal{G})$, $\sigma^2 = \varepsilon$ (the unity permutation). For any $a \in \mathcal{S}(\mathcal{G})$, σ maps both of the pairs (a, a^σ) and (a^σ, a) to the other one which contradicts the condition that $\mathcal{G}$ is a tournament.

PROPOSITION 3: Let $\mathcal{G}$ be a finite tournament with $|\mathcal{S}(\mathcal{G})| = n > 2$ and $\mathcal{A}$ a subgroup of $\mathcal{A}(\mathcal{G})$, acting transitively on $\mathcal{E}(\mathcal{G})$. For $a, b \in \mathcal{S}(\mathcal{G})$ both $(\mathcal{A} : \mathcal{A}_a) = n$ and $(\mathcal{A} : \mathcal{A}_{(a,b)}) = \binom{n}{2}$ are valid.

PROOF: The proposition is a known consequence of $|\mathcal{S}(\mathcal{G})| = n$ and $|\mathcal{E}(\mathcal{G})| = \binom{n}{2}$. (The first statement is not true for $n = 2$.)

THEOREM 1: Let $\mathcal{G}$ be a finite tournament with $|\mathcal{S}(\mathcal{G})| = n > 2$, and $\mathcal{A}$ a subgroup of $\mathcal{A}(\mathcal{G})$, acting transitively on $\mathcal{E}(\mathcal{G})$. Then $n \equiv 3 \pmod 4$.

PROOF: By Proposition 2, $\mathcal{A}$ has no elements of order 2 and so its order is an odd number. Therefore the index of each subgroup of $\mathcal{A}$ is an odd number, too. From this it follows, using Proposition 3., that both n and $\binom{n}{2}$ are odd, i.e. $n \equiv 3 \pmod 4$.

COROLLARY: If $\mathcal{G}$ is a finite homogeneous tournament, then $|\mathcal{S}(\mathcal{G})| \equiv 3 \pmod 4$.

PROOF: It follows from Theorem 1., choosing, especially, $\mathcal{A} = \mathcal{A}(\mathcal{G})$.

Now, let $\mathcal{G}$ be a 1-homogeneous tournament and denote $\mathcal{B}^*$ the complement of the join of all $\mathcal{B}(\mathcal{G})_a$'s in $\mathcal{B}(\mathcal{G})$.

PROPOSITION 4: $|\mathcal{B}^*| = |\mathcal{S}(\mathcal{G})| - 1$.

PROOF: From the 1-homogenity it follows that $\mathcal{B}(\mathcal{G})_{(a,b)} = \{\varepsilon\}$. Therefore, by proposition 3., $|\mathcal{B}(\mathcal{G})| = \binom{n}{2}$ and $|\mathcal{B}(\mathcal{G})_a| = |\mathcal{B}(\mathcal{G})| :$ $: (\mathcal{B}(\mathcal{G}) : \mathcal{B}(\mathcal{G})_a) = \binom{n}{2} : n = \frac{n-1}{2}$. One gets, using $\mathcal{B}(\mathcal{G})_a \cap \mathcal{B}(\mathcal{G})_b = \mathcal{B}(\mathcal{G})_{(a,b)}$, that

$$\Big| \bigcup_{a \in \mathcal{S}(\mathcal{G})} \mathcal{B}_a \Big| = \Big| \sum_{a \in \mathcal{S}(\mathcal{G})} \frac{n-1}{2} - 1 \Big| + 1$$

from which

$$|\mathcal{B}^*| = |\mathcal{B}(\mathcal{G})| - |\bigcup_{a\in\mathcal{S}(\mathcal{G})} \mathcal{B}_a| = \frac{n(n-1)}{2} - \frac{n(n-3)}{2} - 1 = n-1$$

follows.

PROPOSITION 5: The conjugates of $\mathcal{B}^*$ coincide with itself.

PROOF: It is an immediate consequence of the fact that the subgroups $\mathcal{B}(\mathcal{G})_a$ are, for $a\in\mathcal{S}(\mathcal{G})$, just the conjugates to each other.

PROPOSITION 6: $\mathcal{B}^*$ consists of at most 2 classes of conjugates.

PROOF: Let σ be an element of $\mathcal{B}^*$, mapping the element $\underline{a}$ of $\mathcal{S}(\mathcal{G})$ to $\underline{b}$, which is different, by the choice of σ, from $\underline{a}$. The elements $\varrho_1, \dots, \varrho_{\frac{n-1}{2}}$ of $\mathcal{B}(\mathcal{G})_a$ map $\underline{b}$ to different elements, because of 1-homogenity. Then, for the automorphisms $\varrho_i^{-1}\sigma\varrho_i$ $(i = 1, \dots, \frac{n-1}{2})$, the elements

$$a^{\varrho_i^{-1}\sigma\varrho_i} = a^{\sigma\varrho_i} = b^{\varrho_i}$$

are different, proving that the given $\frac{n-1}{2}$ automorphisms are different, too. By Proposition 5., these automorphisms belong to $\mathcal{B}^*$. If $\mathcal{B}^*$ is not one class of conjugates, then it has 2 elements σ_1, σ_2 being not conjugates to each other. Let $[\sigma]$ denote the set of conjugates of σ. It was just proved that both $|[\sigma_1]|$ and $|[\sigma_2]|$ are $\geqq \frac{n-1}{2}$. The disjointness of $[\sigma_1]$ and $[\sigma_2]$, the inclusions $[\sigma_1], [\sigma_2] \subseteqq \mathcal{B}^*$ and $|\mathcal{B}^*| = n-1$ complete the proof.

Now, let $\mathcal{B}$ denote the join $\varepsilon \cup \mathcal{B}^*$.

PROPOSITION 7: $\mathcal{B}$ is a normal subgroup of $\mathcal{B}(\mathcal{G})$.

PROOF: From Theorem 1., it follows that the order of $\mathcal{B}(\mathcal{G})$ is an odd number. Therefore, by the known theorem of Feit and Thompson, $\mathcal{B}(\mathcal{G})$ is solvable. Being $|\mathcal{B}(\mathcal{G})| = n.\frac{n-1}{2}$ and $(n, \frac{n-1}{2}) = 1$, one can use the theorem of Hall, i.e. $\mathcal{B}(\mathcal{G})$ has a subgroup $\bar{\mathcal{B}}$ of order $\underline{n}$. One gets, using again $(n, \frac{n-1}{2}) = 1$, that $\bar{\mathcal{B}} \cap \mathcal{B}(\mathcal{G})_a = \{\varepsilon\}$, for each $a\in\mathcal{S}(\mathcal{G})$. So is $\bar{\mathcal{B}} \subseteqq \mathcal{B}$

and, by $|\bar{\mathcal{B}}| = |\mathcal{B}|$, $\bar{\mathcal{B}} = \mathcal{B}$. Hence, $\mathcal{B}$ is a subgroup of $\mathcal{B}(\mathcal{G})$ and Proposition 5. proves the statement.

PROPOSITION 8: $\mathcal{B}$ has no proper invariant subgroup.

PROOF: Let $\mathcal{C}$ be an invariant subgroup of $\mathcal{B}$, different from the unity one. This means that $\mathcal{C}$ contains, besides the unity, at least one class of conjugates $[\sigma]$ in $\mathcal{B}(\mathcal{G})$. From $|\mathcal{C}| \geqq |\varepsilon| + |[\sigma]| \geqq + \frac{n-1}{2} > \frac{n}{2}$ follows $(\mathcal{B}:\mathcal{C}) < 2$, i.e. $\mathcal{B} = \mathcal{C}$.

PROPOSITION 9: $\mathcal{B}$ is an elementary p-group.

PROOF: Because of the solvability of $\mathcal{B}$, the last but one element of the chain of commutator-subgroups is a commutative invariant subgroup of $\mathcal{B}$, different from the unity one. Then, by Proposition 8, it coincides with $\mathcal{B}$; hence $\mathcal{B}$ is commutative.

For a prime number p, $\mathcal{B}^p$ is also an invariant subgroup of the commutative group $\mathcal{B}$. If p divides n, then $\mathcal{B}^p \neq \mathcal{B}$, i.e., by Proposition 8., $\mathcal{B}^p = \{\varepsilon\}$; hence $\mathcal{B}$ is an elementary p-group, indeed.

THEOREM 2: If $\mathcal{G}$ is a finite 1-homogeneous tournament, then $|\mathcal{S}(\mathcal{G})| = p^k$, with odd k and a prime $p \equiv 3 \pmod 4$.

PROOF: From the definition of the subgroup $\mathcal{B}$ there follow, using Propositions 4 and 9., the equalities $|\mathcal{S}(\mathcal{G})| = |\mathcal{B}| = p^k$. One gets, by the Corollary of Theorem 1., $p^k \equiv 3 \pmod 4$ which is only possible in the case $p \equiv 3 \pmod 4$, and only if k is an odd number.

Now, let $\mathcal{G}$ be a tournament. Let us, further, suppose that there is defined a multiplication on $\mathcal{S}(\mathcal{G})$ which makes it a group. The tournament $\mathcal{G}$ is called a tournament over this group if the multiplications both on the right and on the left belong to $\mathcal{A}(\mathcal{G})$.

THEOREM 3: Any 1-homogeneous tournament is similar to a tournament over a group.

PROOF: Let us, for the 1-homogeneous tournament $\mathcal{G}$, the groups $\mathcal{B}(\mathcal{G})$, $\mathcal{B}(\mathcal{G})_a$ and $\mathcal{B}$ define as above. Let, further, an element $a \in \mathcal{S}(\mathcal{G})$ be fixed. If, for $\sigma, \tau \in \mathcal{B}$, it fulfills $a^\sigma = a^\tau$, then we have $a^{\sigma\tau^{-1}} = a$, i.e. $\sigma\tau^{-1}$ belongs to $\mathcal{B}(\mathcal{G})_a$. On the other hand, $\sigma\tau^{-1} \in \mathcal{B}$ being $\mathcal{B}$ a group; hence $\sigma\tau^{-1} = \varepsilon$, i.e. $\sigma = \tau$, for $\mathcal{G}$ is 1-homogeneous. Thus, the mapping $\sigma \longrightarrow a^\sigma$ is a one-to-one correspondence of $\mathcal{B}$ onto $\mathcal{S}(\mathcal{G})$. Now, $\mathcal{B}$ may be made into a tournament as follows:

$$\text{"}(\sigma,\tau) \in \mathcal{E}(\mathcal{B}) \quad \text{if and only if} \quad (a^\sigma, a^\tau) \in \mathcal{E}(\mathcal{G})\text{".}$$

This tournament is, even by definition, similar to $\mathcal{G}$. Let, finally, $(\sigma,\tau) \in \mathcal{E}(\mathcal{B})$, i.e. $(a^\sigma, a^\tau) \in \mathcal{E}(\mathcal{G})$. From $\varrho \in \mathcal{A}(\mathcal{G})$ it follows that $(a^{\sigma\varrho}, a^{\tau\varrho}) \in \mathcal{E}(\mathcal{G})$, i.e. $(\sigma\varrho, \tau\varrho) \in \mathcal{E}(\mathcal{B})$ holds. The commutativity of $\mathcal{B}$ completes the proof.

THEOREM 4: There exists, for each prime $p \equiv 3 \pmod 4$ and for each odd number k, a 1-homogeneous tournament $\mathcal{G}$, with $|\mathcal{S}(\mathcal{G})| = p^k$.

PROOF: Let $\mathcal{F}_n$ denote the finite field with n elements. One has to consider the field $\mathcal{F} = \mathcal{F}_n$ for the given $n = p^k$. Let $\mathcal{F}^*$ denote the multiplicative group of $\mathcal{F}$, for which $|\mathcal{F}^*| = p^k - 1$ holds. The squares of elements of $\mathcal{F}^*$ form a subgroup $\mathcal{N}$ of it with $(\mathcal{F}^* : \mathcal{N}) = 2$. Because $p^k - 1 \equiv 2 \pmod 4$, $\mathcal{F}^*$ has no element of order 4 from which it follows that -1 does not belong to $\mathcal{N}$. Thus, for each $x \in \mathcal{F}$ exactly one of the relations

$$x = 0 \qquad x \in \mathcal{N} \qquad -x \in \mathcal{N}$$

holds. Therefore, by the definition

$$\text{"}(a,b) \in \mathcal{E}(\mathcal{F}) \quad \text{if and only if} \quad b-a \in \mathcal{N}\text{"}$$

$\mathcal{F}$ becomes a tournament.

Now, let us consider, for $a \in \mathcal{N}$ and $b \in \mathcal{F}$, the mappings $x \rightarrow ax + b$ of $\mathcal{F}$ into itself. These mappings form, clearly, a subgroup $\mathcal{L}$, of the group of linear mappings of the field $\mathcal{F}$. From $(u,v) \in \mathcal{E}(\mathcal{F})$ it follows that $v - u \in \mathcal{N}$, further, $(av+b)-(au+b) = a(v-u) \in \mathcal{N}$, since $\mathcal{N}$ is closed under multiplication. Thus, $(au+b, av+b) \in \mathcal{E}(\mathcal{F})$, i.e. $\mathcal{L}$ is contained in $\mathcal{A}(\mathcal{G})$.

Let, finally, be both (a,b) and (c,d) elements of $\mathcal{E}(\mathcal{F})$. The mapping $x \to (d-c)(b-a)^{-1}x + [c-(d-c)(b-a)^{-1}a]$, belonging to $\mathcal{L}$, maps (a,b) to (c,d), and, clearly, this is the only mapping in $\mathcal{L}$ of this property. Hence, one can choose the group $\mathcal{L}$ for $\mathcal{B}(\mathcal{G})$, i.e. the given tournament is 1-homogeneous.

REMARK: The tournament, constructed in Theorem 4., is not stiff, for $k \neq 1$. Namely, the mapping $x \to x^p$ belongs to $\mathcal{A}(\mathcal{G})$ and it is, for $k \neq 1$, different from the identical one.

THEOREM 5: Let be given a prime number $p \equiv 3 \pmod 4$. Then, there exists, up to similarity, exactly one homogeneous tournament $\mathcal{G}$ with $|\mathcal{S}(\mathcal{G})| = p$, which is, in addition, stiff.

PROOF: As mentioned, $\mathcal{A}(\mathcal{G})$ is solvable and, for $|\mathcal{S}(\mathcal{G})| = p$, it is a solvable subgroup of the symmetrical group of grade p, mereover, by Proposition 1., it is a transitive subgroup of it.

Thus, because of the known result, $\mathcal{A}(\mathcal{G})$ is a subgroup of the linear group $\mathcal{L}_p$. Thus, $|\mathcal{A}(\mathcal{G})|$ is an odd divisor of $|\mathcal{L}_p| = p(p-1)$, i.e. $|\mathcal{A}(\mathcal{G})|$ divides $\frac{p(p-1)}{2}$. On the other hand, using 1-homogeneity, $|\mathcal{B}(\mathcal{G})| = |\mathcal{E}(\mathcal{G})| = \frac{p(p-1)}{2}$. The inequality $|\mathcal{B}(\mathcal{G})| \leqq |\mathcal{A}(\mathcal{G})|$ inplies the equality $\mathcal{B}(\mathcal{G}) = \mathcal{A}(\mathcal{G})$, i.e. $\mathcal{G}$ is a stiff homogeneous tournament. It is possible, as it is known, to denote the elements of $\mathcal{S}(\mathcal{G})$ with the elements of $\mathcal{F}_p$, such that the elements of $\mathcal{L}_p$ act on $\mathcal{S}(\mathcal{G})$ as linear functions. Since $(\mathcal{L}_p : \mathcal{A}(\mathcal{G})) = 2$, the subgroup $\mathcal{A}(\mathcal{G})$ of $\mathcal{L}_p$ is uniquely determined; its elements are the mappings in the form $x \to a^2x + b$.

Now, let us suppose, that for the tournaments $\mathcal{G}$ and $\mathcal{G}'$ one has $(0,u) \in \mathcal{E}(\mathcal{G})$ and $(0,v) \in \mathcal{E}(\mathcal{G}')$. The mapping $\varphi : x \to vu^{-1}x$ is, clearly, a one-to-one mapping of $\mathcal{S}(\mathcal{G})$ to $\mathcal{S}(\mathcal{G}')$, which set is, of course, equal to $\mathcal{S}(\mathcal{G})$. From $(a,b) \in \mathcal{E}(\mathcal{G})$ it follows that $(b-a)u^{-1}$ is a square, for the mapping

$x \to (b-a)u^{-1}x + a$ maps $(0,u)$ to (a,b). Thus, φ maps (a,b) to $(vu^{-1}a, vu^{-1}b) = ((b-a)u^{-1}0 + vu^{-1}a, (b-a)u^{-1}v + vu^{-1}a)$ which belongs to $\mathcal{E}(\mathcal{G}')$. being the image of $(0,v)$ under the mapping $x \to (b-a)u^{-1}x + vu^{-1}a$. Hence, φ gives a similarity, indeed.

REMARK: The fully ordered sets are, clearly, special cases of tournaments. Let $\mathcal{G}$ be a fully ordered set which is, as a tournament, homogeneous. Let, further, $a<b$, which means that $(a,b) \in \mathcal{E}(\mathcal{G})$, and $a<c$. Then there is, by reason of homogeneity, an automorphism $\sigma \in \mathcal{A}(\mathcal{G})$ mapping $\underline{b}$ to $\underline{c}$. It is easy to verify that the mapping

$$x^{\tau} = x \text{ for } x \leqq a \qquad x^{\tau} = x^{\sigma} \text{ for } a < x$$

is also an element of $\mathcal{A}(\mathcal{G})$, different from the unity. If $|\mathcal{S}(\mathcal{G})| > 2$, then there is, by Proposition 1., an automorphism $\varrho \in \mathcal{A}(\mathcal{G})$ mapping $\underline{b}$ to $\underline{a}$ from which it follows that $a^{\varrho} < a$. Clearly, the mapping τ maps (a^{ϱ}, a) to itself, i.e. a fully ordered set with at least 2 elements is never a stiff homogeneous tournament.

PROBLEMS

1) Determine $\mathfrak{A}(\mathcal{G})$ of the 1-homogeneous tournaments $\mathcal{G}$ given in Theorem 4!

2) Are there 1-homogeneous tournaments $\mathcal{G}$ and $\mathcal{G}'$ with $|\mathcal{S}(\mathcal{G})| = |\mathcal{S}(\mathcal{G}')|$, if they are not similar?

3) The same, for homogeneous tournaments.

4) Is there a homogeneous tournament $\mathcal{G}$ for which $|\mathcal{S}(\mathcal{G})|$ is not a power of a prime?

5) Is there a homogeneous tournament $\mathcal{G}$ for which $|\mathcal{S}(\mathcal{G})|$ is divisible by a prime $p \equiv 1 \pmod 4$?

6) Is there a homogeneous tournament $\mathcal{G}$ for which $|\mathcal{S}(\mathcal{G})|$ is equal to a given natural number $n \equiv 3 \pmod 4$?

7) Is there a stiff homogeneous tournament $\mathcal{G}$ for which $|\mathcal{S}(\mathcal{G})|$ is not a prime?

8) Are there infinite stiff homogeneous tournaments?

REFERENCES:

[1] FEIT, W. THOMPSON J.G. Solvability of groups of Odd Order. Pacific J. Math. 13 (1963) 775-1029.

[2] FRIED, E. A generalization of ordered algebraic systems. Acta. Sci. Math.

[3] FUCHS, L. Teilweise geordnete algebraische Strukturen. Akadémiai Kiadó, Budapest 1966.

[4] HALL, Ph. A note on soluble groups, J. London Math. Soc. 3 (1928) 98-105.

[5] HARARY, F.-NORMAN, R.Z. - CARTWRIGHT, D.; Structural models: An introduction to the theory of directed graphs, New York, Wiley (1965)

[6] VAN der WAERDEN B.L.: Algebra Springer Verlag.

On the foundations of combinatorial theory
IV
Finite vector spaces and Eulerian generating functions

by

J. Goldman and G.-C. Rota

Cambridge, U.S.A. Cambridge U.S.A.

1. INTRODUCTION

The purpose of this paper is to carry out a small part of the program that was begun in Foundations I. Our main concern is the study of the combinatorial aspects of the lattice of subspaces of a vector space over a finite field, and its use in deriving various classical and new identities to be found in the literature under various guises and disguises. The central idea is to obtain as systematically as possible a set-theoretic interpretation in terms of enumeration of vector spaces and of linear transformation between vector spaces over finite fields, of various identities classically known as q-identities. These identities have almost universally been studied from different points of view, namely, from the point of view of the theory of partitions of numbers, and from the point of view of the theory of elliptic functions. The analogy between these identities and classical binomial identities has been remarked many times. In fact, it is the theme of the entire work of Jackson and of the small school of English formalists that he left. Unfortunately, Jackson's work is purely analytic, and does not reveal the set-theoretic basis for this analogy. We believe that our systematic attempt at such an interpretation reveals the structure of this analogy, which is the similarity between the lattices of subspaces of a finite vector space and the lattice of sub-sets of a finite set. The numerical analog of this similarity is the fact that as $q \to 1$, every finite identity on vector spaces tends to an identity on a Boolean algebra.

We begin with a brief study of the Gaussian coefficients, namely the number of subspaces of dimension k in a vector space of dimension n, displaying various analogs of binomial identities, which we prove by set-theoretic means. We then proceed to develop the method of Eulerian generating functions, which are derived as a subalgebra of the incidence algebra of the lattice of finite dimensional subspaces of an infinite-dimensional vector space (always over a finite field.) We call this subalgebra the reduced incidence algebra.

Sections 4 and 5 are perhaps the most interesting. At the beginning of Section 4 we give a set-theoretic interpretation of formulas relating Eulerian generating functions to enumeration with the reduced incidence algebra. We then proceed to apply this principle of interpretation to various situations, obtaining enumerations of various quantities connected with vector spaces. Section 5 contains analogs of the binomial theorem for finite vector spaces and applications of the Möbius inversion formula over a finite vector space leading to various classical q-identities. We conclude with a speculative section relating to future work in this field.

We are greatly indebted to the previous work of Philip Hall, who was the first to develop the Möbius inversion formula in the context of p-groups, and to the numerous and profound papers of L. Carlitz, H.W. Gould, Sharma, Chak, Segre and several other authors, whose papers could not be listed in the bibliography because of their number.

2. THE GAUSSIAN COEFFICIENTS

We begin with the q-analog of the binomial coefficients. Just as $\binom{n}{k}$ counts the number of elements of rank (size) k in the lattice of subsets of a set of n elements, we let $\binom{n}{k}_q$ be the number of subspaces of rank (= dimension) k in lattice $L(V_n)$ of subspaces of an n-dimensional vector space over the finite field $GF(q)$. The numbers of elements of rank k in this lattice are called the Gaussian coefficients.

It is easy to derive a formula for the Gaussian coefficients $\binom{n}{k}_q$. First enumerate all ordered bases of k-dimensional subspaces of V_n, as follows. Choose the first vector y_1 in any one of q^n-1 ways (that is, excluding the zero vector). There are q vectors linearly dependent upon y_1, so the next vector y_2 can be chosen in q^n-q ways. y_1 and y_2 span a two-dimensional subspace containing q^2 vectors, so we may choose y_3 in q^n-q^2 way, etc. Thus there are $(q^n-1)(q^n-q)\ \ldots (q^n-q^{k-1})$ linearly ordered sets of k linearly independent vectors in V_n. But each k-dimensional subspace has, by the same argument, (let $n=k$) $\prod_{i=0}^{k-1}(q^k-q^i)$ ordered bases. Thus dividing out the overcount we obtain the well-known expression

$$(1)\qquad \binom{n}{k}_q = \frac{(q^n-1)(q^n-q)\cdots(q^n-q^{k-1})}{(q^k-1)(q^k-q)\ldots(q^k-q^{k-1})} = \frac{(q^n-1)(q^{n-1}-1)\ldots(q^{n-k+1}-1)}{(q^k-1)(q^{k-1}-1)\ldots(q-1)}.$$

Note that as $q\to 1$, $\binom{n}{k}_q \to \binom{n}{k}$, and thus the Gaussian coefficient can be expected to share many of the properties of binomial coefficients. It is a heuristic principle that all identities between Gaussian coefficients yield as corollaries identities between binomial coefficients. Perhaps the simplest example is

$$(2)\qquad \binom{n}{k}_q = \binom{n}{n-k}_q .$$

This can be seen immediately from (1), or by noting that the lattice $L(V_n)$ is selfdual and thus there are as many k spaces of V_n as n-k-spaces. Letting $q \to 1$ we get the familiar identity $\binom{n}{k} = \binom{n}{n-k}$.

Since (2) was derived for q any power of prime and n any positive integer, we can now think of it as an algebraic function of the variable q varying over a wider domain. Similarly we take (1) as a definition when q is a variable.

If we think of q as a real variable $0<q<1$ then as $n \to \infty$,

$$\binom{n}{k}_q \to \frac{1}{(1-q)\cdots(1-q^k)} \tag{3}$$

a fact which will be used later. Identity (3) remain true as q ranges over the q-adic integers.

We shall now use combinatorial arguments on the lattice $L(V_n)$ to derive identities for the Gaussian coefficients by counting sets of subspaces in different ways, in analogy with the counting of subsets in a Boolean algebra. We limit ourselves to a few examples, from which the reader will be able to glean the power and the methods.

Prop. 1. (q-Pascal triangle)

$$\binom{n}{h}_q = \binom{n-1}{h-1}_q + q^h \binom{n-1}{h}_q \tag{4}$$

Note: when $q \to 1$, (4) reduces to the usual Pascal triangle identity.

Proof: Choose a basis $x_1, \ldots, x_n$, and let V_{n-1} be the space spanned by $x_1, \ldots, x_{n-1}$. Now let V_h be an h-dimensional subspace of V_n. There are two possibilites for V_h.

Case 1. V_h includes the whole line spanned by x_n. If so, then $V_h \cap V_{n-1}$ is a subspace of dimension $h-1$, and this intersection can be chosen in $\binom{n-1}{h-1}_q$ ways, accounting for the first term on the right of (4).

Case 2. V_h does not include the vector x_n. But then, the projection of V_h onto V_{n-1} along the line x_n is a subspace of dimension h, call it W_h, of V_{n-1}. One then obtains V_h by choosing such a W_h, and then "lifting it up", that is, choosing a basis $y_1, \ldots, y_h$ of W_h, and adding to each y_i a multiple of x_n. There are altogether q^h ways of performing the latter operation, and $\binom{n-1}{h}_q$ ways of performing the former. This accounts for the second term on the right of (4), and concludes the proof.

The next proposition yields both a q-identity and also enumerates a useful quantity.

Prop. 2. $N_{k,1}$, the number of k-subspaces of V_n containing a fixed one-dimensional subspace, is given by

$$N_{k,1} = \binom{n-1}{n-k}_q = \frac{\binom{n}{k}_q \binom{k}{1}_q}{\binom{n}{1}_q} \,. \tag{5}$$

Proof. $N_{k,1} = \binom{n-1}{n-k}_q$ follows immediately from the self duality of the lattice, as already remarked, see (2). By flipping $L(V_n)$ upside down the number of k-spaces containing a given 1-space is equal to the number of $(n-k)$-spaces contained in a given $n-1$ space which by (1) is $\binom{n-1}{n-k}_q$.

To derive the right side of (5) we look at the bipartite graph whose distinct sets of vertices A and B are the k-spaces and 1-spaces respectively and we connect a k-space $V \in A$ to an 1-space $V' \in B$ by an edge iff $V \supseteq V'$. Now we count the number of edges in this graph in two ways. First there are $\binom{n}{k}_q$ vertices in A and $\binom{k}{1}_q$ edges at each of these vertices. On the other hand there are $\binom{n}{1}_q$ vertices in B and $N_{k,1}$ edges at each vertex. Thus

$$\binom{n}{k}_q \binom{k}{1}_q = N_{k,1} \binom{n}{1}_q \,, \qquad \text{q.e.d.}$$

The next identity is a q-generalization of the binomial theorem, first proved by Cauchy by purely algebraic means. Other q-binomial theorems will be derived in Section 5.

Prop. 3.

$$y^n = \sum_{k=0}^{n} \binom{n}{k}_q (y-1)(y-q)\dots(y-q^{k-1}) . \tag{6}$$

Note. As $q \to 1$ we get $y^n = \sum_{k=0}^{n} \binom{n}{k} (y-1)^k$ or letting $z = y-1$, $(z+1)^n = \sum_{k=0}^{n} \binom{n}{k} z^k$. Thus (6) is a q-analog of the binomial expansion of $(z+1)^m$.

Proof. (6) counts, in two ways, all linear transformations of V_n into a space Y with y vectors. Indeed: let $x_1, \dots, x_n$ be a basis for V_n. Then each of the x_i can map into any of the y vectors of Y and these determine the linear transformation. There are altogether y^n choices for the x_i. This accounts for the left side of the identity.

On the right hand side we enumerate linear transformations by the dimension of their null spaces. Given a subspace V_k of dimension k (and there are $\binom{n}{k}_q$ of these) let $z_1, \dots, z_{n-k}, z_{n-k+1}, \dots, z_n$ be a basis of V_n such that $z_{n-k+1}, \dots, z_n$ generates V_k. A linear transformation has V_k as its null space if and only if it maps $z_{n-k+1}, \dots, z_n$ into zero and the remaining $n-k$ vectors $z_1, \dots, z_{n-k}$ onto an independent set in Y. z_1 can be mapped anywhere into Y except the zero vector i.e. in $y-1$ ways. The vector z_2 can be mapped anywhere except to the line spanned by the image of z_1. Since such a line has q points we have $y-q$ possiblities for z_2. Proceeding in this by now familiar way we find there are

$$(y-1)(y-q)\dots(y-q^{n-k-1})$$

maps whose nullspace is a given k space V_k. Thus, there are $\binom{n}{k}_q (y-1)(y-q)\dots(y-q^{n-k-1}) = \binom{n}{n-k}_q (y-1)(y-q)\dots(y-q^{n-k-1})$ linear transformations whose nullspace has dimension k. Summing over k we get the right side of (6).

Identity (6) can be interpreted as a q-analog of the classical binomial distribution, as follows. In the space Y, choose independently and at random a set of n vectors. What is the probability that they shall span a subspace of dimension k? By the preceding argument, this probability is

$$\binom{n}{k}_q \frac{(y-1)(y-q)\cdots(y-q^{k-1})}{y^n} .$$

As q tends to one, this tends to the classical binomial distribution

$$\binom{n}{k}(1-\frac{1}{y})^k(\frac{1}{y})^{n-k} .$$

3. EULERIAN GENERATING FUNCTIONS

Our chief tool in the study of $L(V_n)$ will be the Möbius function, which in this case gives the q-adic generalization of the principle of inclusion-exclusion. In this section we shall review some of the important points of Foundations I in the context of finite vector spaces and introduce some new concepts.

The Möbius inversion formula in $L(V_n)$ is as follows:

Let $N_=(V)$ and $N_\geq(V)$ be any two functions defined on $L(V_n)$, $V \in L(V_n)$, (with values in a commutative ring which we generally take to be the integers) satisfying the system of equations

(1) $$N_\geq(V) = \sum_{W \geq V} N_=(W) ;$$

then there exists a function $\mu(V,W)$ defined on $L(V_n)$, independent of the functions such that

(2) $$N_=(V) = \sum_{W \geq V} \mu(V,W) N_\geq(W) .$$

The function μ is given by the formula

(3) $$\mu(V,W) = \mu(0,W/V) = (-1)^k q^{\binom{k}{2}}$$

where $k = (\dim W/V)$.

Since $\mu(V,W)$ depends only on the difference of dimensions we set

(4) $$\mu_k = (-1)^k q^{\binom{k}{2}} = (-1)^k q^{\frac{k(k-1)}{2}} .$$

Since $L(V_n)$ is self-dual an equivalent form of Möbius inversion is

(1a) $$N_{\leq}(V) = \sum_{W \leq V} N_{=}(W) \quad \text{implies}$$

(2a) $$N_{=}(V) = \sum_{W \leq V} N_{\leq}(W)\, \mu(W,V).$$

If both $N_{=}(V)$ and $N_{\leq}(V)$ depend only on the dimension of V, that is, $N_{=}(V) = a_k$ if $\dim V = k$ and similarly $N_{\geq}(V) = b_k$, then collecting terms of the same dimension (1a)-(3) gives at once the beautiful numerical inversion formula

(5) $$b_n = \sum_{k=0}^{n} \binom{n}{k}_q a_k ; \quad a_n = \sum_{k=0}^{n} (-1)^k q^{\binom{k}{2}} \binom{n}{k}_q b_k .$$

This is the q-analog (let $q \to 1$) of the classical inversion formula (see Riordan)

(6) $$b_n = \sum_{k=0}^{n} \binom{n}{k} a_k ; \quad a_k = \sum_{k=0}^{n} (-1)^k \binom{n}{k} b_k$$

which arises from μ-inversion over the lattice of subsets of an n-element set.

But let us put the μ function in a more general setting. It is just one of many functions of two variables in $L(V_n)$ which form an interesting structure.

Def.: The incidence algebra $I(V_n)$ of V_n is the set of all functions $F(V,W)$ of two variables defined on $L(V_n)$, which take values in a commutative ring R, (which we generally take to be the integers) and such that $F(V,W)=0$ unless $V \leq W$, together with the following operations:

a) Addition: if $f,g \in I(V_n)$
let $h(V,W) = (f+g)(V,W) \equiv f(V,W) + g(V,W)$ be their sum,

b) if $c \in R$ and $f \in I(V_n)$
let $(cf)(V,W) \equiv c(f(V,W))$.

c) if $f,g \in I(V_n)$ their convolution (or product) is given by
$L(V,W) = f*g\,(V,W) \equiv \sum_{V \leq Z \leq W} f(V,Z)g(Z,W)$.

It is easily verified that $I(V_n)$ is an algebra over R.

If one embeds the partial order of $L(V_n)$ in a linear order and lists the subspaces in the order $W_1, W_2, \ldots,$ then a typical element f of $I(V_n)$ can be thought of as a matrix $A_{ij} = f(W_i, W_j)$ and $I(V_n)$ is isomorphic to an algebra of upper triangular matrices.

The zeta function, $\zeta(V,W) = 1$ if $V \leq W$, otherwise, belongs to $I(V_n)$ and its inverse is the Möbius function.

If one translates this to the isomorphism with upper triangular matrices then it is seen that Möbius inversion is a special matrix inversion. (See Foundation I for further details.)

It turns out in practice that one often doesn't have to study the full incidence algebra but a special subalgebra of particular combinatorial interest.

For this we recall that an interval or segment $[V,W]$ in $L(V_n)$ is given by

$$[V,W] = \{Z \mid V \leq Z \leq W\}.$$

Def. The reduced incidence algebra $R(V_n)$ of $L(V_n)$ is the subalgebra of $I(V_n)$ consisting of all function $f \in I(V_n)$ s.t.

if $[V,W]$ is isomorphic to $[V',W']$ then $f(V,W) = f(V',W')$ i.e. those functions constant on isomorphism classes of intervals. Isomorphism is taken in the sense of partially ordered sets.

It is easily verified that $R(V_n)$ is a subalgebra.

We next determine the reduced incidence algebra of $L(V_\infty)$, the lattice of finite dimensional subspaces of a countably infinite dimensional vector space over $GF(q)$. All definitions and results of this section hold for $L(V_\infty)$ and it proves more convenient to work with this lattice which contains as sublattices $L(V_n)$ for all n.

Let the height of a segment $[V,W]$ be $(\dim W - \dim V) = \dim(W/V)$. Then any two segments are isomorphic if they have the same height.

In the convolution sum

$$\sum_{V \le W \le U} f(V,W) g(W,U) = L(V,U) \tag{7}$$

there occur as many segments $[V,W]$ of height k as there are subspaces of dimension k of the quotient space U/V. Thus if $f,g \in R(V_n)$ i.e. $f(V,W) = a_k$ whenever $\dim(W/V) = k$ and $g(W,U) = b_{n-k}$ whenever $d(U/W) = n-k$, then equation (7) simplifies to

$$c_n = \sum_{k=0}^{n} \binom{n}{k}_q a_k b_{n-k} \tag{8}$$

where $c_n = h(V,U)$ and $d(U/V) = n$. $\{c_n\}$ is called the Gaussian convolution of $\{a_n\}$ and $\{b_n\}$. When $q \to 1$ (8) reduces to the binomial convolution $c_n = \sum_{k=0}^{n} \binom{n}{k}_q a_k b_{n-k}$. Since

$$\binom{n}{k}_q = \frac{(1-q)(1-q^2)\dots(1-q^n)}{(1-q)(1-q^2)\dots(1-q^k)(1-q)(1-q^2)\dots(1-q^{n-k})}$$

we can rewrite (8) as

(9) $$\sum_{k=0}^{n} \frac{a_k}{(1-q)(1-q^2)\dots(1-q^k)} \; \frac{b_{n-k}}{(1-q)(1-q^2)\dots(1-q^{n-k})} = \frac{c_n}{(1-q)(1-q^2)\dots(1-q^n)} .$$

Defining an Eulerian series to be a series of the form

$$\sum_{n=0}^{\infty} \frac{a_n x^n}{(1-q)(1-q^2)\dots(1-q^n)}$$

then by (8) we have proved the following:

Theorem 1: The reduced incidence algebra of the lattice $L(V_\infty)$ of all finite dimensional subspaces of a countable infinite dimensional vector space over a finite field with q elements is isomoprhic to the algebra of Eulerian series where multiplication is defined as formal multiplication of power series. The isomorphism maps the Eulerian series

$$\sum_{n=0}^{\infty} \frac{a_n}{(1-q)(1-q^2)\dots(1-q^n)}$$

to the element f of $L(V_\infty)$ defined by $f(V,W) = a_k$ if $d(W/V) = k$ and $f(V,W) = 0$ if $V \nsubseteq W$.

In particular the zeta function corresponds to the Eulerian series

(10) $$E_q(x) = \sum_{k=0}^{\infty} \frac{x^k}{(1-q)(1-q^2)\dots(1-q^k)}$$

and the Möbius function is given by the Eulerian series

(11) $$e_q(x) = \sum_{k=0}^{\infty} \frac{(-1)^k q^{\frac{k(k-1)}{2}} x^k}{(1-q)(1-q^2)\dots(1-q^k)}$$

and since ζ is the inverse of μ in $I(V_n)$

(12) $$E_q(x)\, e_q(x) \equiv 1 .$$

In analogy to the above, the reduced incidence algebra of the lattice of finite subsets of a countable set is isomoprhic to the algebra of exponential series, i.e. series of the form

$$\sum_{n=0}^{\infty} a_n \frac{x^n}{n!} .$$

Note however that the Eulerian series $\sum_{n=0}^{\infty} \frac{a_n x^n}{(1-q)\dots(1-q^n)}$ does not converge to the exponential series $\sum \frac{a_n x^n}{n!}$ as one might expect from previous discussion. The difficulty lies in the fact that Eulerian series as we write them refer to affine spaces,not projective spaces, i.e. $(1-q)\dots(1-q^n)$ refers to enumeration of affine points (vectors) not projective points (affine lines). If we renormalize our series and write them in the form

$$A((1-q)x) = \sum \frac{a_n x^n (1-q)^n}{(1-q)\dots(1-q^n)} = \sum \frac{a_n x^n}{\binom{1}{1}_q \binom{2}{1}_q \cdots \binom{n}{1}_q}$$

then this latter series converges to $\sum \frac{a_n x^n}{n!}$ as $q \to 1$. Since the correspondence

$$A(x) = \sum \frac{a_n x^n}{(1-q)\dots(1-q^n)} \longleftrightarrow \sum \frac{a_n x^n}{\binom{1}{1}_q \cdots \binom{n}{1}_q} = A((1-q)x)$$

is an automorphism of the algebra of Eulerian series it makes no real difference if we use the affine or projective form. At present we find it more convenient to use the affine form. It should be noted, however, that it always seems to be the projective form of equation that converge to the corresponding results in sets as $q \to 1$.

As we previously remarked, there are two versions of the Möbius inversion formula, according as we sum upwards" or "downwards". One of these remains unchanged whether the dimension is finite or infinite. The other instead becomes an infinite sum, and questions of convergence become relevant

We shall see that, in contrast to other instances of Möbius inversion (see for example Hille for a discussion of the difficult convergence questions associated with the classical Möbius inversion formula), all convergence questions here can be easily resolved by use of the q-adic norm.

In the reduced incidence algebra, the "upwards" inversion formula become:

(13) $$g_n = \sum_{k \geq n} \binom{n}{k}_q f_k$$

(14) $$f_n = \sum_{k \geq n} \binom{k}{n}_q \mu_{k-n} g_k .$$

Proposition. A necessary and sufficient condition that either - and hence both - of the series (13) and (14) converge in the q-adic field is that

$$\sum_{n \geq 0} f_n$$

converges q-adically.

Proof. Recall that a series converges q-adically if and only if the nth term converges to zero q-adically.

A straightforward computation gives $\| \binom{n}{k}_q \|_q = 1$.

Suppose $\sum f_n < \infty$ so that $\|f_n\|_q \to 0$. Then $\| \binom{k}{n}_q f_k \|_q = \|f_k\|_q \to 0$ and $q_n = \sum_{k \geq n} \binom{k}{n}_q f_k < \infty$. But since g_k and $\binom{k}{n}_q$ are bounded and $\|\mu_i\|_q = \frac{1}{q^{\binom{i}{n}}} \to 0$ we also have $f_n = \sum_{k \geq n} \binom{n}{k}_q \mu_{k-n} g_k < \infty$.

Conversely, suppose that the right side of (14) converges for some n, say $n = 1$, then it converges for all n, and the partial sums of the tail end on the right side of (14) must tend to zero. Thus, $f_n \to 0$, and hence $\sum_{n \geq 0} f_n$ converges, q.e.d.

4. THE INCIDENCE COALGEBRA

In the preceding section we studied the incidence algebra of the lattice of subspaces of a vector space as an algebra of operators acting on functions from the lattice to a commutative ring. In this section we introduce an entirely different interpretation of the incidence algebra, which will lead to an combinatorial interpretation of the convolution of two elements in the incidence algebra.

We begin with some very general notions applying to every locally finite partially ordered set, but quickly specialize to the reduced incidence algebra of $L(V_n)$. It is suggested that the reader of this section refer to the notion of coalgebra, as is found for example in MacLane (page 197), or in the recent survey work of Heinemann and Sweedler.

Let P be a locally finite partially ordered set. Let $V(P)$ (abbreviated V) be the module, over any ring R, spanned by the intervals $[x,y]$, where $x \leq y$. We introduce a comultiplication in V as follows. It is a function

(1) $$\psi : V \to V \otimes V,$$

where the right side is the tensor product taken relative to the ring R (which we may as well assume to be commutative), defined on the basis elements on V as follows:

(2) $$[x,y] = \sum_{x \leq z \leq y} [x,z] \otimes [z,y],$$

where the summation ranges over the variable z. It is easily verified that this comultiplication is coassociative. This means that the iteration of the comultiplication leads to summations of the form

(3) $$[x,y] = \sum [x,z_1] \otimes [z_1,z_2] \otimes \cdots \otimes [z_n,y],$$

where the summation on the right ranges over all sequences $z_1, z_2, \ldots, z_n$ such that

(4) $$x \le z_1 \le z_2 \le \ldots \le z_n \le y.$$

The counit ϵ is defined by mapping

(5) $$\epsilon([x,y]) = \begin{cases} 1 & \text{if} \quad x = y \\ 0 & \text{otherwise.} \end{cases}$$

With this definition we obtain a coalgebra, as is easily verified. We call this the incidence coalgebra $C(P)$ of the partially ordered set P, over the ring R. Most combinatorial operations on the incidence algebra really refer to the incidence coalgebra, and in fact, when reinterpreted in terms of the incidence coalgebra, they reveal their combinatorial meaning. We shall see in a moment how this is the case. Before that, let us formally recall the relationship between the incidence coalgebra and the incidence algebra. This comes from the well-known fact that the set V^* of all linear functionals on $C(V)$ with values in R has the structure of an algebra, which is precisely the incidence algebra of the partially ordered set P.

To obtain a combinatorial interpretation of the incidence algebra, we only have to closely inspect formula (3) above. Let us call a typical summand on the right hand side of (3), that is an expression

(6) $$[x,z_1] \otimes [z_1,z_2] \otimes \ldots \otimes [z_n,y], \quad x \le z_1 \le \ldots \le z_n \le y$$

a multichain (or chain) of a partially ordered set P. The single entries in the multichains will be called the links, and the chain as displayed in (6) will be said to be of length $n+2$. We can and will now consider a multichain of length $n+2$ as the underlying set of $n+1$ links. Note that the trivial link $[z,z]$ is also allowed. We are led to the following

Main Problem.

To every interval $[x,y]$ associate an element of a finite set $C(x,y)$. We are to enumerate the functions from the multichains between x and y to the set $\bigcup_{x\leq y} C(x,y)$ with the property that the first link $[x,z_1]$ is mapped into the set $C(x,z_1)$, the second link $[z_1,z_2]$ is assigned to the set $C(z_1,z_2)$, and so on.

The problem is easily visualized if one interprets the sets $C(x,y)$ as "colors" to be assigned to each link of the chain. In this way, one asks for the number of colored multichains between x and y with the property that the first link is assigned a color from a given set, the second link is assigned a color from another given set, etc.

The solution of the problem is given in the following

Proposition 1.

Let $f(x,y)$ be the number of elements of the set $C(x,y)$, for $x\leq y$. Then the solution of the Main Problem is given by the convolution, in the sense of the incidence algebra

$$\sum f(x,z_1)f(z_1,z_2)\cdots f(z_n,y), \tag{7}$$

where the summation ranges as in (4). This gives the number of colored chains of length $n+2$.

The proof is immediate, since the Proposition is a restating of the fact that the incidence algebra is obtained as the dual of the incidence coalgebra, when the ring R is taken to be the integers.

Thus we see that the elements of the incidence algebra, which are linear functionals on the incidence coalgebra, can be considered as the solution of problems of enumeration. This justifies the contention, first advanced in

Foundations I, that the incidence algebra generalizes the notion of generating function.

In a similar vein, we can interpret the convolution of different elements of the incidence algebra. Here we need assignments $C_k(x,y)$, $k = 1, 2, \ldots$ of "colors" to segments. Letting $f_k(x,y)$ be the size of the set $C_k(x,y)$ we obtain the convolution of $f_1, f_2, \ldots, f_{n+1}$ as the number of colored chains in which the i-th link is assigned a color from one of the sets $C_i(x,y)$. We spare the reader the obvious details, moving instead to more concrete applications, in the incidence algebra $L(V_\infty)$ of all finite-dimensional subspaces of an infinite-dimensional space over a field with q elements. For simplicity consider the reduced incidence algebra, where convolution in the avove sense reduces to an Eulerian convolution as studied in the preceding section. This amounts to studying the Main Problem in the special case when $f(x,y) = f(u,v)$ whenever $\dim y/x = \dim v/u$.

For example, let $a_i^{(2)}$ = the number of colored chains with 2 links connecting x and y, where the height of $[x,y]$ is i, that is $\dim y - \dim x = i$. Then we want to color chains of the form $[x,A] \otimes [A,y]$. Since $[x,y] \cong L(V_i)$, there are $\binom{i}{k}_q$ chains such that $[x,A]$ has height k and $[A,y]$ height $i-k$. The first link can be colored in $a_k = f(x,A)$ ways and the second in $a_{n-k} = f(A,y)$ ways. Thus

$$a_i^{(2)} = \sum_{k=0}^{i} \binom{i}{k}_q a_k a_{i-k} .$$

In other words, the sequence $a_i^{(2)}$ is the Eulerian convolution of the sequence a_i with itself. If $A(x)$, $A_2(x)$ are the Eulerian generating functions of a_i and $a_i^{(2)}$ respectively, then $A_2(x) = (A(x))^2$. Similarly, if $a_i^{(k)}$ is the number of chains with k links, between x and y, where the height of $[x,y]$ is i, and $A_n(x)$ is its generating function, then $A_n(x) = (A(x))^n$.

Example 1.

Let $a_i = 1$ for all i, and let us count all multichains of length 2. Setting $x = 0$ (the null space), we obtain the Eulerian convolution

$$\sum_{k \geq 0} \binom{n}{k}_q = G_n , \tag{8}$$

which gives the number of subspaces of the vector space V_n of dimension n (see Goldman-Rota for a study of these numbers, called the Galois numbers).

Example 2.

Let $f(u,v) = 1$ if $\dim k - \dim u = 1$ and $f(u,v) = 0$ otherwise: in other words, set $a_1 = 1$, and $a_i = 0$ if $i \neq 1$. The number of chains of length $n+2$ with this restriction is simply the total number of maximal chains with $n+1$ non-trivial links connecting x to y. Its Eulerian generating function is

$$A_n(x) = \left(\frac{x}{1-q}\right)^n , \tag{9}$$

where n is the difference of dimensions between the subspaces x and y.

Summing of all n, we obtain

$$\bar{A}(x) = \sum_{n=1}^{\infty} A_n(x) = \sum_{n=1}^{\infty} \left(\frac{x}{1-q}\right)^n = \frac{1}{1 - \frac{x}{1-q}} \tag{10}$$

where the coefficients of x^i counts all maximal chains in a segment $[x, y]$ where $\dim y/x = i$.

Next we consider some examples of convolutions of distinct functions.

Example 3.

Let $C_1(x,z)$ be the family of all sets of vectors in the quotient space z/x which span the space z/x. Let $C_2(z,y)$ be a set with one element. Then obtain

(11) $$\sum_{k=0}^{n} \binom{n}{k}_q D_k = 2^{q^n},$$

where the right hand side is the total number of subsets of a vector space of dimension n, and where D_k is the number of spanning subsets for a vector subspace of dimension k. But equation (11) states that the sequence $(2^{q^0}, 2^{q^1}, \ldots)$ is the Eulerian convolution of the sequences $(1,1,1,\ldots)$ and $(D_0, D_1, D_2, \ldots)$. Translating this into Eulerian generating functions we get

(12) $$E_q(x)D(x) = S(x)$$

where $E_q(x)$ is the zeta-function of Section 3,

$$D(x) = \sum_{n=0}^{\infty} \frac{D_n x^n}{(1-q)\ldots(1-q^n)} \quad \text{and} \quad S(x) = \sum_{n=0}^{\infty} \frac{2^{q^n} x^n}{(1-q)\ldots(1-q^n)}.$$

Solving for $D(x)$ we get

(13) $$D(x) = \frac{1}{E_q(x)} S(x) = e_q(x)S(x)$$

where $e_q(x)$ is the Eulerian generating function of the Möbius function. Equating coefficients in (13) we get

$$D_n = \sum_{n=0}^{\infty} \binom{n}{k}_q (-1)^k q^{\binom{k}{2}} 2^{q^{n-k}}$$

a result which we could also get directly by Möbius inversion.

Example 4.

We wish to count the set of all pairs (A_1, A_2), where A_1 is an atom of the lattice, A_2 is a coatom of the lattice that is, a line and a hyperplane respectively, and A_1 contained in A_2.

This is achieved by taking an Eulerian convolution according to the above prescriptions, as follows: set $a_1 = 1$, $a_i = 0$ for $i \neq 0$, and set $b_i = 1$ for all i. Then the desired number is the coefficient of

$$\frac{x^n}{(1-q)(1-q^2)\dots(1-q^n)}$$

in the Eulerian generating function $A(x)\,B(x)\,A(x)$, where $A(x)$ is the Eulerian generating function of a_i and $B(x)$ is the Eulerian generating function of b_i.

Example 5.

Suppose we have a store of a_i colors, and we wish to count the number of colored maximal chains with no trivial links between x and y. Assume again that the dimension of y/x is n. Let $f_i(n)$ be such a number. Now, the last link of such a chain is of the form $[A_{n-1}, y]$ where A_{n-1} is any $(n-1)$-dimensional subspace and each such link can receive any of the a_1 colors. Thus, we are led to the recursion

$$f_1(n) = \binom{n}{n-1}_q a_1 f_1(n-1).$$

Taking the Eulerian generating function of the sequence $f_1(n)$ we obtain for the Eulerian generating function $F_1(x)$ the recursion

$$F_1(x) = \frac{1}{1 - \frac{a_1 x}{1-q}}, \tag{14}$$

which gives the explicity form. When $a_1 = 1$ this gives the result of the Example 2, giving $\bar{A}(x)$ in (10).

Example 6.

Generalizing the preceding Example let $f_k(n)$ be the number of colored chains between x and y such that links of sizes $1,2,\ldots,k$ are allowed. The links of size i can be colored in a_i colors. Repeating the preceding argument we are led to the recursion

$$f_k(n) = \sum_{i=1}^{k} a_i \binom{n}{n-i}_q f_k(n-i) \tag{15}$$

which is the general q-difference equation with constant coefficients. Again, taking Eulerian generating functions $F_k(x)$ we find

$$F_k(x) = \frac{1}{1 - \frac{a_1 x}{1-q} - \frac{a_2 x^2}{(1-q)(1-q^2)} - \cdots - \frac{a_k x^k}{(1-q)\ldots(1-q^k)}} = \tag{16}$$

$$= \sum_{n=0}^{\infty} \frac{f_k(n) x^n}{(1-q)(1-q^2)\ldots(1-q^n)} .$$

Thus, we see that the general q-difference equation has a combinatorial interpretation in terms of enumeration of multichains in the lattices of subspaces of a vector space.

We hope these few examples have given the reader an idea of the scope of the method.

5. SOME EULERIAN IDENTITIES

We shall now derive by combinatorial arguments some identities, which have traditionally been associated with the theory of partitions of a number. We use two methods: direct enumeration and Möbius inversion on the lattice of subspaces.

If x and z are subspaces of a vector space, we write $X \div Y = (X-Y) \cup \{0\}$ for simplicity.

We begin by giving a combinatorial derivation of a very general q-analog of the binomial theorem. It includes the q-binomial theorem of Section 2 as a special case and has other cases of particular combinatorial significance.

Theorem. (q-binomial theorem) Let $P_k(x,y) = (x-y)(x-qy)\dots$ $\dots(x-q^{k-1}y)$ then

(1) $$P_n(x,z) = \sum_{k=0}^{n} \binom{n}{k}_q P_k(x,y)\, P_{n-k}(y,z).$$

Note: Although the variable y appears only on the right side, and cancels out when the right side is expanded, it nevertheless proves very useful to write the identity in this form. As $q \to 1$ the identity reduces to the trivial identity

$$(x-z)^n = \sum \binom{n}{k}(x-y)^k (y-z)^{n-k}.$$

Proof. Let V_n, X, Y, Z be vector spaces such that $Z \subset Y \subset X$, $\dim V_n = n$, and $\dim V_n < \dim Z$. Say X, Y, Z have x, y, z vectors respectively. Equation (1) count in two ways the set of all one-to-one linear transformations $f: V_n \to X$ such that $f^{-1}(Z) = 0$ (or equivalently $f(V_n) \cap Z = 0$).

Indeed, let $v_1, \dots, v_n$ be a basis for V_n. We count the ways of mapping this basis into a set of n independent vectors in X whose span intersects Z in $\{0\}$. The vector v_1 can be mapped into any vector in X not in Z i.e. in $x-z$ ways; the vector v_2 can be mapped into any of $x-qz$ vectors, namely, all vectors in X except those lying in the subspace spanned by Z together with the image of V_1; similarly, for the vector v_3 there are $x-q^2z$ choices, all vectors in X except the members of the space space spanned by the images of V_1 and V_2 together with Z, and so on. Thus, the number of one-to-one linear transformations whose image doesn't intersect Z is $(y-z)(y-qz)\cdots(y-q^{n-1}z) = P_n(x,z)$.

Next, we again count the set of all one-to-one linear transformations whose image is disjoint from Z, according to the position of the image of V_n relative to Y. Let f be such a transformation. Then $f(V_n) \cap Y$ is a subspace of some dimension, say k; hence it is the image of a k-dimensional subspace of V_n. Thus we can construct such an f by first choosing an arbitrary subspace U of V_n, next mapping U into Y but outside $Z - \{0\}$, and mapping $V_n - U$ into $X \div Y$. This leads to the following enumeration: let $v_1, \ldots, v_n$ be a basis for V_n such that $v_1, \ldots, v_k$ is a basis for U. Then, as in the first part of the proof, the number of one-to-one linear maps of U into $Y \div Z$ is $P_k(y,z) = (y-z)(y-qz)\cdots(y-q^{k-1}z)$. The remaining basis vectors must map into $X \div Y$ in such a way that f is one-to-one. As above, this can be done in $P_{n-k}(x,y)$ ways since any set of independent vectors in $X - Y$ is independent of any set of vectors in Y. Thus, for every k-dimensional subspaces U of V_n there are $P_k(y,z)P_{n-k}(x,y)$ one-to-one linear tranformations of U_n into X whose image intersected with $Y \div Z$ is the image of U. Hence there are $\sum_{k=0}^{n} \binom{n}{k}_q P_k(y,z) P_{n-k}(x,y)$ one-to-one linear maps of V_n into $X \div Z$. We conclude that

$$P_n(x,z) = \sum_{k=0}^{n} \binom{n}{k}_q P_k(y,z) P_{n-k}(x,y),$$

which is the desired result.

Several special cases of (1) are worth remarking.

Corollary 1. Setting $z = 0$, $y = 1$ in (1), we obtain

$$x^n = \sum \binom{n}{k}_q (x-1)(x-q)\ldots(x-q^{k-1})$$

which is Prop. 3 of Section 2.

Corollary 2. Set $z = 1$, $y = 0$ then

$$(2)\qquad (x-1)(x-q)\dots(x-q^{n-1}) = \sum_{k=0}^{n} \binom{n}{k}_q x^k q^{\binom{n-k}{2}} (-1)^{n-k} =$$

$$= \sum_{k=0}^{n} \binom{n}{k}_q (-1)^k q^{\binom{k}{2}} x^{n-k} .$$

This identity goes back to Cauchy (and probably even earlier).

(2) can also be derived directly by a Möbius inversion argument as follows: count all one-to-one linear transformations from V_n into X. The left side clearly counts this directly by the same method used in proving (1). To derive the right side let $N_=(W)$ be the number of linear transformations from V_n to X whose null space equals W and let $N_{\geq}(W)$ be the number of linear transformation from V_n to X whose null space contains W. Clearly

$$N_{\geq}(U) = \sum_{W \geq U} N_=(W)$$

for every subspace U of V_n. Hence by Möbius inversion

$$N_=(U) = \sum_{W \geq U} \mu(U,W) N_{\geq}(W)$$

and setting $U = 0$ (0 is the zero subspace)

$$N_=(0) = \sum_{W \subseteq V_n} \mu(0,W) N_{\geq}(W) .$$

But $N_=(0)$ counts all one-to-one linear transformations since a linear transformation is one-to-one iff its nullspace is 0. Thus we count the one-to-one maps by "sieving" through all maps. We now have the identity

$$(3)\qquad (x-1)(x-q)\dots(x-q^{n-1}) = \sum_{W \subseteq V_n} \mu(0,W) N_{\geq}(W) .$$

To compute $N_{\geq}(W)$, the set of all linear transformations that send W into 0, let $v_1, \dots, v_n$ be a basis for V_n such that $v_1, \dots, v_k$, $k = \dim W$, are a basis for V_n. Then each of $v_1 \dots v_k$ must map into zero and the remaining $v_{k+1}, \dots, v_n$ can map into any vector in X and we can choose the images in x^{n-k} ways, i.e. $N_{\geq}(W) = x^{n-\dim W}$. Substituting this result and the value of μ into (3) we prove our identity (2).

However (3) also gives us a new method of computing the Möbius function. If we substitute the value of $N_{\geq}(W)$ into (3), observe that $\mu(0, W)$ depends only on the dimension of W (i.e. the isomorphism type of $(0, W)$) [see Rota (1964)], and equate the right hand side of equations (2) and (3), then equating coefficients of x^k we get

$$\mu(0,W) = (-1)^k q^{\binom{k}{2}} \qquad \text{where} \qquad k = \dim W .$$

It is now easy to derive from (2) a famous identity due to Euler. Replace y by x^{-1} and multiple both sides by x^n, obtaining

$$(1-x)(1-qx)(1-q^2x)\dots(1-q^{n-1}x) = \sum_{k=0}^{n} (-1)^k \binom{n}{k}_q q^{k(k-1)/2} x^k .$$

This is a polynomial identity holding for all x and q. We may therefore let n tend to infinity, and obtain Euler's identity (convergence obtains in the q-adic norm, trivially):

$$\prod_{n=0}^{\infty} (1-xq^n) = \sum_{k=0}^{\infty} \frac{(-1)^k q^{k(k-1)/2} x^k}{(1-q)(1-q^2)\dots(1-q^k)} = e_q(x) \tag{4}$$

where we have used the fact that, as $n \to \infty$, in the q-adic norm

$$\begin{bmatrix} n \\ k \end{bmatrix} \to \frac{1}{(1-q)(1-q^2)\dots(1-q^k)} ,$$

and where $e_q(x)$ is the generating function of μ introduced in Section 3.

Therefore the general fact that the Möbius function is the inverse of the zeta function yields at once another famous identity of Euler, namely

$$(5)\qquad \prod_{n=0}^{\infty} \frac{1}{(1-xq^n)} = \sum_{k=0}^{\infty} \frac{x^k}{(1-q)(1-q^2)\dots(1-q^k)} = E_q(x).$$

Corollary 3. Set $y = 0$ in (1), then

$$(6)\qquad (x-z)\dots(x-q^{n-1}z) = \sum_{k=0}^{n} \binom{n}{k}_q x^k (-1)^{n-k} q^{\binom{n-k}{2}} z^{n-k} =$$

$$= \sum_{k=0}^{n} \binom{n}{k}_q (-1)^k q^{\binom{k}{2}} z^k x^{n-k}.$$

This identity seems a little more general than (2) but can actually be derived from it by setting $x = \frac{x}{z}$. We can again derive this directly by counting all one-to-one linear transformations of V_n into X such that $F(V_n) \cap Z = 0$. The left side has been proved in (2). The right side comes from setting

$N_=(W) =$ number of linear transformation f such that $f^{-1}(Z) = W$.

$N_\geq(W) =$ number of linear transformations f such that $f^{-1}(Z) \geq W$,

and proceeding by Möbius inversion as in Corollary 2.

It is interesting to note that setting $z = 1$ in (6) yields (2) algebraically. Geometrically, it corresponds to shrinking the subspace Z to 0. It would be of the utmost interest to extend this correspondence between the general algebraic and geometric theory.

To conclude, we derive an identity which does not appear to be a corollary of (1).

Proposition 1.

(7) $$(x-z)(x-qz)\dots(x-q^{n-1}z) =$$

$$= \sum_{k\geq 0} \binom{n}{k}_q (-1) q^{\binom{k}{2}} \prod_{i=0}^{k-1} (z-q^i) \prod_{i=0}^{n-k-1} (x-q^{k+i}) .$$

Proof: As in (2) the left side counts the set of all one-to-one linear transformations f from V_n to X s.t. $f(V_n)\cap Z = 0$. We derive the right side as in (2) by Möbius inversion, but this time we sieve only through the one-to-one transformations instead of all linear transformations. Let

$N_=(W)$ be the number of one-to-one linear transformations $f: V_n \to X$ such that $f^{-1}(Z) = W$

$N_\geq(W)$ equal the number of one-to-one linear transformations $f: V_n \to X$ such that $F^{-1}(Z) \supseteq W$.

Then $N_\geq(U) = \sum_{W\supseteq U} N_=(W)$. Inverting and setting $U=0$ we get $N_=(0) = \sum_{W\subseteq V_n} \mu(0,W)N_\geq(W)$ which is the desired identity.

We next compute $N_\geq(W)$. Let $k = \dim W$. Then a linear transformation is counted by $N_\geq(W)$ whenever it is one-to-one and it sends W into Z. Therefore,

$$N_\geq(W) = \prod_{i=0}^{k-1} (z-q^i) \prod_{i=0}^{n-k-1} (x-q^{k+i}) .$$

Substituting this in the formula for $N_=(0)$ we get (7), q.e.d.

Letting $n\to\infty$ in (7) obtain the following infinite q-identity

(8) $$\prod_{i=0}^{\infty} (x-q^i z) = \sum_{k=0}^{\infty} \frac{(-1)^k q^{\binom{k}{2}} \prod_{i=0}^{k-1} (z-q^i) \prod_{i=0}^{\infty} (y-q^{k+1})}{(1-q)(1-q^2)\dots(1-q^k)}$$

The particular interest case of this comes from setting $x = 1$, $z = 0$. This yields

$$1 = \sum_{k=0}^{\infty} \frac{(-1)^k q^{\binom{k}{2}} \prod_{i=0}^{k} (-q^i) \prod_{i=0}^{\infty} (1-q^{k+i})}{(1-q)\dots(1-q^k)} =$$

$$= \sum_{k=0}^{\infty} \frac{q^{2\binom{k}{2}} \prod_{i=0}^{\infty} (1-q^{k+i})}{(1-q)\ \dots\ (1-q^k)} \ \frac{\prod_{i=1}^{k} (1-q^i)}{\prod_{i=1}^{k} (1-q^i)} =$$

$$= \sum_{k\geq 0} \frac{q^{2\binom{k}{2}} (1-q^k) \prod_{i=0}^{\infty} (1-q^i)}{(1-q)^2 \dots (1-q^k)^2} .$$

From this we arrive at the

Corollary

$$\frac{1}{\prod_{i=0}^{\infty} (1-q^i)} = \sum_{k=0}^{\infty} \frac{q^{2\binom{k}{2}} (1-q^k)}{(1-q)^2 \dots (1-q^k)^2} .$$

This generalizes Durfee's identity (Hardy and Wright pg. 281) in the theory of partitions of a number, by looking at the largest $k \times k-1$ rectangle in the Ferrars diagram of a partition instead of the largest (Durfee) square.

6. FURTHER WORK AND OPEN PROBLEMS

The present paper does little more than scratch the surface of a field of research that may prove fertile in connecting various branches of mathematics. In closing, we should like to outline some of the directions in which further work might proceed, and some of the research problems that appear at present to us to be most promising.

Perhaps the most tantalizing open problem is a construction of a q-probabilistic setup within which formulas such as the Eulerian expansions (4) and (5) of the preceding section can be justified. One can see heuristically that these formulas should be related to some sort of q-Poisson distribution, as follows. From formula (2) of the preceding section we see that the following is true: Given a vector space X with x vectors, and a linear transformation from V_n to X picked at random, the probability that this linear transformation shall be one-to-one is

$$\sum_{k=0}^{n} \binom{n}{k} (-1)^k q^{\binom{k}{2}} x^{-k} . \tag{1}$$

This is simply the right side of (2) divided by the total number of linear transformations from the space V_n to the space X, which is x^n.

Now let $n \to \infty$ in (1). The series converges in the q-adic norm and yields precisely the Eulerian formula (4) of the preceding section, with x^{-1} replacing x. It is therefore reasonable to surmise that this should be a candidate for the probability stated above.

A deeper argument would yield the q-analog of a Poisson distribution along the same lines. However, a set-theoretic justification of this heuristic limit-taking is much more difficult to obtain. We have obtained some results in this direction using some methods drawn from Von Neumann's theory of continuous geometries, specifically, using as the analog of probability the dimension function in a continuous geometry constructed by an inductive limit of finite-dimensional vector spaces over a field with q elements. In a certain sense, this continuous geometry is the q-analog of a nonatomic probability space. The difficulty is that this geometry is not represented by subspaces of a vector space (only by ideals in a regular ring), and the development of a probability theory becomes exceedingly delicate. Nevertheless, we think that it is both possible and desirable.

In the same vein, we surmise that most of the identities that heretofore have been proved using the theory of partitions and elliptic functions are interpretable set-theoretically in terms of enumeration in finite or in infinite-dimensional vector spaces. A simple example is the theta series

$$(2) \qquad \sum_{k=0}^{\infty} (-1)^n q^{n(n-1)/2} x^n ,$$

which is the Eulerian generating function of the number of automorphisms of a vector spaces of dimension n over a field with q elements. The ultimate goal is a set-theoretic proof and interpretation of the Rogers-Ramanujan identities.

In this connection, one of the mysteries is the dual interpretation of various Eulerian formulas both as enumeration of subspaces or of linear transformations in vector spaces with given properties, and on the other hand as partitions of a number with given properties. A connection between these two approaches would be very revealing.

In another direction, the interpretation of linear q-difference equations with constant coefficients derived in Section 4 could be extended to more complex q-difference equations with nonconstant coefficients (for example, the q-difference equations for the q-hypergeometric function of Jackson).

The Theorem of Section 5 can be made the basis for q-analogs of the theories of Appell polynomials and basic polynomials. The first such theory would be concerned with polynomials $r_n(x,y)$ satisfying the identities

$$(3) \qquad r_n(x,z) = \sum_{k=0}^{n} \binom{n}{k}_q r_k(x,y)\, P_{n-k}(y,z) .$$

where P_n is defined in Section 5. These are the analogs of Appell polynomials. We know of several examples of polynomials in the literature that satisfy these identities. For example, the q-Hermite polynomials introduced by Carlitz.

On the other hand, a system of basic polynomials $b_n(x,z)$ satisfies the identities

(4) $$b_n(x,y) = \sum_{k=0}^{n} \binom{n}{k}_q b_k(x,y) b_{n-k}(y,z),$$

an analogy with the basic polynomials such as developed for example in Foundations III. Such systems of polynomials seem to be rare. In the same vein, one can introduce the analog of shift operators and the notion of a shift-invariant operator, again in an analogy with Foundations III.

The analogy between q-identities and binomial identities can be carried further to a straight set-theoretic analogy between problems and concepts for subspaces of vector spaces and on the other hand for subsets of a set. One of the most intriguing notions to be developed is the q-analog of the notion of a partition of a set. In this connection see the recent work of Bender-Goldman.

A particularly simple analog is the classical result of Sperner regarding the maximal antichain in the lattice of subsets of a set. This theorem, as well as the well-known proof given by Lubell, carry over almost without change to the lattice of subspaces of a vector space (see Lubell, or Harper-Rota). A more intriguing conjecture is the analog of Ramsey's theorem, which has recently been studied by Graham and Rothschild. Finally, one of us has begun a study of q-analogs of the Euler characteristic, q-analogs of simplicial complexes, and, in a more speculative vein, q-analogs of homology.

All these questions point to a major problem, which is of a somewhat philosophical nature; this is the problem of explaining why q-identities relating to vector spaces tend to ordinary identitites for binomial coefficients as q tends to 1. A purely set-theoretic explanation of this fact would be of great significance.

REFERENCES

This bibliography does not repeat any references listed under Foundations I, II, III.

[1] BENDER, E. and GOLDMAN J., "The Enumerative Uses of Generating Functions", Journal of Mathematics and Mechanics (to appear).

[2] CARLITZ, L., "Some Polynomials Related to Theta Functions", Ann. Math. Pura App 11. (1956), 359-373.

[3] CARLITZ, L., "Some Polynomials Related to Theta Functions", Duke Math. J. 24 (1957), 521-527.

[4] CRAPO, H., ROTA, G.-C., "On the Foundations of Combinatorial Theory II, Combinatorial Geometries" (to appear).

[5] GAUSS, C.G., "Summation Quarundam Serierum Signularium", Werke 2 (1876), 9-45.

[6] GOLDMAN, J., "The Method of Special Cases in Combinatorial Analysis and Probability", (to appear).

[7] GOLDMAN, J. and ROTA, G.-C., "The Number of Subspaces of a Vector Space" in Recent Progress in Combinatorics edited by W. Tutte, Academic Press (1969), 75-83.

[8] GOULD, H.W., "Combinatorial Identities, A Standardized Set of Tables Listing 500 Binomial Coefficient Summations", West Virginia University (Nov. 1959).

[9] GRAHAM,R.and ROTHSCHILD, B., "Ramsey's Theorem for n-Dimensional Arrays", Bulletin of American Math. Soc., 75 (1969) 418-422.

[10] HARPER, L.H. and ROTA, G.-C., "Matching Theory, An Introduction, to appear in Studies in Probability ed. by P. Ney, North Holland.

[11] HEYNEMAN, R., SWEELER, M., "Affine Hopf Algebras, I", Journal of Algebra 13 (1969) 192-241.

[12] JACKSON, F.H., "Basic Intergration", Quarterly J. of Math. (Oxford) (2), p. 1-6 (1951).

[13] MACLANE, S., 'Homology", Academic Press and Springer Verlag (1963).

[14] MULLIN, R., ROTA, G.-C., "On the Foundations of Combinatorial Theory III, Theory of Binomial Enumeration", to appear (Academic Press).

[15] ROTA, G.-C., "On the Foundations of Combinatorial Theory, I", Zeitschrift für Wahrscheinlichkeitstheorie, Band 2, Heft 4, 5. (1964) 340-368.

[16] ROTA, G.-C., "The Number of Partitions of a Set", Amer. Math. Monthly 71, No. 5 (1964), 498-504.

[17] ROTA, G.-C., "Möbius Functions and the Euler Characteristic", Rado Festschrifte, Academic Press, (London) 1970.

Combinatorics and graph theory of abundance and stability of chemical species

by

M. Gordon
Essex, England

ABSTRACT

PÓLYA stressed the importance of the hierarchy of groups of a chemical structure considered as a graph, as a three-dimensional object, etc. Group-theoretical invariants occur as factors in molecular partition functions, especially the order of the molecular rotation group. Accordingly, purely enumerative theories (Cayley, Pólya) can be generalised to treat stability and relative abundance of chemicals. I. J. Good speculated that the number T of distinct rooted ordered trees isomorphic with a given tree-like molecula was a measure of its abundance in nature. This view has been confirmed for certain polymers, but the statistical mechanical significance of T, which also features prominently in applications of cascade (branching) theory to synthetic polymers, needed clarification. Harmony is restored (Gordon & Temple) by a theorem valid for all finite vertex-labelled trees and generalisable to vertex-labelled graphs:

$$T = \underline{f}! \sum f_i / |G| \prod f_j$$

Thus T is inversely proportional to the order $|G|$ of the automorphism group. (f_i the degree of the ith vertex, and $\underline{f}! \equiv f_1! f_2! \ldots$).

The power of cascade theory for chemical combinatorics lies in the way it maps algebraic functions into trees. It leads to new results for enumeration of Cayley trees under restrictions.

1. STATISTICAL FORESTS (s.f.)

The simplest kinds of cascade (or "branching") processes [1, 2a, 2b] lead to statistical forests (s.f.) of family trees, defined in terms of the pgf's $F_0(\Theta)$, $F_1(\Theta)$, ... for the number of offsprings of members of generations $g_0, g_1, \ldots$ etc. We shall mainly be concerned with forests of $(1,f)$-trees, i.e. trees containing only two kinds of vertices: those of degree f (a constant ≥ 2) which are called nodes, and those of degree unity, called 'terminals'. The most random s.f. of node-rooted $(1,f)$-trees is generated thus:

(1) $$F_0(\Theta) = (1-\alpha+\alpha\Theta)^f$$

(2) $$F_n(\Theta) = (1-\alpha+\alpha\Theta)^{f-1} \qquad n = 1, 2, \ldots$$

The parameter α is the fraction of vertices on generation g_1 which are nodes.* By repeated use of a simply modified cascade substitution [2a] one obtains the g.f. for the probability w_x of a tree in the s.f. being an x-mer (i.e. having exactly x nodes):

(3) $$W(\Theta) = \sum w_x \Theta^x = \Theta F_0(\Theta F_1(\Theta F_1(\Theta F_1(\ldots)))).$$

*Chemists study the statistical forest defined by eq. (1) and (2) in connection with f-functional random polycondensation; α is called the degree of advancement of the reaction, and w_x the weight fraction of x-mer, i.e. of trees having exactly x nodes.

The same function $W(\Theta)$ may also serve as an enumeration [2a] function of node-rooted ordered x-mers, e.g. by putting $\alpha = 1/2$. Two non-isomorphic finite x-mer trees are called 'isomers' [3] and we shall refer to the jth of such isomers (in any arbitrary fixed system of enumeration) simply as the j th x -mer. The forest generated by (3) is shown in this paper to contain every isomer with a probability proportional to the number of distinct ordered rooted trees [4] with which it is isomorphic. The typographical trick of writing a product n factors $a_1 a_2 \cdots a_n$ so as to emphasise a_1:

$$(4) \qquad a_1 a_2 \cdots a_n \equiv a_1 \begin{cases} a_2 \\ a_3 \\ \vdots \\ a_n \end{cases}$$

allows us to display the connection of the combinatorial formula (3) with graph theory. Substituting (1) and (2) in (3), for example, for the simple case $f = 3$, we obtain fig. 1. The Θ-symbols in fig. 1 clearly stand in one-one correspondence to the nodes in the infinite lattice upon which the rooted trees of the s.f. can be arranged. Any single term in Θ^x $(x = 1,2,\ldots)$ in the expansion of $W(\Theta)$ can only correspond to a singly-connected tree which is part of this lattice. Moreover, this tree must be rooted, in that one of the x Θ-factors of such a term must be the one lying on g_0. Moreover, every distinc ordered rooted x-mer tree on the lattice of fig. 1 contributes an equal term $\alpha^{x-1}(1-\alpha)^{fx-2x+2}$ to the probability w_x of an x-mer tree in the s.f. This term corresponds to the probability α^{x-1} of finding the $x-1$ links between pairs of nodes, times the independent probability $(1-\alpha)^{fx-2x+2}$ of finding the $fx-2x+2$ terminals which occur in any x-mer tree. The number of such equal terms, moreover, is the number of distinct node-rooted ordered $(1,f)$-trees, i.e. the number $T_{nx}(f)$ of distinct node-rooted trees of x nodes which can be drawn on a lattice such as fig. 1. It is shown [5] by Lagrange expansion, that

(5) $$w_x = \frac{(fx-x)!\,f}{(x-1)!\,(fx-2x+2)!}\,\alpha^{x-1}(1-\alpha)^{fx-2x+2},$$

a classical result previously obtained by various ad hoc combinatorial methods. It follows that

(6) $$T_{nx}(f) = (fx-x)!\,f/(x-1)!\,(fx-2x+2)!$$

a result which can be established directly (cf. Good [2b]). Examining the tree-distribution in finer detail than merely in terms of the number x of nodes in a tree, one sees that

(7) $$w_{xj} = T_{xj}(f)\,(1-\alpha)^{fx-2x+2}\,\alpha^{x-1}$$

where w_{xj} is the chance of a jth x-mer $(1-f)$-tree in the s.f., and $T_{xj}(f)$ the number of distinct ordered rooted trees isomorphic with the jth x-mer.

2. GROUPS, GRAPHS, AND ABUNDANCE OF CHEMICALS

As long ago as 1960, I. J. Good [2b] speculated that the number T of rooted ordered trees isomorphic with a given isomer tree was some kind of measure of the abundance of that isomer in nature. The development of the cascade formalism (in conjunction with chemical experiments) provided a quantitative support for this view as regards abundance in synthetic inorganic [6] and organic [7] polymer systems, at least when allowance is made for a small amount of cyclic graphs and other disturbances. Recently, Burchard and his coworkers [8] have found convincing evidence - based on interpreting light-scattering data in terms of cascade-theoretical formulae - that the natural polymer amylopectin contains isomeric structures in proportion to T. The cascade theory approach seemed, however, to be of an ad hoc nature and not properly integrated with statistical mechanical theory, [9] which should of course be able to predict the abundance of isomers on its own terms. Thus in the petrol industry, for instance, abundance of the various octane isomers in cracking mixtures would be calculated from statistical mechanics through the partition functions of these gases, which make no mention of T. Instead, when properly formulated, the corresponding integral invariant which appears there is the reciprocal of the order $|R|$ of the rotation group of the molecule. Complete harmony between cascade theory and statistical mechanics was recently achieved, [10] however; the essential link was the proof [10, 11] of eq. (9) below. This shows that for any finite tree, the number T of distinct rooted ordered representations is inversely proportional to the order $|G|$ of its automorphism group, of which the group R is always (cf. Pólya [12]) a subgroup. Eq. (9) is valid also for vertex-labelled trees, and, after extending appropriately [11] the definition of ordering a tree, to vertex-labelled graphs containing cycles.

The vertices of a given tree fall into topological equivalence classes under the automorphism group G; let s_a be the order of the ath equivalence class. Let T_a be the number of distinct ordered tree, rooted on a vertex of the ath equivalence class, which are isomorphic with the given tree. By the technique of labelling and delabelling, putting distinct labels on all the $x-1$ edges, one readily shows (cf. the proof of eq. (18) below) that

(8) $$T_a = s_a f_a \underline{f}! / |G| \prod f_j$$

(because $|G|/s_a$ is the order of the stabiliser group of a vertex of the ath equivalence class!).

If we sum (8) over all equivalence classes we obtain the total number of distinct rooted ordered trees isomorphic with the given tree:

(9) $$T = (\sum f_i)\, \underline{f}! / |G| \prod f_j$$

Restriction to $(1, f)$-trees yields:

(10) $$T_{xj} = 2(fx - x + 1)((f-1)!)^{x} / |G_{xj}|$$

If we sum (8) merely over the equivalence classes which comprise the nodes (leaving out the terminals), we obtain the number of distinct node-rooted ordered trees isomorphic with the given tree; for $(1,f)$-trees we find

(11) $$T_{n\varkappa j} = f\varkappa((f-1)!)^{\varkappa}/|G_{\varkappa j}|$$

From (10) and (11) follows:

(12) $$T_{\varkappa j} = 2T_{n\varkappa j}(f\varkappa-\varkappa+1)/f\varkappa$$

By summation and comparison with (6), we see that

(13) $$\sum_j T_{\varkappa j} \equiv T_{\varkappa} = 2T_{n\varkappa}(f\varkappa-\varkappa+1)/f\varkappa = $$
$$= 2(f\varkappa-\varkappa)!(f\varkappa-\varkappa+1)/(f\varkappa-2\varkappa+2)!\varkappa!$$

Likewise summing (11) over j, and comparing with (6), we, discover the interesting relation:

(14) $$\sum_j |G_{\varkappa j}|^{-1} = (f\varkappa-\varkappa)!/\varkappa!(f\varkappa-2\varkappa+2)![(f-1)!]^{\varkappa}$$

Using Stirling's approximation yields

(15) $$\lim_{\varkappa\to\infty} \sum_j |G_{\varkappa j}|^{-1} = (e/(f-2))^{(f-2)\varkappa}$$

PÓLYA [12] has given an asymptotic formula for the number $\varrho_{\varkappa}$ of distinct $\varkappa$-mers for $f=4$ (the case of saturated acyclic hydrocarbons with $\varkappa$ carbons):

$$\lim_{\varkappa\to\infty} \varrho_{\varkappa} = \varkappa^{-5/2}\varrho^{-\varkappa}$$

with

(16) $$0.35 < \varrho < 0.36$$

Hence the average reciprocal automorphism group order of such hydrocarbons behaves asympotically as

(17) $$\langle |G_{xj}|^{-1} \rangle \sim x^{5/2} (\varrho^{1/2} e/2)^{2x} \to 0$$

The implication that $\langle |G_{xj}| \rangle$ diverges rather strongly as $x \to \infty$ is somewhat disappointing for theories of physical properties of randomly branched polymers. In these the assumption that symmetry (due to operations of G_{xj}) can be neglected for sufficiently large structures would often lead to useful simplificiations (especially in connection with eigenvalue problems).

3. NUMBER OF CAYLEY TREES WITH RESTRICTIONS

Because of the important relation of the automorphism group order to the integral invariant in the partition function of a (perfectly flexible) molecule, combinatorial formulae for various quantities, which carry $|G|$ in the denominator, just like Pólya's cycle index, [12] are of interest in chemical physics. To the various examples just given we add another one, due to Pólya himself,[12] which again is immediately obvious using the technique of labelling and delabelling:

(18) $$\mathfrak{J} = y^{y-2} = y! \sum |G_i|^{-1}$$

Here $\mathfrak{J}$ is the number of distinct trees of order y, whose y vertices are distinctly coloured (with y fixed colours). The first equality of (18) is, of course, due to Cayley (cf. Moon in reference [3]); Pólya's result is the second equality, where the summation is over all topologically distinc uncoloured trees.

Proof: The ith topological tree can have its vertices coloured (labelled) in $y!$ ways, which would all be distinct, were it not for the fact that they fall into classes of $|G_i|$ mutually indistinguishable members each. The members of each class are related to each other under mappings of the automorphism group of the uncoloured (delabelled) tree.

A useful route can be discerned to combinatorial formulae for quantities defined with (often irksome) inbuilt restrictions. The route consists in eliminating $|G|^{-1}$ from two relations of the kind just discussed. The first example is the equation for the relative number of orderings of a given tree when roots are restricted to two different vertices:

(19) $$T_a / T_b = s_a f_a / s_b f_b$$

obtained by eliminating $|G|$ between the applications of eq. (8) to vertex a and to vertex b. Eq. (19) like (9) itself, has been generalised as far as graphs with cycles whose vertices may be coloured. Eq. (19) is deceptively simple; the first proof found [11] was of a probabilistic nature and rested on the theory of cascade processes. The same "vanishing trick" of the group-theoretical content, by cancelling $|G|$, may be applied to (18) as a second example. From the proof given, (18) may immediately be restricted to x-meric $(1,f)$-trees in the following form:

(20) $$\mathcal{F}_{(1,f)} = y! \sum_j |G_{x_j}|^{-1}$$

with the summation restricted to automorphism groups of all topologically distinct $(1,f)$-trees. We obtain the analogue of Cayley's result (first equality of (18)), subject to this restriction, by eliminating $\sum_j |G_{x_j}|^{-1}$ between (20) and (14). Since the total number y of vertices is related to the number x of nodes of any $(1,f)$-tree thus:

(21) $$y = fx - x + 2$$

the result may be written:

(22) $$\mathcal{F}_{(1,f)} = \binom{fx-x+2}{x} (fx-x)! [(f-1)!]^{-x}$$

The loss in generality and elegance in passing from $\mathcal{F}_4$ to $\mathcal{F}_{(1,f)}$ is compensated by the fact that forests of Cayley trees restricted to $(1,f)$-type are, unlike unrestricted ones, of practical interest to chemists. Cascade theory in principle gives access to combinatorial formulae analogous to (14) in which further restrictions are placed on the kinds of trees (apart from restriction to $(1,f)$-trees). This will be reported in detail elsewhere. Accordingly, there are prospects of solving problems on the number of Cayley trees under more restrictive conditions.

The asymptotics (as $x \to \infty$) of the ratio R_f of the numbers of restricted (i.e. $(1-f)$-) to unrestricted Cayley trees are found to be as follows (using eq. (21) and (22)):

(23) $$R_2 = \mathcal{T}_{(1,2)} / \mathcal{T}_y \sim (\pi/2)^{1/2} x^2 e^{2-x}$$

while for $f > 2$:

(24) $$R_f \sim (f-1)^3/(f-2)^{5/2} [(f-2)^{f-2} e^{f-1} (f-1)!]^x; \quad f > 2$$

If $1 \ll f \ll x$, Stirling's approximation may be used on $(f-1)!$, thus

(25) $$R_f \sim f^{1/2}/[\sqrt{2\pi}\ f^{2f-(5/2)}]^x \qquad 1 \ll f \ll x \to \infty$$

Formulae for enumerating somewhat similarly restricted Cayley trees by elegant methods were recently published by Rényi. [13]

4. SUMMARY OF CONTRIBUTION OF CASCADE THEORY TO CHEMICAL COMBINATORICS

Eqs. (1) and (2) are merely the simplest (binomial) prototypes of generating functions for polymer systems. Short-range correlations which destroy the completely random behaviour inherent in the binomal distribution, have been used to attain greater chemical realism. The moments of the resulting w_x-distributions, or expected values of various other graph-theoretical parameters obtained from w_x, have physical meanings and can be checked against experiments. Thus cascade theory has been used to interpret measurements of equilibrium swelling and elastic moduli, [14] of configurational statistics [15, 16] of particle scattering functions, [17] and of diffusion constants [18] in polymer systems. The method of Lagrange expansion of eq. (3), mentioned above for the simple 'random' s.f. of eq. (5), has been extended to a case with local correlations in the trees, [4] and to the more complex bivariate case. [14]

ACKNOWLEDGEMENT:

I thank the Department of Statistics, University of North Carolina, Chapel Hill, for hospitality under the auspices of the Combinatorial Mathematics Year for the month of June 1969, when the work on enumerating Cayley trees was done.

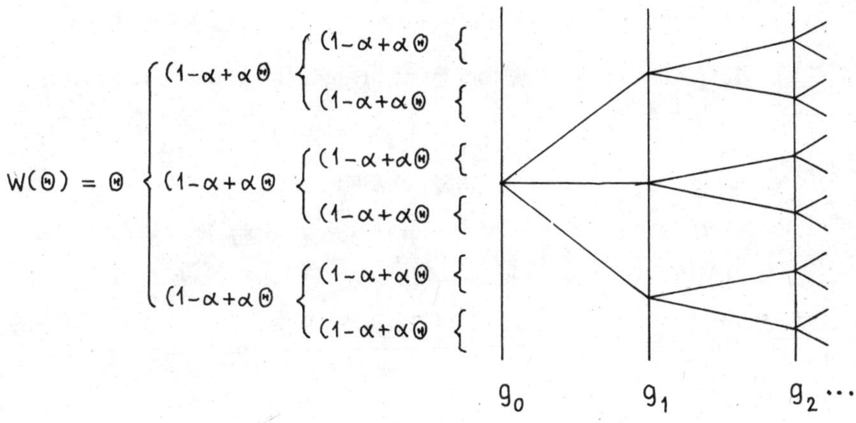

FIGURE 1. Relation of the function $W(\theta)$ to a lattice. The lattice (shown on the right) may be regarded as the intersection of all finite rooted ordered (1, 3)-trees.

ROOTING means associating one of the vertices of a given (free) tree with the root vertex of the lattice on generation g_o.

ORDERING of a rooted tree means associating one-to-one the remaining vertices and the edges of the tree with a set of vertices and edges of the lattice, while preserving all incidence relations (connectivity).

REFERENCES

[1] T.E. HARRIS, 1963, The theory of branching processes, Berlin, Springer.

[2a] I.J. GOOD, Proc. Camb. Phil. Soc., 1955, 51, 240 (reprinted in Proc. Roy. Soc., London, 1962, A268, 256)

[2b] I.J. GOOD, 1960, Proc. Camb. Phil. Soc., 1960, 56, 367.

[3] F. HARARY, 1967, A Seminar on graph Theory, New York.

[4] M. GORDON and G. R. SCANTLEBURY, 1964, Trans. Faraday Soc., 60, 604.

[5] I.J. GOOD, 1963, Proc. Roy. Soc. (London) A272, 54.

[6] M. GORDON and G.R. SCANTLEBURY, Proc. Roy. Soc. (London) 1966, A292, 380.

[7] M. GORDON and G.R. SCANTLEBURY, 1967, J. Chem. Soc., B, 1.

[8] K. KAJIWARA, B. PFANNEMÜLLER and W. BURCHARD, to be published.

[9] P. WHITTLE, 1965, Proc. Roy. Soc. (London) A285, 501.

[10] M. GORDON and W. B. TEMPLE, to be published.

[11] M. GORDON, T.G. PARKER and W.B. TEMPLE, submitted to J. Combinatorial Theory.

[12] G. PÓLYA, Acta. math., 1937, 68, 145.

[13] A. RÉNYI, 'Lectures on the theory of search', Mimeo 600.7, Department of Statistics, UNC, Chapel Hill, N.C., U.S.A.

[14] G. R. DOBSON and M. GORDON, 1965, J. Chem. Phys. 43, 705, reprinted Rubb. Chem. Technol., 1966, 39, 1472; M. Gordon, S. Kuchárik and T. C. Ward, to be published; K. Dusek and W. Prins, Adv. Polymer Sci., 1969, 6, 1.

[15] G. R. DOBSON and M. GORDON, 1964, J. Chem. Phys. 41, 2389.

[16] D. S. BUTLER, M. GORDON and G. N. MALCOLM, 1966, Proc. Roy. Soc. (London) A295, 29.

[17] K. KAJIWARA, W. BURCHARD and M. GORDON, submitted to Brit. Polymer J.

[18] M. GORDON and R. ASHLEY, to be published.

On a close relation between Sperner collections and certain chance games

by

F. Göbel

Enschede, The Netherlands

1. SPERNER COLLECTIONS

Let Ω be a set of n elements, to be indicated by the first n natural numbers: $\Omega = \{1, 2, \dots, n\}$. Subsets of Ω will be denoted by $\Lambda, \Lambda', \Lambda_1, \dots$ A collection $\mathcal{S}$ of subsets $\Lambda_1, \dots, \Lambda_k$ of Ω, to be denoted $\mathcal{S} = \{\Lambda_1 \vee \dots \vee \Lambda_k\}$, is called a <u>SPERNER collection</u> if $\Lambda_i \subseteq \Lambda_j$ implies $i = j$. The empty collection is denoted by $\emptyset$, the collection consisting of the empty set by $\{\emptyset\}$.

A collection $\mathcal{F}$ of subsets of Ω is called a <u>SPERNER family</u> if $\Lambda \in \mathcal{F}, \Lambda' \supseteq \Lambda$ imply $\Lambda' \in \mathcal{F}$.

The minimal elements (in the set theoretic sense) of a Sperner family form a Sperner collection. Conversely, to each Sperner collection we can assign a Sperner family in an obvious way.

Collections $\mathcal{S}$ and $\mathcal{S}'$ are called <u>equivalent</u> if there exists a permutation σ of Ω such that the "induced" permutation of $\mathcal{S}$ yields $\mathcal{S}'$.

If $\mathcal{S}$ and $\mathcal{S}'$ are Sperner collections, then $\mathcal{S} \succ \mathcal{S}'$ ($\mathcal{S}$ <u>dominates</u> $\mathcal{S}'$) means by definition: the family of $\mathcal{S}$ is included in the family of $\mathcal{S}'$.

The <u>dual</u> $\hat{\mathcal{S}}$ of a Sperner collection $\mathcal{S}$ is formed as follows:

a) form the corresponding family,

b) replace each member of the family with its complement with respect to Ω,

c) take the complement with respect to 2^{Ω},

d) take the minimal elements.

It is easily shown that $\hat{\mathcal{S}}$ is a Sperner collection, and that it has the same size (number of elements of Ω in $\cup \Lambda_j$) as $\mathcal{S}$.

2. SPERNER STRATEGIES IN CHANCE GAMES

2.1 Description of a game

A player plays n games of chance in succession. The probability that he wins the j-th game is $p_j > 0$. The games are stochastically independent.

The initial capital of the player is 1. Money is assumed to be infinitely divisible, and the player can stake any real number between 0 and his momentary capital.

When he wins, he obtains his stake plus the same amount; when he loses, he forfeits his stake. His final capital is denoted by Z. A strategy (i.e. a rule according to which the player stakes) is τ-optimal when $P\{Z \geq \tau\}$ is maximum.

In his paper "Optimal Gambling Systems for Favorable Games" (Proc. 4th Berkeley Symp. <u>1</u> (1961), 65-78), L. Breiman describes an optimal strategy in the case $p_1 = \dots = p_n = p > \frac{1}{2}$.

In this note, we present a class of strategies, the Sperner strategies, which have a number of desirable properties, to be discribed below.

2.2 Sperner strategies, definition

Let a Sperner collection $\mathcal{S} = \{A_1 \cup \dots \cup A_k\}$ on Ω be given. We will construct a strategy $\Sigma(\mathcal{S})$, and a target $\tau(\mathcal{S})$. We number the games backwards: $n, n-1, \dots, 2, 1$.

Initial step. When $n=1$, there are 3 cases: $\mathcal{S} = \{\phi\}$ or $\{\{1\}\}$ or ϕ.

When $\mathcal{S} = \{\phi\}$, we define Σ: "stake 0", $\tau = 1$.

When $\mathcal{S} = \{\{1\}\}$, we define Σ: "stake 1", $\tau = 2$.

When $\mathcal{S} = \phi$, we define Σ: "stake 0", $\tau = \infty$.

Induction step. Suppose a strategy and a target have been assigned to all Sperner collections on $n-1$ elements $(n \geq 2)$. It is convenient to distinguish two cases.

A) $n \notin \bigcap A_j$,

B) $n \in \bigcap A_j$.

A) Suppose we stake x on game n.

When we win, we use* $\Sigma(\mathcal{S}_1)$, where $\mathcal{S}_1$ is obtained from $\mathcal{S}$ as follows: omit all n's from $A_1, \dots, A_k$ and then take the minimal elements. When we lose game n, we use $\Sigma(\mathcal{S}_2)$ where $\mathcal{S}_2$ is obtained from $\mathcal{S}$ by omitting all A's containing n. We will use the notation $\tau_j = \tau(\mathcal{S}_j)$, $j = 1, 2$. $\tau(\mathcal{S})$ is now defined by

$$\tau(\mathcal{S}) = \max_x \min \{\tau_1(1+x), \tau_2(1-x)\},$$

that is:

$$\tau(\mathcal{S}) = \frac{2\tau_1\tau_2}{\tau_1 + \tau_2},$$

* Here and in the sequel, all stakes in a strategy should be multiplied by the new initial capital, if this differs from 1.

and the maximum is achieved for

$$x_0 = \frac{\tau_2 - \tau_1}{\tau_1 + \tau_2}$$

$\Sigma(\mathcal{S})$ is defined as: "Stake x_0 on game n; when you win, use $\Sigma(\mathcal{S}_1)$; when you lose, use $\Sigma(\mathcal{S}_2)$".

B) We stake 1 on game n. When we win, we use $\Sigma(\mathcal{S}_1)$ as defined under A). τ is defined as $2\tau_1$. When we lose, our capital is 0, and the target will not be reached.

This completes the construction.

To each $\mathcal{S}$ we can assign a probability of success $\pi(\mathcal{S})$ namely the probability of reaching the target $\tau(\mathcal{S})$ when $\Sigma(\mathcal{S})$ is applied. Between $\mathcal{S}$ and $\Sigma(\mathcal{S})$ we have the relation: the strategy $\Sigma(\mathcal{S})$ increases the capital to $\tau(\mathcal{S})$ if and only if there exists $\Lambda_j \in \mathcal{S}$ such that all games in Λ_j are won.

2.3 THEOREMS

THEOREM 1: If $\mathcal{S}$ dominates $\mathcal{S}'$, then $\pi(\mathcal{S}) \leq \pi(\mathcal{S}')$, and $\tau(\mathcal{S}) \geq \tau(\mathcal{S}')$.

THEOREM 2: Let $\tau_0 \leq 2^n$ be given; then the class of Sperner strategies contains a τ_0-optimal strategy.

Proof: Let Σ_0 be given with probability of success p_0. Define a Sperner family $\mathcal{S}'$ as follows: $\Lambda \in \mathcal{S}'$ if and only if in the case of a win in all games of Λ, τ_0 is reached (at least). Let $\mathcal{S}$ be the collection of minimal elements of $\mathcal{S}'$. Now obviously $\pi(\mathcal{S}) = p_0$, and from the construction of $\Sigma(\mathcal{S})$ we see that $\tau(\mathcal{S}) \geq \tau_0$.

Notation: If all $p_j = p$ then $\pi(\mathcal{S})$ will possibly be denoted by $\pi(p,\mathcal{S})$; similarly for τ.

THEOREM 3: $\pi(\mathcal{S}, \frac{1}{2})\, \tau(\mathcal{S}, \frac{1}{2}) = 1$.

Proof: Since the game is fair, the expected final capital $\mathcal{E}Z$ is 1, From the construction it follows that Z is either τ or 0. Hence $1 = \mathcal{E}Z = \tau \cdot P\{Z = \tau\} + 0 \cdot P\{Z = 0\}$, etc.

THEOREM 4: If $\mathcal{S}$ is equivalent so $\mathcal{S}'$, then $\tau(\mathcal{S}) = \tau(\mathcal{S}')$.

Proof: $\tau(\mathcal{S}) = \tau(\mathcal{S}, \frac{1}{2})$ because τ is independent of $p_1, \dots, p_n$ provided $p_j > 0$. According to Theorem 3, $\tau(\mathcal{S}) = (\pi(\mathcal{S}, \frac{1}{2}))^{-1}$, and since $\pi(\mathcal{S}, p) = \pi(\mathcal{S}', p)$, we have $\tau(\mathcal{S}) = \tau(\mathcal{S}')$.

Remark: The strategies by which the common target of equivalent Sperner collections is reached, are entirely different in general.

THEOREM 5: $\pi(\mathcal{S}, p) + \pi(\hat{\mathcal{S}}, 1-p) = 1$; $\frac{1}{\tau(\mathcal{S})} + \frac{1}{\tau(\hat{\mathcal{S}})} = 1$.

Proof: Let $\mathcal{S}$ be given, suppose $\Sigma(\mathcal{S})$ leads to a success, i.e. $\exists \Lambda \in \mathcal{S}$ such that all games g in Λ are won. A careful examination of the definition of $\hat{\mathcal{S}}$ tells us that this is equivalent to: all $\Lambda \in \hat{\mathcal{S}}$ contain at least one success. Now let us call a failure a "success", and conversely. Thus, all $\Lambda \in \hat{\mathcal{S}}$ contain at least one "failure". The complementary event is: $\exists \Lambda \in \hat{\mathcal{S}}$, such that Λ contains all "successes".

Since $P\{\text{"success"}\} = 1-p$, the first half of the theorem follows. The second half is an easy consequence of Theorem 3.

2.31 Numerical example

Let $n = 4$. The first of the table below gives the possible values of $16/\tau(\mathcal{S})$ when $\mathcal{S}$ ranges through all Sperner collections on $\{1, 2, 3, 4\}$. The second row gives the number of $\mathcal{S}$ for which the values in the first row occur.

0	1	2	3	4	5	6	7	8	9	10	11	12	13	14	15	16
1	1	4	6	10	13	18	19	24	19	18	13	10	6	4	1	1

Ramsey's theorem for N-parameter sets: an outline

by

R. L. Graham and B. L. Rotschild
Murray Hill, USA Cambridge, USA

INTRODUCTION

In 1930, F. P. Ramsey [7, 9] proved the following theorem:

Theorem [Ramsey]: Let ℓ, k, r be positive integers. Then there is a number $N = N(\ell, k, r)$ depending only on ℓ, k and r with the following property: If S is a set with at least N elements, and if all the subsets of S with k elements are divided into r classes in any way, then there is some subset of ℓ elements with all of its subsets of k elements in a single class.

Since this theorem appeared there has been interest in finding generalizations, applications and analogues of it. The work presented here was motivated by a conjecture made by Gian-Carlo Rota, a geometric analogue to Ramsey's theorem, which can be stated as follows:

Conjecture [Rota]: Let ℓ, k, r be nonnegative integers, and F a field of q elements. Then there is a number $N = N(q, r, \ell, k)$ depending only on q, r, ℓ and k with the following property: If V is a vector space over F of dimension at least N, and if all the k-dimensional subspaces of V are divided into r classes in any way, then there is some ℓ-dimensional subspace with all of its k-dimensional subspaces in a single class.

This conjecture is obtained from the statement of Ramsey's theorem essentially by replacing the notions of set and cardinality by those of vector space and dimension, respectively, if we replace the notion of vector space with that of affine space, then we obtain another conjecture. This conjecture is actually equivalent to Rota's conjecture [2, 8]. In this paper we outline the proof of another analogue to Ramsey's theorem, in which we replace the notion of n-dimensional affine space by the notion of n-parameter set, which we define later. The n-parameter sets are similar to n-dimensional affine spaces in certain ways, and, in fact, by appropriate choice of certain variables we can obtain results for vector and affine spaces. In particular, the affine conjecture is shown to be true for the cases of $k = 0$ and $k = 1$, with any choice for ℓ, r and q. This implies that Rota's conjecture is true for $k = 1$ and $k = 2$ [2, 8]. Some other interesting results which follow from the n-parameter set analogue are presented as corollaries to the main result.

In general, we shall not present the details of the proofs of various assertions since they are rather long and will appear elsewhere. What we shall attempt to do instead is to give a rough indication of the proofs and to show how the results may be applied.

k-PARAMETER SETS

All of the aforementioned analogues to Ramsey's theorem are just statements about some special kinds of subsets of certain sets and their inclusion relationships.

Ramsey's theorem itself can be thought of thus as a statement about the lattices of subsets of finite sets; Rota's conjecture refers to the lattices of subspaces of finite vector spaces; the affine analogue concerns the partially-ordered sets of the subspaces of finite affine spaces. So also is the n-parameter set analogue a statement about partially ordered sets of special subsets of certain sets. We give here a precise definition of a k-parameter set followed by some (less formal) convenient notation with which the concepts involved can be more readily digested.

Let $A = \{a_1, a_2, \dots, a_t\}$ be a finite set with $t \geq 2$. Let $H : A \to A$ be a permutation group acting on A. For $a \in A$, $\sigma \in H$ the action is denoted by $a \to a^{\sigma}$. Also, for $\sigma_1, \sigma_2 \in H$, $\sigma_1 \cdot \sigma_2 \in H$ is defined by $a^{\sigma_1 \cdot \sigma_2} = (a^{\sigma_1})^{\sigma_2}$ for all $a \in A$. For a nonempty subset $B \subseteq A$, let $\bar{B} = \{\bar{b} : b \in B\}$ be the set of constant maps of A into A given by $x^{\bar{b}} = b$ for $x \in A$, $\bar{b} \in \bar{B}$. A^t denotes the cartesian product $A \times A \times \dots \times A$ (t factors) which is just $\{(x_1, \dots, x_t) : x_i \in A,\ 1 \leq i \leq t\}$.

For $x = (x_1, \dots, x_t) \in A^t$, $\sigma \in H$, we define an action of $H : A^t \to A^t$ by

$$x^{\sigma} = (x_1, \dots, x_t)^{\sigma} = (x_1^{\sigma}, \dots, x_t^{\sigma}) \in A^t .$$

Similarly $\bar{B}$ acts on A^t by

$$x^{\bar{b}} = (x_1, \dots, x_t)^{\bar{b}} = (x_1^{\bar{b}}, \dots, x_t^{\bar{b}}) = (b, \dots, b) \in A^t$$

for $x \in A^t$, $\bar{b} \in \bar{B}$.

For fixed integers $n > 0$ and $0 \leq k \leq n$, let $\Pi = \{S_0, S_1, \dots, S_k\}$ be a partition of the set $I_n = \{1, 2, \dots, n\}$ with $S_i \neq \emptyset$ for $1 \leq i \leq k$. $S_0 = \emptyset$ is possible. Let $f : I_n \to H \cup \bar{B}$ be a mapping with the property:

$$f(i) \in \bar{B} \quad \text{if} \quad i \in S_0$$

$$f(i) \in H \quad \text{if} \quad i \in I_n - S_0 .$$

The set $P = (A, \bar{B}, H, \Pi, f, n, k) = P$ is defined by

$$P = \bigcup_{1 \le i_0, \ldots, i_k \le t} \left\{ (x_1, \ldots, x_n) : x_j = a_{i_y}^{f(j)} \quad \text{if} \quad j \in S_y \right\} \subseteq A^n .$$

Definition: A subset $P \subseteq A^n$ is said to be a k-parameter set in A^n if $P = P(A, \bar{B}, H, \Pi, f, n, k)$ for some meaningful choice of these variables. What this means is the following. Let us write Π symbolically as:

$$\overbrace{\qquad\qquad}^{S_0} \ \overbrace{\qquad\qquad}^{S_1} \qquad\qquad \overbrace{\qquad\qquad}^{S_k}$$

We imagine that we have bunched together the elements in the blocks of the partition Π. With each $i \in I_n$ we associate an element $f(i) \in \bar{B} \cup H$. We can write this as

$$[\overbrace{\bar{a} \ \ldots \ \bar{b}}^{S_0} \ \overbrace{\pi_1 \ \ldots \ \delta_1}^{S_1} \ \ldots \ \overbrace{\pi_k \ \ldots \ \delta_k}^{S_k}]$$

where $\bar{a}, \ldots, \bar{b} \in \bar{B}$, $\pi_1, \ldots, \delta_k, \ldots, \pi_k, \ldots, \delta_k \in H$. With ℓ_0 defined by

$$\ell_0 = \begin{pmatrix} a_1 \\ a_2 \\ \vdots \\ a_t \end{pmatrix} \in A^t$$

(we occasionally write elements of A^t as column vectors when this is useful for our purposes), the preceding is shorthand notation for

$$\left[\overbrace{\ell_0^{\bar{a}} \ \ldots \ \ell_0^{\bar{b}}}^{S_0} \quad \overbrace{\ell_0^{\pi_1} \ \ldots \ \ell_0^{\delta_1}}^{S_1} \quad \ldots \quad \overbrace{\ell_0^{\pi_k} \ \ldots \ \ell_0^{\delta_k}}^{S_k}\right]$$

which we can write as .

$$\begin{bmatrix} \overbrace{a_1^{\bar{a}} \cdots a_1^{\bar{b}}}^{S_0} & \overbrace{a_1^{\pi_1} \cdots a_1^{\delta_1}}^{S_1} & \cdots & \overbrace{a_1^{\pi_k} \cdots a_1^{\delta_k}}^{S_k} \\ a_2^{\bar{a}} \cdots a_2^{\bar{b}} & a_2^{\pi_1} \cdots a_2^{\delta_1} & \cdots & a_2^{\pi_k} \cdots a_2^{\delta_k} \\ \vdots \qquad \vdots & \vdots \qquad \vdots & & \vdots \qquad \vdots \\ a_t^{\bar{a}} \cdots a_t^{\bar{b}} & a_t^{\pi_1} \cdots a_t^{\delta_1} & \cdots & a_t^{\pi_k} \cdots a_t^{\delta_k} \end{bmatrix}$$

which, of course, is just

$$\begin{bmatrix} \overbrace{a \cdots b}^{S_0} & \overbrace{a_1^{\pi_1} \cdots a_1^{\delta_1}}^{S_1} & \cdots & \overbrace{a_1^{\pi_k} \cdots a_1^{\delta_k}}^{S_k} \\ a \cdots b & a_2^{\pi_1} \cdots a_2^{\delta_1} & \cdots & a_2^{\pi_k} \cdots a_2^{\delta_k} \\ \vdots \qquad \vdots & \vdots \qquad \vdots & & \vdots \qquad \vdots \\ a \cdots b & a_t^{\pi_1} \cdots a_t^{\delta_1} & \cdots & a_t^{\pi_k} \cdots a_t^{\delta_k} \end{bmatrix}$$

Now, consider an n-tuple $x = (x_1, \ldots, x_n) \in A^n$ formed in the following way:

$$x = (\overbrace{a, \ldots, b}^{S_0}, \overbrace{a_{i_1}^{\pi_1}, \ldots, a_{i_1}^{\delta_1}}^{S_1}, \ldots, \overbrace{a_{i_k}^{\pi_k}, \ldots, a_{i_k}^{\delta_k}}^{S_k})$$

where $1 \leq i_1, i_2, \ldots, i_k \leq t$. We can think of x as being formed by taking "row cross-sections" under the various S_i and independently piecing these together. The set of all such x forms the set P. Since each $\pi_i, \ldots, \delta_i$ is a permutation of A, then a different choice of "row cross-sections" results in a different n-tuple x. Hence, $|P| = t^k$.

Thus, P is a k-parameter set in A^n iff P can be generated by some expression of the form

(1)
$$[\overbrace{a \cdots b}^{S_0} \overbrace{\pi_1 \cdots \delta_1}^{S_1} \cdots \overbrace{\pi_k \cdots \delta_k}^{S_k}]$$

If P_ℓ is an ℓ-parameter set in A^n, we say that P_k is a k-parameter subset of P_ℓ if P_k is a k-paramter set in A^n and P_k is a subset of P_ℓ (with the same $A, \bar{B}, H, n$).

We point out here that a set of 2^k points of A^n may possibly have many representations of the form (1). It is a k-parameter set, however, iff there is at least one such representation.

For example, for any choice of $\sigma_1, \sigma_2, \ldots, \sigma_n \in H$ the set denoted by $[\overbrace{\sigma_1}^{S_1} \overbrace{\sigma_2}^{S_2} \cdots \overbrace{\sigma_n}^{S_n}]$ is just A^n, which is an n-parameter subset of itself.

We next state several facts about k-parameter sets whose proofs we omit.

(i) Let $P = P(A, \bar{B}, H, \Pi, f, n, k)$ be a k-parameter set in A^n and $\beta_i \in H$, $1 \leq i \leq k$. Define $f' : I_n \to H \cup \bar{B}$ by

$$f'(j) = \begin{cases} \beta_i f(j) & \text{for} \quad j \in S_i, \quad 1 \le i \le k, \\ f(j) & \text{for} \quad j \in S_0 \end{cases}$$

Then $P' = P(A, \bar{B}, H, \Pi, f', n, k) = P$.

(ii) Let $P = P(A, \bar{B}, H, \Pi; f, n, \ell)$ be an ℓ-parameter set in A^n. The general k-parameter subset $P_k \subseteq P_\ell$ is formed as follows: Choose an unrefinement Π' of Π, say $\Pi' = \{S'_0, S'_1, \ldots, S'_k\}$ with $S_0 \subseteq S'_0$ and $S'_i \neq \phi$, $i > 0$. For each $S_i \subseteq S'_0$, $i > 0$, choose $\tau_i \in \bar{B}$; for each $S_i \nsubseteq S'_0$, choose $\tau_i \in H$.

Define $I_n \to H \cup \bar{B}$ by $f'(j) = \begin{cases} \tau_i f(j), & j \in S_i, \ i > 0, \\ f(j), & j \in S_0. \end{cases}$

Then $P_k = P(A, \bar{B}, H, \Pi', f', n, k)$ is a k-parameter set in A^n, $P_k \subseteq P_\ell$ and all k-parameter subsets of P_ℓ can be obtained this way.

CONSTRUCTION OF $*$-SETS

We now give a new construction which will be essential in the remainder of the paper. We retain the notation of the preceding section. Define

$$L_{\bar{A}} = \{\ell_0^{\bar{a}} : a \in A\} = \{(a, \ldots, a) : a \in A\} \subseteq A^t,$$

$$L_{\bar{B}} = \{\ell_0^{\bar{b}} : b \in B\},$$

$$L_H = \{\ell_0^{\sigma} : \sigma \in H\},$$

$$L = L_{\bar{A}} \cup L_H = \{\ell_1, \ldots, \ell_u\} \subseteq A^t.$$

For $x = (x_1, \ldots, x_u) \in L^u$, $\sigma \in H$, we define an action of $H : L^u \to L^u$ by

$$x^{\sigma} = (x_1^{\sigma}, \dots, x_u^{\sigma}).$$

Similarly, define $\bar{B} : L^u \to L^u$ by

$$x^{\bar{b}} = (x_1^{\bar{b}}, \dots, x_u^{\bar{b}}).$$

For all $\ell, m \in L$, define the map $\bar{\ell} : L \to L$ by

$$m^{\bar{\ell}} = \ell .$$

This induces a map $\bar{\ell} : L^u \to L^u$ by

$$x^{\bar{\ell}} = (x_1^{\bar{\ell}}, \dots, x_u^{\bar{\ell}}) = (\ell, \dots, \ell) \in L^u .$$

Finally we make the following definitions:

$$\ell_0^* = (\ell_0^{\bar{a}_1}, \dots, \ell_0^{\bar{a}_t}, \ell_0^{\sigma_1}, \dots, \ell_0^{\sigma_h}) \in L^u ,$$

$$C = L_H \cup L_{\bar{B}}, \quad \bar{C} = \{\bar{c} : c \in C\} = \overline{L_H} \cup \overline{L_{\bar{B}}} ,$$

$$L_H^* = \{\ell_0^{*\sigma} : \sigma \in H\}, \quad L_{\bar{C}}^* = \{\ell_0^{*\bar{c}} : \bar{c} \in \bar{C}\} .$$

As before, we have the notion of k-parameter sets in L^n.

For L^n we modify the notation slightly by writing a k-parameter set $P_k^* = P(L, \bar{C}, H, \pi^*, g, n, k)$ as

$$\left[\overbrace{\underbrace{\overline{\ell_0^{b}} \dots \overline{\ell_0^{d}}}_{T_0^*} \; \underbrace{\overline{\ell_0^{\pi_0}} \dots \overline{\ell_0^{\delta_0}}}_{V_0^*}}^{S_0^*} \; \underbrace{\pi_1 \dots \delta_1}_{S_1^*} \; \dots \; \underbrace{\pi_k \dots \delta_k}_{S_k^*}\right]$$

where $\ell_0^{\bar b}, \ldots, \ell_0^{\bar d} \in L_{\bar B}$ and $\ell_0^{\pi_0}, \ldots, \ell_0^{\delta_0} \in L_H$ (i.e., $\pi_0, \ldots, \delta_0 \in H$). Slightly expanded, this is

$$u\left\{\begin{array}{cccc}
\overbrace{\overbrace{\ell_0^{\bar b}\cdots\ell_0^{\bar d}}^{T_0^*}\;\;\overbrace{\ell_0^{\pi_0}\cdots\ell_0^{\delta_0}}^{V_0^*}}^{S_0^*} & \overbrace{(\ell_0^{\bar a_1})^{\pi_1}\cdots(\ell_0^{\bar a_1})^{\delta_1}}^{S_1^*} & \cdots & \overbrace{(\ell_0^{\bar a_1})^{\pi_k}\cdots(\ell_0^{\bar a_1})^{\delta_k}}^{S_k^*} \\
\vdots \quad \vdots \quad \vdots \quad \vdots & \vdots \quad \vdots & & \vdots \quad \vdots \\
\ell_0^{\bar b}\cdots\ell_0^{\bar d}\;\;\ell_0^{\pi_0}\cdots\ell_0^{\delta_0} & (\ell_0^{\sigma_h})^{\pi_1}\cdots(\ell_0^{\sigma_h})^{\delta_1} & \cdots & (\ell_0^{\sigma_h})^{\pi_k}\cdots(\ell_0^{\sigma_h})^{\delta_k}
\end{array}\right.$$

THE M MAP

We define a map $M : L^n \longrightarrow 2^{A^n}$ as follows: For $x = (x_1, \ldots, x_n) \in L^n$, $x_i = (x_{i1}, \ldots, x_{it}) \in L \subseteq A^t$, $1 \le i \le n$, let

$$M(x) = \left\{\begin{array}{l}
(x_{11}, x_{21}, \ldots, x_{n1}), \\
(x_{12}, x_{22}, \ldots, x_{n2}), \\
\quad\vdots \qquad\quad \vdots \\
(x_{1t}, x_{2t}, \ldots, x_{nt})
\end{array}\right\} \subseteq A^n$$

For $S \subseteq L^n$ we define $M(S)$ to be $\bigcup_{s \in S} M(S)$.

Suppose $P_\ell^* = P^*(L, \bar{C}, H, \Pi^*, g, n, \ell)$ is a ℓ-parameter set in L^n. It can be shown that if $V_0^* \neq \emptyset$ then $M(P_\ell^*)$ is just an $(\ell+1)$-parameter set $P_{\ell+1}$ in A^n. Furthermore if P_{k+1} is any $(k+1)$-parameter subset of $P_{\ell+1}$ in which T_0^* and V_0^* are not in the same block of the partition for P_{k+1} then there exists a k-parameter set P_k^* in L^n for which the following diagram is commutative:

$$\begin{array}{ccc} \exists\ P_k^* & \subseteq & P_\ell^* \\ M \downarrow & & \downarrow M \\ P_{k+1} & \subseteq & P_{\ell+1} \end{array}$$

Before proceeding to the outline of the proof we make a remark on terminology. By an r-coloring of a set X we just mean a partition of X into r disjoint (possibly empty) classes. Of course, the "r colors" correspond to the r classes into which X is partitioned. In general, we shall use this "chromatic" terminology in preference to that of partitions and classes.

THE MAIN RESULT

Theorem: Given A, B, H and integers $k, r, t_1, \cdots, t_r$, there exists an $N = N(A, \bar{B}, H, k, r, t_1, \ldots, t_r)$ such that if $n \geq N$ and $P_n = P(A, \bar{B}, H, \Pi, f, w, n)$ is any fixed n-parameter set in A^w, then for any r-coloring of the k-parameter subsets of P_n there is an i, $1 \leq i \leq r$, such that some t_i-parameter subset of P_n has all its k-parameter subsets the ith color.

Proof: The proof will proceed basically by double induction on k and $t_1 + \cdots + t_k$. The proof for $k = 0$ and all t_i is relatively straightforward once certain notational difficulties have been overcome. For a fixed integer $k \geq 0$ assume the theorem has been established for this k and all values of $r, t_1, \cdots, t_r$. We prove the theorem for $k+1$. Of course, the theorem is immediate for $r = 1$, and it is true vacuously for

$t_1 + \dots + t_n \le (k+1)r - 1$ (since in this case for some i, $t_i < k+1$). Henceforth we assume that $r \ge 2$, $t_i \ge k+1$, and that for some $p \ge (k+1)r-1$ the theorem holds for all $t_1 + \dots + t_r \le p$ We must prove the theorem with the additional assumption, which we now make, that $t_1 + \dots + t_r = p+1$.

Definition: Let $P_m = P(X, \bar{Y}, G, \Pi, f, w, m)$ be an m-parameter set in X^w, where Π is the partition $\{S_0, S_1, \dots, S_m\}$. Then for $k \le m$ and $1 \le i \le m$, an S_i-crossing k-parameter subset of P_m is a k-parameter subset $P_k = P(X, \bar{Y}, G, \Pi', f', w, k)$ with the partition $\Pi' = \{S'_0, S'_1, \dots, S'_k\}$, and $S_i \not\subseteq S'_0$.

Now let L, C and the map M be as before. We state a lemma whose proof we omit.

Lemma 1: Let $P_{m+1} = P(A, \bar{B}, H, \Pi, f, w, m+1)$ be an $(m+1)$-parameter set in A^w with partition $\Pi = \{S_0, S_1, \dots, S_{m+1}\}$. Let $\ell \ge 0$ be an integer. If $m \ge N(L, C, H, k, r, \underbrace{\ell, \dots, \ell}_{r})$ (which is meanigful by the induction hypothesis), then for any fixed i, $1 \le i \le t$, and for any r-coloring of the $(k+1)$-parameter subsets of P_{m+1}, there is an S_i-crossing $(\ell+1)$-parameter subset $P_{\ell+1}$ of P_{m+1} such that for some j, $1 \le j \le r$, all the S_i-crossing $(k+1)$-parameter subsets of $P_{\ell+1}$ have the jth color.

Let A, B, C, H, L and M be as before. Let $P^*_m = P(L, \bar{C}, H, \Pi^*, f, w, m)$ be an m-parameter set in L^w with partition $\Pi^* = \{V^*_0 \cup T^*_0 = S^*_0, S^*_1, \dots, S^*_m\}$, $V^*_0 \neq \phi$. Then $P_{m+1} = M(P^*_m)$ is an $(m+1)$-parameter set in A^w. Let $P_{\ell+2}$ be a V^*_0-crossing $(\ell+2)$-parameter subset of P_{m+1},

$$P_{\ell+2} = [\overbrace{\bar{b} \dots \bar{d}}^{S_0}\ \overbrace{\overbrace{\pi_0 \dots \delta_0}^{V^*_0} \ \dots \ \overbrace{\sigma_j \pi_j \dots \sigma_j \delta_j}^{S^*_j}}^{S_1}\ \overbrace{\pi \dots}^{S_2} \ \dots \ \overbrace{\dots \delta}^{S_{\ell+2}}].$$

Then $P_{\ell+2}$ is the disjoint union of t $(\ell+1)$-parameter subsets $P^i_{\ell+1}$, $1 \le i \le t$, none of which are V_0^*-crossing subsets, defined by

$$P^i_{\ell+1} = \Bigg[\overbrace{\overbrace{\bar{b} \ldots \bar{d}}^{S_0} \quad \overbrace{\overbrace{\overline{a_i^{\pi_0}} \ldots \overline{a_i^{\delta_0}}}^{V_0^*} \quad \ldots \quad \overbrace{\overline{a_i^{\sigma_j\pi_j}} \ldots \overline{a_i^{\sigma_j\delta_j}}}^{S_j^*}}^{S_1}}^{S'_0} \quad \overbrace{\pi \ldots}^{S_2} \quad \ldots \quad \overbrace{\ldots \delta}^{S_{\ell+2}}\Bigg].$$

Definition: The $P^i_{\ell+1}$ are called V_0^*-translates of each other in $P_{\ell+2}$ (or just translates when no confusion arises).

Remark 1: Let $P_{\ell+2}$ be a V_0^*-crossing $(\ell+2)$-parameter subset of P_{m+1}, as above, with V_0^*-translates $P^i_{\ell+1}$, and let P_{k+2} be a V_0^*-crossing $(k+2)$-parameter subset of $P_{\ell+2}$. Then

$$P_{k+2} = \Big[\overbrace{\overbrace{\bar{b} \ldots \bar{d}}^{S_0} \;\ldots\; \overbrace{\overbrace{\pi_0 \ldots \delta_0}^{V_0^*} \;\ldots\; \overbrace{\sigma_j\pi_j \ldots \sigma_j\delta_j}^{S_j^*}}^{S_1}}^{S'_1} \;\ldots\; \overbrace{\tau \ldots \eta}^{S_e} \; \overbrace{\pi' \ldots}^{S'_2} \;\ldots\; \overbrace{\ldots \delta'}^{S'_{k+2}}\Big].$$

P_{k+2} is the disjoint union of the t V_0^*-translates P^i_{k+1}, where

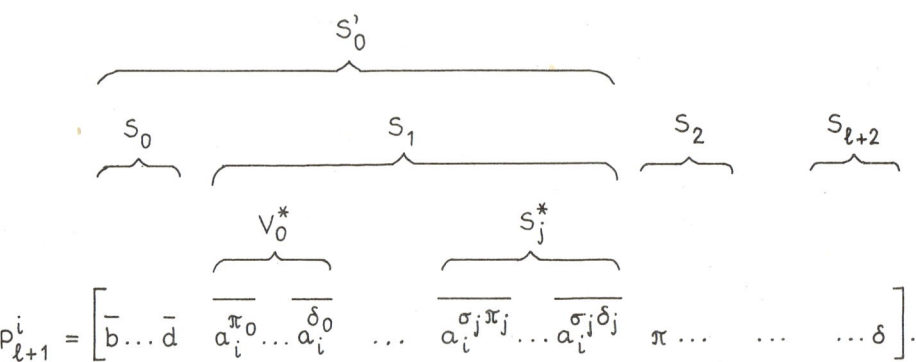

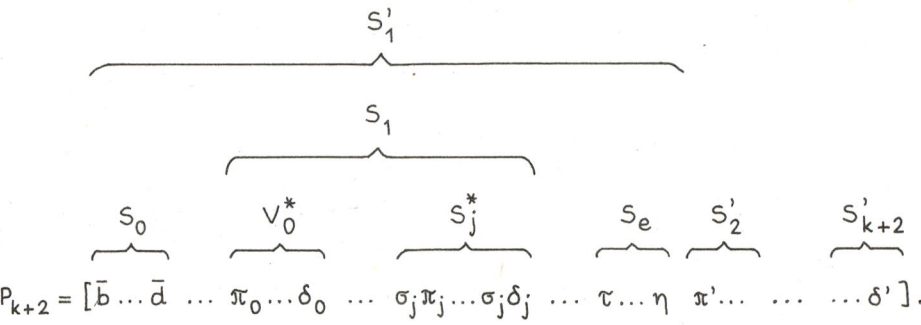

$$P^{i}_{k+1} = \Big[\overbrace{\overbrace{\overbrace{\bar{b} \ldots \bar{d}}^{S_0}}^{S_0} \ldots \overbrace{\overbrace{\overbrace{\overline{a_i^{\pi_0}} \ldots \overline{a_i^{\delta_0}}}^{V_0^*} \ldots \overbrace{\overline{a_i^{\sigma_j \pi_j}} \ldots \overline{a_i^{\sigma_j \delta_j}}}^{S_j^*}}^{S_1} \ldots \overbrace{\overline{a_i^{\tau}} \ldots \overline{a_i^{\eta}}}^{S_e}}^{S_1'}}^{S_0''} \overbrace{\pi' \ldots}^{S_2'} \ldots \overbrace{\ldots \delta'}^{S_{k+2}'} \Big]$$

We see that $P^{i}_{k+1} = P^{i}_{\ell+1} \cap P_{k+2}$ because $P^{i}_{k+1} \subseteq P^{i}_{\ell+1}$ and $P_{k+2} \subseteq P_{\ell+2}$, and on the other hand, any point in $P^{i}_{\ell+1}$ and P_{k+2} must be in P^{i}_{k+1} as can be checked by verifying the inclusion properties of parameter sets.

Remark 2: If P_{k+1} is any $(k+1)$-parameter subset of $P^{i}_{\ell+1}$, then there is some V_0^*-crossing $(k+2)$-parameter subset of $P_{\ell+2}$ with $P_{k+2} = \bigcup_{j=1}^{t} P^{j}_{k+1}$, the P^{j}_{k+1} being V_0^*-translates, such that $P_{k+1} = P^{i}_{k+1}$. In particular, taking $P^{i}_{\ell+1}$ to be as in the definition, $P_{k+1} \subseteq P^{i}_{\ell+1}$ must look like

$$P_{k+1} = \Big[\overbrace{\overbrace{\bar{b} \ldots \bar{d}}^{S_0} \ldots \overbrace{\overbrace{\overline{a_i^{\pi_0}} \ldots \overline{a_i^{\delta_0}}}^{V_0^*} \ldots \overbrace{\overline{a_i^{\sigma_j \pi_j}} \ldots \overline{a_i^{\sigma_j \delta_j}}}^{S_j^*}}^{S_1} \ldots \overbrace{\overline{a_x^{\tau}} \ldots \overline{a_x^{\eta}}}^{S_g}}^{S_0''} \overbrace{\pi' \ldots}^{S_2'} \ldots \overbrace{\ldots \delta'}^{S_{k+2}'} \Big]$$

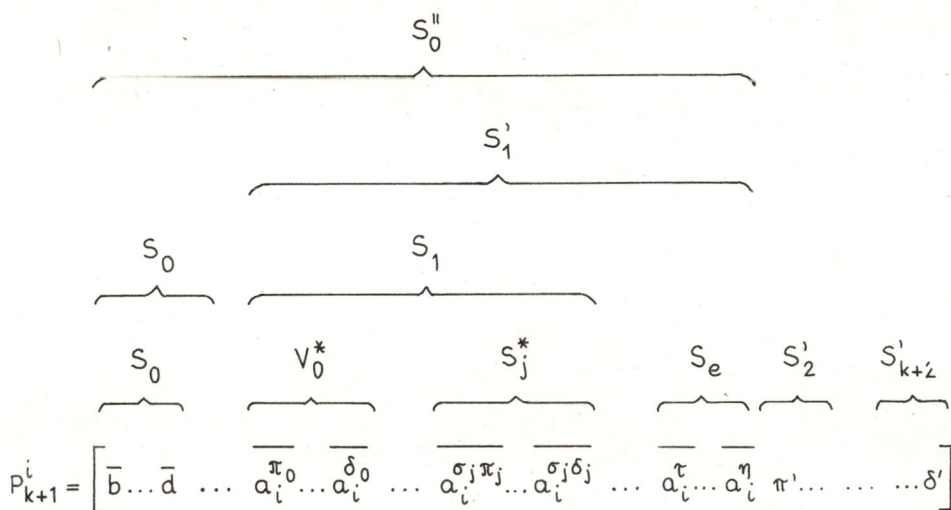

Then we can take

$$P_{k+2}=\Big[\overbrace{\overbrace{\bar{b}\ldots\bar{d}}^{S_0}\;\ldots\;\overbrace{\bar{a}_x^{\tau}\ldots\bar{a}_x^{\eta}}^{S_g}}^{S_0'}\;\overbrace{\overbrace{\pi_0\ldots\delta_0}^{V_0^*}\;\ldots\;\overbrace{\sigma_j\pi_j\ldots\sigma_j\delta_j}^{S_j^*}}^{S_1'=S_1}\;\overbrace{\pi'\ldots}^{S_2'}\;\ldots\;\overbrace{\ldots\delta'}^{S'_{k+2}}\Big]$$

This choice of P_{k+2} is well-defined. That is, $S_1' = S_1$ is the smallest set we can choose from S_0'' to generate a $(k+2)$-parameter set which is V_0^*-crossing and is contained in $P_{\ell+2}$ (since any such S_1' must contain S_1). We shall refer to this particular P_{k+2} as the V_0^*-expansion of P_{k+1} in $P_{\ell+2}$.

Remark 3: It should be noted that if P_{k+1} is any $(k+1)$-parameter subset of $P_{\ell+2}$, then either P_{k+1} is a V_0^*-crossing $(k+1)$-parameter set, or $P_{k+1} \subseteq P_{\ell+1}^{(i)}$ for some i.

This follows from the way in which the $(k+1)$-parameter subsets of $P_{\ell+2}$ must be formed.

Definition: Let A, B, H be as above. Let P_{m+v} be an $(m+v)$-parameter subset of A^W with partition $\{S_0, S_1, \ldots, S_v, V_1, \ldots, V_m)$. For each $i = 1, 2, \ldots, m$, P_{m+v} is the union of t disjoint $(m+v-1)$-parameter subsets $P^j_{(m+v-1),i}$, $1 \le j \le t$, which are V_i-translates of each other. Let P_{k+1} be a $(k+1)$-parameter subset of P_{m+v} which is V_i-crossing for at least one i. Let $\ell = m - \max\{i : P_{k+1}$ is V_i crossing $\}$. Then we associate with P_{k+1} the $(\ell+1)$-tuple $(\ell; j_m, j_{m-1}, \ldots, j_{m-\ell+1})$, where for $m-\ell < i \le m$ we define j_i by: $P_{k+1} \subseteq P^{j_i}_{m+v-1,i}$. (For $\ell = 0$ we get merely (0)). We call this the signature of P_{k+1} in P_{m+v} with respect to $(V_1, V_2, \ldots, V_m)$. An r-coloring of the $(k+1)$-parameter subsets of P_{m+v} will be called a $(V_1, V_2, \ldots, V_m)$-coloring if the colors of all $(k+1)$-parameter subsets with the same signature are the same.

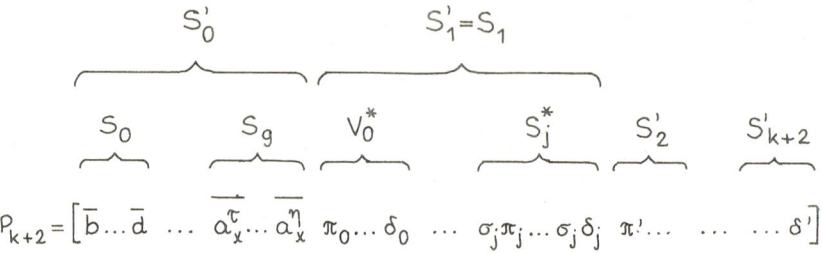

We next present an iterated form of Lemma 1. For arbitrary positive integers m and v, define the integers v_i, $1 \le i \le m$ as follows (where the various values of the function N exist by the induction hypothesis of the theorem):

$$v_1 = N(L, \bar{C}, H, k, r^{t^{m-1}}, v, \ldots, v),$$

$$v_2 = N(L, \bar{C}, H, k, r^{t^{m-2}}, v_1 + 1, \ldots, v_1 + 1),$$

$$\vdots$$

$$v_{i+1} = N(L, \bar{C}, H, k, r^{t^{m-i-1}}, v_i + 1, \ldots, v_i + 1),$$

$$\vdots$$

$$v_m = N(L, \bar{C}, H, k, r^{t^0}, v_{m-1} + 1, \ldots, v_{m-1} + 1).$$

Lemma 2: Let m and v be positive integers. Let $P_x = P(A, B, H, \Pi, f, w, x)$ be an x-parameter set in A^w with $x \ge v_m$. Suppose the $(k+1)$-parameter subsets of P_x are r-colored.

Then P_x contains an $(m+v)$-parameter subset P_{m+v}, with partition $\{S_0, S_1, \ldots, S_v, V_1, \ldots, V_m)$, such that the r-coloring restricted to P_{m+v} is a $(V_m, V_{m-1}, \ldots, V_1)$-coloring.

With the help of Lemma 2, the theorem can now be proved. We define $v = \max_{1 \le i \le r} N(A, \bar{B}, H, k+1, r, t_1, \ldots, t_i - 1, \ldots, t_r)$, $m = N(A, B, H, 0, r^K, 1, 1, \ldots, 1)$, $K = \binom{t^v}{t^{k+1}}$, and we let $v_1, \ldots, v_m$

be as previously defined. The induction step can be shown to hold for the choice $N(A, \bar{B}, H, k+1, r, t_1, \ldots, t_r) = v_m$.

A very informal sketch of the remainder of the proof can be given as follows. The original r-coloring of the $(k+1)$-parameter subsets of an N-parameter set P (with $N \geq v_m$) induces an r-coloring of the k-parameter subsets of the induced $(N-1)$-parameter set P^*. By the induction hypothesis we can find a "large" parameter set $Q^* \subseteq P^*$ with all k-parameter subsets of Q^* the same color.

Using the map M, we can show that this induces a large parameter set $Q \subseteq P$ such for some "direction" all the crossing $(k+1)$-parameter subsets of Q are one color. The remaining $(k+1)$-parameter subsets of Q fall into t classes which are translates of one another such that each class consists of the set of $(k+1)$-parameter subsets of a parameter subset Q' of Q. We choose one of these t classes, say, the class of $(k+1)$-parameter subsets of Q', and recolor each of these $(k+1)$-parameter subsets using r^t colors according to the way in which the corresponding translates of the particular $(k+1)$-parameter subset were r-colored. We again go up into the $*$-sets, use the induction hypothesis and the map M, and obtain a large parameter subset R of Q' such that for some new direction all crossing $(k+1)$-parameter subsets of R are one color (probably a different color than that of the first class of $(k+1)$-parameter subsets). As before, all the remaining $(k+1)$-parameter subsets of R fall into t classes which are translates of one another such that each class consists of the set of $(k+1)$-parameter subsets of a parameter subset R' of R. We choose one of these t classes and recolor the $(k+1)$-parameter subsets in this class using r^{t^2} colors according to the way that the corresponding translates are r^t-colored.

We iterate this process for a large number of steps (the exact numbers need not concern us here) and we then are faced with the following interesting configuration. We have a very large parameter set S and a large set of directions so that the color of any $(k+1)$-parameter set which crosses in any of the specified directions depends only on the directions in which it crosses. The $(k+1)$-parameter sets which do not cross in any of these directions naturally fall into parallel classes, each class being the set of $(k+1)$-parameter subsets of some (still large) parameter subset T of S. There is a natural correspondence we can make between this configuration and a large parameter set $\hat{S}$ in which the "points" of the $\hat{S}$ correspond to these various parameter subsets T. These "points" of $\hat{S}$ may be colored according to the way in which the (ordered) set of $(k+1)$-parameter sets in the corresponding subset T are colored (using a number of colors depending only on r, t and the number of parameters in T). By the Theorem for the case $k = 0$, $\ell = 1$ we can extract a 1-parameter subset $\hat{P}_1$ of $\hat{S}$ all of whose 0-parameter subsets are the same color.

This corresponds in S to a set of t "parallel" large parameter subsets T_i, $1 \leq i \leq t$, all of whose corresponding $(k+1)$-parameter subsets have the same color and such that (by the iterative construction) all $(k+1)$-parameter sets of $\bigcup_i T_i$ which intersect more than one T_i have the same color, say, the ith color. By the choice of v_m, the T_i have so many parameters that either for some $j \neq i$ one of them (and therefore all of them) contains a t_j-parameter subset all of whose $(k+1)$-parameter subsets have the jth color and we are done or each contains a (t_i-1)-parameter subset all of whose $(k+1)$-parameter subsets have the ith color.

However, the union of these t (t_i-1)-parameter sets exactly forms a t_i-parameter set all of whose $(k+1)$-parameter subsets have the ith color and we are also done. This completes the proof of the Theorem.

SOME APPLICATIONS

In this section we present several corollaries to the Theorem, the most well-known of these perhaps being the theorem of van der Waerden on arithmetic progressions (Corollary 7). Other corollaries are new, in particular, the results for affine and vector spaces, which we present first.

The way in which most of the results follow from the Theorem is relatively straightforward and the actual values of the assorted variables needed for the derivations will not be given.

Corollary 1: Let ℓ, r be positive integers, $k = 0$ or 1, and $F = GF(q)$ a finite field. Then there is an integer $N = N(q, r, \ell, k)$, depending only on q, r, ℓ and k with the following property: If A is an affine space over F of dimension $n \geq N$, and if all the k-dimensional affine subspaces of A are r-colored in any way, then there is some ℓ-dimensional affine subspace of A with all of its k-dimensional affine subspaces the same color.

Corollary 2: Let ℓ, r be positive integers, $F = GF(q)$ a finite field and $k = 0$ or 1. Then there is a number $N = N(q, r, \ell, k)$, depending only on q, r, ℓ and k with the following property: If V is an n-dimensional vector space with $n \geq N$, and if the k-dimensional vector subspaces of V are r-colored in any way, then there is an ℓ-dimensional vector subspace of V with all of its k-dimensional vector subspaces one color.

Remark: The last corollary (Rota's conjecture for $k = 0,1$) is also true for $k = 2$. This result is not a direct corollary of the Theorem, but follows from Corollary 1 by an inductive argument, which can be found in [8].

This argument, in fact, shows that if the affine analogue is true for some fixed k, and all q, r, ℓ, then Rota's conjecture is true for $k + 1$, and all q, r, ℓ.

Corollary 3: Given integers ℓ and r, there exists an integer $N(\ell,r)$ such that if S is a finite set with $|S| \geq N(\ell,r)$ and the subsets of S are r-colored, then there exist ℓ disjoint nonempty subsets $S_1, \ldots, S_\ell$ of S such that all $2^\ell - 1$ unions $\bigcup_{j \in J} S_j$, $\emptyset \neq J \subseteq \{1, 2, \ldots, \ell\}$, are one color.

Corollary 4: (J. Folkman, J. Sanders [10], R. Rado [6]) Given integers ℓ and r, there exists an integer $N'(\ell,r)$ such that if $n \geq N(\ell,r)$ and the positive integers $\leq n$ are r-colored then there exists ℓ integers

$a_1, \ldots, a_\ell$ such that all the sums $\{\sum_{i=1}^{\ell} \varepsilon_i a_i : \varepsilon_i = 0$ or 1, not all $\varepsilon_i = 0\}$ are one color.

The case $\ell = 2$ or Corollary 4 was first proved by Schur [11]. Corollary 4 is actually a special case of the following corollary, which is contained in a result of R. Rado [5].

Corollary 5: Let $\mathcal{L} = L_i(x_1, \ldots, x_m)$, $1 \le i \le n$, be a system of homogeneous linear equations with real coefficients with the property that for each j, $1 \le j \le m$, there exists a solution $(\varepsilon_1, \ldots, \varepsilon_m)$ to the system $\mathcal{L}$ with $\varepsilon_i = 0$ or 1 and $\varepsilon_j = 1$. Then given an integer r there exists an integer $N(r)$ such that if $n \ge N(r)$ and the positive integers $\le n$ are r-colored, then $\mathcal{L}$ can be solved with integers of one color.

By a multigrade of order m we mean two disjoint sets of integers $\{c_i\}$, $\{d_i\}$, $1 \le i \le n$, such that

$$\sum_{i=1}^{n} c_i^k = \sum_{i=1}^{n} d_i^k, \quad \text{for} \quad k = 1, 2, \ldots, m.$$

We denote this by

$$c_1, \ldots, c_n \overset{m}{=} d_1, \ldots, d_n$$

Since $\{ac_i + b\}$, $\{ad_i + b\}$, $1 \le i \le m$, is a multigrade of order n if $\{c_i\}$, $\{d_i\}$, $1 \le i \le n$, is, then a straightforward application of the Theorem along the lines used in the preceding corollaries yields

Corollary 6: If the multigrade equations

$$(*) \qquad x_1, \ldots, x_n \overset{m}{=} y_1, \ldots, y_n$$

have any integer solution (which always happens, for example, if $n \ge 2^{m-1}$), then for any r-coloring of the positive integers, $(*)$ always has a solution in integers of one color.

Corollary 7: (van der Waerden [4, 12]) Given integers t and r, there exists an integer $M(t,r)$ such that if $n \geq M(t,r)$ and the non-negative integers $< n$ are arbitrarily r-colored, then there must exist a monochromatic arithmetic progression of length t.

This result is implied by the stronger

Corollary 8: (Hales-Jewett [3]) Let $A = \{a_1, \ldots, a_t\}$ be a finite set. Given an integer r there exists an integer $N(r,t)$ such that if $n \geq N(r,t)$ and the set A^n is r-colored then there exists a set of t elements of A^n of the form

$$X_i = (x_{11}, \ldots, x_{1u}, a_i, x_{21}, \ldots, x_{2v}, a_i, \ldots, a_i, x_{d1}, \ldots, x_{dz}) \in A^n, \quad 1 \leq i \leq t,$$

all of which have the same color.

We conclude with a final (stronger) application of the Theorem.

Let $C_n = \{(x_1, \ldots, x_n) : x_i = 0 \text{ or } 1\}$ be the set of 2^n vertices of a unit n-cube in $\mathbb{R}^n$. Let us call a subset $Q_k \subseteq C_n$ a k-subspace of C_n if $|Q_k| = 2^k$ and Q_k is contained in some k-dimensional euclidean subspace of $\mathbb{R}^n$.

Corollary 9: Given integers k, ℓ, r, there exists an integer $N(k,\ell,r)$ such that if $n \geq N(k,\ell,r)$ and the k-subspaces of C_n are r-colored, then there exists an ℓ-subspace of C_n all of whose k-subspaces are one color.

CONCLUDING REMARKS

Several questions come to mind at this point.

(i) In the corollaries of the Theorem listed, we never really make much use of the freedom we have in choosing B and H. What are some interesting applications for some less trivial choices of B and H?

(ii) Are the various infinite versions of certain of the corollaries valid? A specific simple case would be: If the positive integers are 2-colored, is it true that there always exists an infinite subset A such that all sums

$$\sum_{b \in B} b, \qquad \emptyset \neq B \subseteq A, \qquad B\text{-finite},$$

are one color?

(iii) The reader will notice that the original theorem of Ramsey (for subsets of finite set) does not appear as a corollary to the Theorem. Is there a more general theorem which includes both of these results in a natural way? In a certain sense, Ramsey's theorem for sets corresponds to taking $t = 1$ in the Theorem (something which we are prohibited from doing), much in the spirit found in the paper of Goldman and Rota [1] on finite vector spaces.

(iv) With respect to the corollaries, the upper bounds given by the Theorem on the various N's are rather crude, to say the least. Is it possible to improve the estimates of these numbers? For example, in Corollary 9, the upper bound on $N(1,2,2)$ given by the Theorem is truly enormous, where, in fact, the exact bound is probably < 10.

(v) It was suggested by M. Simonovits that perhaps it would be possible to give an intrinsic definition of k-parameter sets, i.e., one which does not depend on coordinates. If this is possible then conceivably the corresponding proofs might become simpler.

(vi) Our particular definition of k-parameter set was chosen, to a certain extent, because a Ramsey theorem for them could be proved. What other definitions will have this property? In particular, can a suitable one be found which will establish Rota's original conjecture for k-subspaces of finite vector space, $k \geq 3$?

REFERENCES

[1] GOLDMAN, J. and ROTA, G. C., Combinatorial theory of finite vector spaces, (to appear).

[2] GRAHAM, R. L. and ROTHSCHILD, B. L., Ramsey's Theorem for n-dimensional arrays, Bull. A.M.S. 75 (1969) 418-422.

[3] HALES, A. W. and JEWETT, R. I., Regularity and positional games, Trans. A.M.S. 106 (1963) 222-229.

[4] KHINCHIN, A. Y., Three pearls of number theory, Graylock Press, Rochester, 1952.

[5] RADO, R., Note on combinatorial analysis, Proc. London Math. Soc. (2) 48 (1943) 122-160.

[6] RADO, R., see paper in this Proceedings.

[7] RAMSEY, F. P., On a problem of formal logic, Proc. London Math. Soc., 2nd ser., 30 (1930) 264-286.

[8] ROTHSCHILD, B. L., A generalization of Ramsey's theorem and a conjecture of Rota, Doctoral Dissertation, Yale University, New Haven, Connecticut, 1967.

[9] RYSER, H. J., Combinatorial Mathematics, Wiley, New York, 1963.

[10] SANDERS, J., A generalization of a theorem of Schur, Doctoral Dissertation, Yale University, New Haven, Connecticut, 1968.

[11] SCHUR, I., Über die Kongruenz $x^m + y^m \equiv z^m \pmod{p}$, Jahr. Deutsch. Math. - Verein. 25 (1916), 114.

[12] VAN DER WAERDEN, B. L., Beweis einer Baudetschen Vermutung, Nieuw Arch. Wiskunde 15 (1927) 212-216.

Sequences associated with a problem of Turán and other problems

by

R. K. Guy
Calgary, Canada

INTRODUCTION

In obtaining better bounds than those already given in the literature for the Turán numbers and for various crossing numbers, certain sequences arise naturally which could be studied for their own sake.

A CONJECTURE OF TURÁN

In [11] TURÁN proved the well-known theorem that given integers $n > k > 1$, then a graph on n points with more than

$$(1) \qquad \frac{k-2}{2k-2}(n^2-r^2)+\binom{r}{2}$$

edges must contain a complete subgraph K_k on k points, where r is the least non-negative residue of n, modulo $k-1$. The critical graph is the complete $(k-1)$-partite graph on points partitioned into $k-1$ parts of (nearly) equal size. For proofs see also [1, 3, 10].

TURÁN also conjectured [12, 13] that a generalization of the theorem to hypergraphs is also true. The following is the simplest case. If a 3-graph on n points contains more than e_n 3-edges (triangles) then it contains a complete subgraph K_4^3 on four points (tetrahedron).

(2) $$54 e_n = 5 n^3 - 9 n^2 - an + b ,$$

where $a = 0$ or 6 according as $n \equiv 0$ or not, modulo 3, and $b = 0, 10$ or 8 according as $n \equiv 0, 1$, or 2, modulo 3. A critical graph appears to be obtained by partitioning the n vertices into sets A, B, C with $|A| = [n/3]$, $|B| = [(n+1)/3]$, $|C| = [(n+2)/3]$ and taking all triangles whose vertices are in the sets AAB, BBC, CCA and ABC. A major difficulty in establishing the conjecture is that the critical graph is not unique. In [8] Katona et al. give another proof of Turán's theorem and establish his conjecture for $n \leq 9$. If we denote by t_n the maximum number of 3-edges a 3-graph may contain without there being a complete graph on four points, they also prove that

(3) $$(e_n \leq)\, t_n \leq \frac{9}{14} \binom{n}{3} ,$$

and that $t_n / \binom{n}{3}$ tends to a limit, which we will denote by τ.

Thus $\tau \leq 9/14$ and the truth of Turán's conjecture would imply that $\tau = 5/9$.

Recently Vera Sós Turán and M. Simonovits (oral and written communications) independently confirmed the conjecture for $n \leq 12$, so $t_{12} = e_{12} = 136$. The methods of [8] now show that $\tau \leq 34/55$. We are able to improve this to $\tau < 0.608143\ldots$ (with an error of at most 8 in the seventh decimal place) and prove

THEOREM 1. If t_n is the maximum number of edges that a 3-graph may contain without its containing a complete subgraph on 4 points, then

(4) $$23t_n \le 14\binom{n}{3} - \binom{n}{2} + 43n - 944,$$

the inequality being strict except for $n = 66$.

Suppose we have a 3-graph on 13 points containing 177 triangles. These have 531 corners, so that there is some point with at most $40 = [531/13]$ triangles. The remaining 12 points contain $177 - 40 = 137 > 136$ triangles, so by the result of Mrs. TURÁN and SIMONOVITS, there is a complete subgraph on 4 points. Hence $t_{13} \le 176$.

More generally define u_n inductively as the least integer such that

(5) $$u_n + 1 - [3(u_n + 1)/n] = u_{n-1} + 1,$$

where $u_{12} = t_{12} = 136$. Then $t_n \le u_n$. Note the word "least" in the definition; for example (5) is satisfied by both $u_{13} = 176$ and $u_{13} = 177$. Table 1 gives the values of u_n for $12 \le n \le 131$; and (4) may be verified in these cases.

n	u_n	u_{n+20}	u_{n+40}	u_{n+60}	u_{n+80}	u_{n+100}
12	136	3024	13452	36286	76392	138635
13	176	3326	14259	37481	78938	142415
14	224	3647	15097	39439	81540	146264
15	280	3988	15967	41082	84198	150181
16	344	4350	16870	42770	86914	154168
17	417	4733	17807	44503	89687	158225
18	500	5138	18778	46283	92519	162352
19	593	5566	19783	48109	95410	166550
20	697	6017	20824	49983	98360	170820
21	813	6492	21901	51905	101371	175162
22	941	6991	23014	53876	104442	179577
23	1082	7515	24164	55896	107575	184066
24	1236	8064	25352	57966	110770	188629
25	1404	8640	26578	60086	114027	193267
26	1587	9242	27843	62257	117348	197980
27	1785	9872	29148	64480	120733	202769
28	1999	10530	30493	66755	124182	207635
29	2229	11216	31879	69083	127696	212578
30	2476	11931	33306	71465	131276	217599
31	2741	12676	34775	73901	134922	222698

Table 1. Values of u_n defined by (5)

We may rewrite (5) as

$$(u_n + 1)\frac{n-3}{n} + \varepsilon_n = u_{n-1} + 1 \tag{6}$$

where

$$3 \le n\varepsilon_n \le n-3 \tag{7}$$

or as

$$(n-3)u_n + n\varepsilon_n - 3 = nu_{n-1}$$

If we write $u_n = \beta_n \binom{n}{3}$ we obtain

$$\beta_n + \frac{6n\varepsilon_n - 18}{n(n-1)(n-2)(n-3)} = \beta_{n-1}, \tag{8}$$

so β_n is monotonic decreasing and tends to a limit, say β. $\tau \le \beta$. If we sum (8) from $n = k+1$ we obtain

$$\beta_n = \beta_k - \sum_{i=k+1}^{n} \frac{6i\varepsilon_i - 18}{i(i-1)(i-2)(i-3)}. \tag{9}$$

The terms of the sum are non-negative and bounded by (7) so that

$$\beta_k \ge \beta_n \ge \beta_k - \frac{3(k-4)}{k(k-1)(k-2)}.$$

A calculation with $u_{434} = 8228429$ establishes that $\beta = 0.608143$ to the accuracy mentioned above. Alternate convergents (those $> \beta_{434}$) to $\beta_{434} = 8228419/13530384$ are $1/1$, $2/3$, $14/23$, $194/319$, $1150/1891, \ldots$, the last-mentioned of which is within the limits of accuracy of our calculations, so we can state that

$$t_n \le u_n \le \frac{1150}{1891}\binom{n}{3} \tag{10}$$

for sufficiently large n. To prove Theorem 1 we note that $t_n \leq u_n$, and assume inductively that

$$(11) \qquad 23 u_{n-1} \leq 14\binom{n-1}{3} - \binom{n-1}{2} + 43(n-1) - 944,$$

then (6) and (11) give

$$23 u_n + 23 \leq \frac{n}{n-3}\left(14\binom{n-1}{3} - \binom{n-1}{2} + 43(n-1) - 944 + 23\right)$$

so that

$$23 u_n \leq 14\binom{n}{3} - \binom{n}{2} + 43n - 944 - \frac{(2n-173)^2 - 7825}{8(n-3)} \leq$$

$$\leq 14\binom{n}{3} - \binom{n}{2} + 43n - 944$$

provided $n \geq 131$. Theorem 1 is proved, since (4) follows for smaller values of n from Table 1.

It does not seem likely that β is rational; a fortiori u_n is unlikely to be expressible as a cubic with rational coefficients, or as one of a finite number of such cubics. However M.J.T. GUY [2, and see the conclusion of this paper] discovered a sequence of this type, obtained by rounding up successive fractional values, which can be represented by the rounded up value of a cubic with rational coefficients. This arose from a problem of H.T. CROFT: given n points in the plane, what is the minimum number of obtuse angles they form? This problem is closely related to the one presently under consideration, since 4 points determine 4 triangles, at least one of which is obtuse-angled.

THE CROSSING NUMBER OF THE COMPLETE GRAPH

This is another unsolved problem which gives rise to a similar sequence, this time one which it is appropariate to approximate by a fourth degree polynomial. The complete graph K_n consists of n points, each joined to every other by an edge. If one maps this into a plane (or sphere) one can ask for the minimum number of crossings, points of (just two) edges which are mapped into the same point of the plane. Denoting this by $cr(K_n)$ it is known [4] that

$$(12) \qquad cr(K_n) \le \frac{1}{4}\left[\frac{1}{2}n\right]\left[\frac{1}{2}(n-1)\right]\left[\frac{1}{2}(n-2)\right]\left[\frac{1}{2}(n-3)\right]$$

and it is conjectured that equality holds. The conjecture is true for $n \le 10$, and folklore has it for some higher values. We give in Theorem 2 below what we believe to be the best reliable lower bound for $cr(K_n)$. We write

$$cr(K_n) = c_n\binom{n}{4}$$

It is known that c_n tends to a limit, c say, as n tends to infinity, and (12) and the fact that $cr(K_{10}) = 60$ establish that

$$(13) \qquad \frac{2}{7} \le c \le \frac{3}{8}.$$

We shall show that $c > 0.294866$. Consider the complete graph on 11 points. This contains 11 complete subgraphs on 10 points, each of which contains at least 60 crossings when mapped on the plane. However each crossing arises from 4 points, so we have counted each crossing $11-4$ times. The crossing number of K_{11} is thus $\ge 11 \times 60/7 = 94\frac{2}{7}$, i.e. ≥ 95. Moreover it is known from a parity argument that $cr(K_n)$ is even if $n \equiv 1$ or 3, and odd if $n \equiv 5$ or 7, modulo 8. So $cr(K_{11}) \ge 96$. More generally we inductively define

$$(14) \qquad v_n = \left\{\frac{n v_{n-1}}{n-4}\right\}$$

where braces denote the "post-office" function, least integer not less than, and v_n further exceeds this value by one if n is odd, and the parity condition mentioned above demands it. If we take $v_{10} = 60$, then we have

$$cr(K_n) \geq v_n .$$

n	v_n	v_{n+20}	v_{n+40}	v_{n+60}	v_{n+80}	v_{n+100}
11	96	9273	73676	286487	788064	1765937
12	144	10598	79816	303340	823886	1831343
13	209	12060	86333	320926	860915	1898550
14	293	13668	93240	339265	899178	1967589
15	401	15432	100553	358380	938703	2038494
16	535	17361	108288	378290	979517	2111298
17	700	19467	116462	399019	1021648	2186035
18	900	21758	125089	420588	1065123	2262738
19	1140	24245	134188	443021	1109972	2341443
20	1425	26939	143773	466338	1156221	2422183
21	1761	29852	153863	490564	1203901	2504994
22	2153	32995	164475	515722	1253040	2589910
23	2607	36380	175627	541836	1303669	2676966
24	3129	40018	187336	568928	1355816	2766199
25	3726	43923	199622	597025	1409512	2857645
26	4404	48107	212501	626149	1464787	2951339
27	5170	52583	225994	656325	1521672	3047319
28	6032	57364	240119	687579	1580198	3145620
29	6999	62464	254897	719936	1640397	3246280
30	8076	67896	270346	753422	1702299	3349337

Table 2. Values of v_n defined by (14).

Table 2 gives the values of v_n for $11 \leq n \leq 130$. Much as before we may write (14) as

$$(15) \qquad (n-4)v_n = nv_{n-1} + (n-4)\delta_n$$

where the last term is non-negative and bounded above by $n-5$ or $2n-9$ according as n is even or odd. If we put $v_n = \alpha_n \binom{n}{4}$, (15) becomes

$$(16) \qquad \alpha_n = \alpha_{n-1} + 24\delta_n / n(n-1)(n-2)(n-3),$$

showing that α_n is an increasing function of n. Since it is bounded, it tends to a limit, α say, as n tends to infinity.

If we sum (16) and use the bounds on δ_n, we have

$$\alpha_{2k} \leq \alpha_{2n} \leq \alpha_{2k} + \frac{12k-15}{k(2k-1)(2k-2)(2k-3)}$$

which gives limits of error for α. We may take $k = 76$, $v_{152} = 6302482$, $\alpha_{152} = 3151241/10687025$ and α exceeds this by less than 4 in the sixth decimal place, i.e. $\alpha = 0.294868 \pm 2$ and we have proved

THEOREM 2. For sufficiently large n,

$$(17) \qquad cr(K_n) \geq \frac{649}{2201}\binom{n}{4}.$$

TURÁN'S BRICK-FACTORY PROBLEM

Another outstanding problem is to find the crossing number of the complete bipartite graph, $K_{m,n}$. For some years it was believed that Zarankiewicz [14] had established this, but in fact [5] all that was known was

$$(18) \qquad cr(K_{m,n}) \leq \left[\tfrac{1}{2}m\right]\left[\tfrac{1}{2}(m-1)\right]\left[\tfrac{1}{2}n\right]\left[\tfrac{1}{2}(n-1)\right],$$

with equality for m (or n) ≤ 4 . Recently Kleitman [9] has established equality for $m=5$ and hence for $m=6$, i.e.

(19) $$cr(K_{6,n}) = 6\left[\frac{1}{2}n\right]\left[\frac{1}{2}(n-1)\right].$$

Now we may obtain a lower bound for $cr(K_{7,n})$ by counting the crossings in the subgraphs $K_{6,n}$ and $K_{7,n-1}$ of $K_{7,n}$, and taking the maximum of the rounded up values, i.e.

(20) $$cr(K_{7,n}) \ge \max\left(\left\{\frac{7}{5}cr(K_{6,n})\right\}, \left\{\frac{n}{n-2}cr(K_{7,n-1})\right\}\right).$$

This again inductively defines a sequence

(21) $$w_n = \max\left(\left\{\frac{7}{5}cr(K_{6,n})\right\}, \left\{\frac{nw_{n-1}}{n-2}\right\}\right)$$

and if we take $w_6 = 54 = cr(K_{7,6})$ then $cr(K_{7,n}) \ge w_n$.

n	w_n	$2.1(n-1)^2$	f_n
6	54	52.5	36
7	76	75.6	25
8	102	102.9	12
9	135	134.4	27
10	169	170.1	10
11	210	210	21
12	252	254.1	0
13	303	302.4	27
14	354	354.9	12
15	412	411.6	25
16	471	472.5	6
17	538	537.6	25
18	606	606.9	12
19	681	680.4	27
20	757	758.1	10
21	840	840	21
22	924	926.1	0
23	1017	1016.4	27
24	1110	1110.9	12
25	1210	1209.6	25
26	1311	1312.5	6
27	1420	1419.6	25
28	1530	1530.9	12
29	1647	1646.4	27
30	1765	1766.1	10

Table 3. Values of w_n defined by (21)

Table 3 exhibits the values of w_n for $6 \le n \le 30$. This time the sequence is quadratic, and it seems reasonable to expect periodicity in such a case. The present sequence has period 10 in the sense that

$$w_n = (21n^2 - 42n + f_n)/10 \tag{22}$$

where, for $n \ge 7$, f_n depends only on the residue class to which n belongs, modulo 10. Its values are given in Table 3.

To establish (22) note that when n is odd, $\frac{7}{5} cr(K_{6,n}) = 2.1(n-1)^2$ and $21 \le f_n \le 27$, while for n even,

$$\left\{ \frac{n}{n-2} cr(K_{7,n-1}) \right\} \ge \left\{ \frac{n}{n-2} \left(2.1(n-2)^2 + \frac{f_{n-1} - 21}{10}\right) \right\} \ge 2.1\,n(n-2)$$

and the last inequality is strict except when $n \equiv 2$, modulo 10, $f_{n-1} = 21$, $f_n = 0$. Otherwise $0 < f_n \le 12$. We have proved

THEOREM 3. $cr(K_{7,n}) \ge 2.1(n-1)^2$, n odd,

$\ge 2.1\,n(n-2)$, n even,

from which we can immediately deduce

THEOREM 4. For $5 \le m \le n$

$$cr(K_{m,n}) \ge m(m-1)n(n-2)/20. \tag{23}$$

We may also define sequences associated with $cr(K_{8,n})$, $cr(K_{9,n}), \ldots$. These also exhibit periodicity but we have been unable to improve the leading coefficient in (23).

TOROIDAL CROSSING NUMBERS

Crossing numbers can also be defined when the graph is mapped into the torus. They have already been investigated; in [7] it is shown that

$$cr_1(K_n) \geq \frac{4}{35}\binom{n}{4}, \qquad n \geq 15,$$

and the present methods can improve this slightly.

THEOREM 5.

$$cr_1(K_n) \geq \frac{184}{1599}\binom{n}{4}, \qquad n \geq 42.$$

In [6] the methods have already been used to give the following theorems.

THEOREM 6.

$$cr_1(K_{7,n}) \geq 9(n-2)(3n+1)/38, \qquad n \geq 15.$$

THEOREM 7.

$$cr_1(K_{8,n}) \geq 6n(3n-5)/19, \qquad n \geq 15.$$

THEOREM 8. For sufficiently large m and n

$$cr_1(K_{m,n}) \geq \frac{9}{133}\binom{m}{2}\binom{n}{2}.$$

CONCLUSION

It may be of interest to study the sequences defined by

(24) $$x_n = [(n+k)x_{n-1}/n],$$

(25) $$y_n = \{(n+k)y_{n-1}/n\},$$

for their own sake. For $k = 2$ they are probably always periodic in the sense that they can be represented by a finite number of quadratic polynomials, and that these differ only in their absolute terms. For $k > 2$, a similar phenomenon will not occur in general, but J.H. Conway [2] has asked for the values of and the initial conditions which lead to an explicit solution of (25).

The example already mentioned arises from Croft's problem [2]: given n points P_j in a plane, there are $\frac{1}{2}n(n-1)(n-2)$ angles of the type $P_i\hat{P}_j P_k$ formed. What proportion of them must necessarily be obtuse or right angles? M.J.T. Guy has shown that this proportion has a limit, and that it lies between $1/9$ and $4/27$ by the following arguments.

For the upper bound place about one-third of the points on each of three small circular arcs, situated near the vertices of a just acute-angled triangle, of different degrees of smallness, and with each having its centre at the next vertex. By suitable choice of the relative degree of smallness, the limiting proportion $4/27$ is found.

For the lower bound consider all subsets of $n-1$ points, then if z_n is the number of obtuse angles,

(26) $$z_n \geq \left\{ \frac{n z_{n-1}}{n-3} \right\} .$$

Omission of the braces gives rather a poor bound, but (26) with equality and initial condition $z_4 = 1$ has an exact solution which may be established by induction. This leads to

(27) $$z_n \geq \{ n^2(n-3)/18 \}$$

and the lower bound $1/9$ for the proportion of obtuse angles.

REFERENCES

[1] ANDRÁSFAI, B., Neuer Beweis eines graphentheoretischen Satzes von P. Turán, Publ. Math. Inst. Hungar. Acad. Sci., (A), 6 (1962) 193-196. M.R. 27 (1964), # 745.

[2] CROFT, H.T., Research problems (mimeographed 1967 edition), p. 20, Problem V, 8.

[3] DIRAC, G., Extensions of Turán's theorem on graphs, Acta Math. Acad. Sci. Hungar. 14 (1963), 417-422. M.R., 28 (1964), # 605.

[4] GUY, R.K., A combinatorial problem, Nable (Bull. Malayan Math. Soc.), 7 (1960), 68-72. University of Calgary Research Paper # 8, Jan. 1967.

[5] GUY, R.K., The decline and fall of Zarankiewicz's theorem, in Harary, F. (ed) Methods of proof in graph theory, Proc. Ann Arbor Graph Theory Conference, 1968, Academic Press, 1969.

[6] GUY, R.K. and JENKYNS, T.A., The toroidal crossing number of $K_{m,n}$, J. Combinatorial Theory, 6 (1969), 235-250.

[7] GUY, R.K., JENKYNS, T.A. and SCHAER, J.: The toroidal crossing number of the complete graph, J. Combinatorial Theory, 4 (1968), 376-390.

[8] KATONA, G., NEMETZ, T. and SIMONOVITS, M., Ujabb bizonyitás a Turán-féle gráftételre és megjegyzések bizonyos általánositásaira, Mat. Lapok, 15 (1964), 228-237. M.R. 30 (1965), # 2483.

[9] KLEITMAN, D.J., The crossing number of $K_{5,n}$. J. Combinatorial Theory (to appear), University of Calgary Research Paper # 65, Nov., 1968.

[10] MOTZKIN, T.S. and STRAUS, E.G., Maxima for graphs and a new proof of a theorem of Turán, Canad. J. Math., 17 (1965), 533-540, M.R. 31 (1966), # 89.

[11] TURÁN, P., Egy gráfelméleti szélsőértékfeladatról, Mat. Fiz. Lapok, 48 (1941), 436-452. M.R. 8 (1947), 284.

[12] TURÁN, P., On the theory of graphs, Colloq. Math., 8 (1954), 19-30. M.R. 15 (1954), 976.

[13] TURÁN, P., Research problems, Publ. Math. Inst. Hungar. Acad. Sci., (A), 6 (1961), 417-423. M.R. 31 (1966), # 2107.

[14] ZARANKIEWICZ, K., On a problem of P. Turán concerning graphs, Fund. Math., 41 (1954), 137-145. M.R. 16 (1955), 156.

A Helly-type problem in trees

by

A. Gyárfás and J. Lehel
Budapest, Hungary

In our paper we present some results in connection with a problem of T. GALLAI.

$R_1, R_2, \dots, R_c$ will denote c distinct parallel lines in the plane. Let I_k be a closed interval of R_k, then the set $A = \bigcup_{k=1}^{c} I_k$ is said to be a c-interval. I_k is the k-th component of A.

Let $\mathcal{A} = \{A^\nu\}$ be an arbitrary finite family of c-intervals, any two of them having common points. It is well known that in case $c=1$ the set $\bigcap_{A^\nu \in \mathcal{A}} A^\nu$ is non-empty, or which means the same, there exists a point p so that $p \cap A^\nu \neq \phi$ $(A^\nu \in \mathcal{A})$. (This is Helly's theorem in one-dimension.)

T. GALLAI has posed the problem for c-intervals: to find the least integer $\ell(c)$ for which there is a set $P \subset \bigcup_{k=1}^{c} R_k$ of $\ell(c)$ points that $P \cap A^\nu \neq \phi$ $(A^\nu \in \mathcal{A})$. We may assume that P consists of endpoints of the components of A^ν-s.

In the first part we prove the existence of $\ell(c)$, (Theorem 1.) and we show that $\ell(2)=2, \ell(3)=4$ (Theorem 2. and Theorem 4.) Theorem 2. was proved by J. Surányi and L. Surányi, independent from us.

Now we formulate the problem which we are dealing with in the second part: instead of distinct lines we suppose that $R_1, R_2, \dots, R_c$ coincide. We define $\ell^*(c)$ in this case on the analogy of $\ell(c)$. It is clear that $\ell(c) \leq \ell^*(c)$. We will prove that $\ell^*(c)$ exists, (Theorem 5.) and $\ell^*(2) = 3$ (Theorem 6.)

Replacing the lines by trees (a tree is a connected graph without circuits) and the intervals by subtrees, all the theorems mentioned above remain true. In particular, if every tree is a path, our problem is equivalent to the original one. This generalization was suggested by L. Lovász, who proved Theorem 2. in this form.

For the sake of simplicity we only sketch the proofs for trees unless they demand different methods as in case of Theorem 6.

I.

Without restriction of generality we may assume that the c-intervals have no common endpoints.

A family of sets is said to be t-independent if the maximal number of its pairwise disjoint sets is t. A family of pairwise intersecting sets is clearly 1-independent. For t-independent c-intervals $\ell_t(c)$ is defined similar to $\ell(c)$. It is obvious that $\ell_1(c) = \ell(c)$. $\ell_t(1) = t$ according to a theorem of Hajnal and Surányi. [1]

The existence of $\ell_t(c)$ follows from

THEOREM 1. $\ell_t(c) \leq \ell_T(c-1) + t$ where $T = [(t+1)^{c-1} - 1]$

In the proof we use the following

LEMMA: let $\mathfrak{K}_1 = \{X_j^1\}_{j=1}^{t^c}$, $\mathfrak{K}_2 = \{X_j^2\}_{j=1}^{t^c}$, ..., $\mathfrak{K}_t = \{X_j^t\}_{j=1}^{t^c}$ be systems of pairwise disjoint c-intervals. Then we can choose from $\mathfrak{K}_1, \mathfrak{K}_2, \dots, \mathfrak{K}_t$ the c-intervals $X_{j_1}^1, X_{j_2}^2, \dots, X_{j_t}^t$ which are also pairwise disjoint.

PROOF: we prove by induction on c.

(i) In case of $c=1$ we may assume the disjoint (1-)intervals to follow one another from left to the right. Let us choose the index $i_1 \in \{1,2,\dots,t\}$ for which the right endpoint of $X_1^{i_1}$ is to the extreme left. We continue the process by choosing an $i_2 \in \{1,2,\dots,t\} - \{i_1\}$ for which the right endpoint of $X_2^{i_2}$ is to the extreme left, and so on. This procedure obviously leads to an interval-system of the desired property.

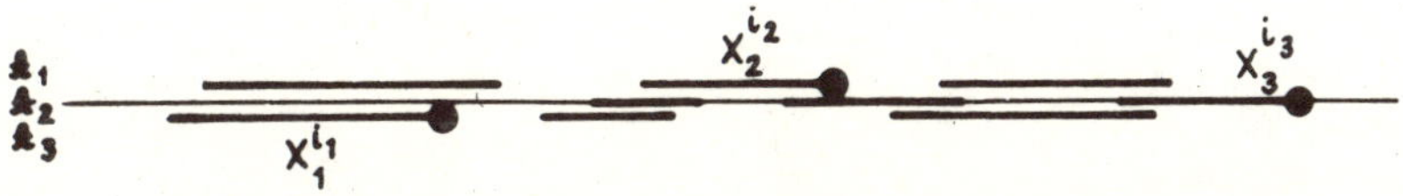

(ii) Supposing the Lemma to be true for $c-1$, let $\{X_j^i\}_{j=1}^{t^c}$ be a system of c-intervals satisfying the assumptions of the Lemma. $(i=1,2,\dots,t)$. We may suppose that for every i the first components of $X_1^i, X_2^i, \dots, X_{t^c}^i$ follow one another from left to the right on R_1. Let Y_k^i be the convex hull of the union of the first components of $X_{(k-1)t^{c-1}+1}^i$, $X_{(k-1)t^{c-1}+2}^i, \dots, X_{k\cdot t^{c-1}}^i$.

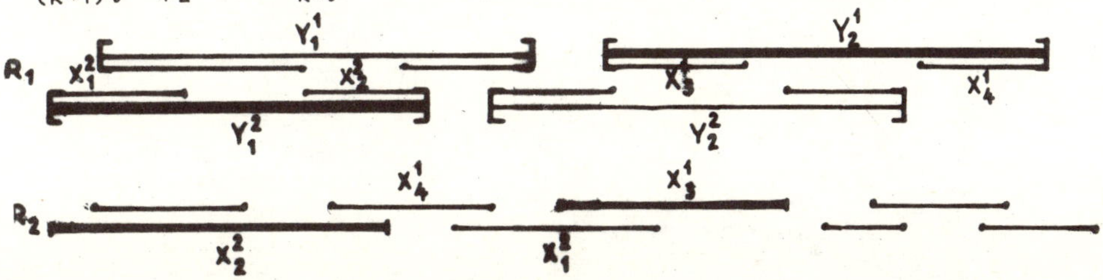

The intervals Y_k^i are pairwise disjoint for all i, so because of (i) we can choose the pairwise disjoint intervals $Y_{j_1}^1, \dots, Y_{j_t}^t$. We apply the inductive hypothesis for the last components of the system $\{X_\nu^i\}_{\nu=(j_i-1)t^{c-1}+1}^{j_i t^{c-1}}$.

The t pairwise disjoint c-intervals with their first components are t pairwise disjoint c-intervals, which completes the proof.

Proof of Theorem 1.: let $\{A^\nu\}$ be a t-independent system of c-intervals, $A^\nu = I^\nu \cup B^\nu$ where I^ν is the first component of A^ν (B^ν is a $(c-1)$-

interval). We define a sequence of points of R_1 as follows: $P_0 = -\infty$ and P_r is to the extreme right with the property: the system $\{B^\nu : I^\nu \subset (P_{r-1}, P_r)\}$ is $[(t+1)^{c-1}-1]$-independent. ((a,b) and $(a,b]$ mean open and half-closed intervals). As a consequence of the definition of P_r, there exists an $I^{\nu_r} \subset (P_{r-1}, P_r)$ so that the right endpoint of I^{ν_r} is P_r, moreover the system $\{B^\nu : I^\nu \subset (P_{r-1}, P_r]\}$ is $(t+1)^{c-1}$-independent, so it contains a pairwise disjoint subsystem $X_1^r, \dots, X_{(t+1)^{c-1}}^r$. It follows from the Lemma (applying to $t+1$ and $c-1$) that $P_{t+1} = +\infty$. We decompose the system $\{A^\nu\}$ into two parts:

$$\{A^\nu\} = \bigcup_{r=1}^{t+1} \{A^\nu : I^\nu \subset (P_{r-1}, P_r)\} \cup \{A^\nu : P_s \in I^\nu \text{ for some } s \quad (1 \leqq s \leqq t)\}$$

It is clear that the system of (c-1)-intervals B^ν belonging to the A^ν-s of the first part of the decomposition is $t[(t+1)^{c-1}-1]$ -independent and the theorem follows. The simple consequence of this theorem is

THEOREM 2. $\ell(2) = 2$

PROOF: because of $\ell(2) = \ell_1(2) \leqq \ell_1(1) + 1 = 2$ we only have to prove $\ell(2) \geqq 2$ which is obvious from the following simple example:

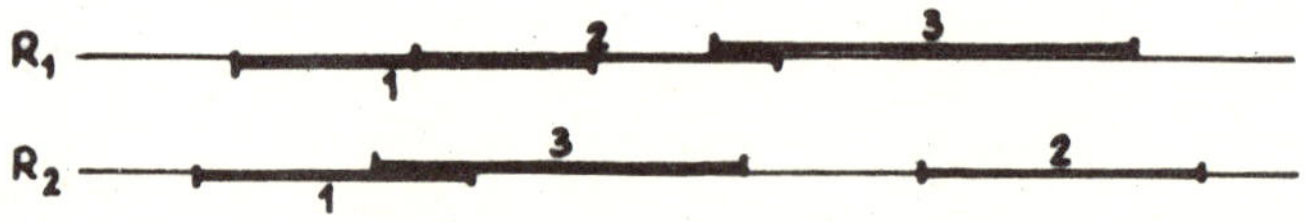

We will prove that Theorem 2. also holds under more general assumptions. For this purpose we need the notion of the interval-graph. A finite graph G is called an interval-graph if its vertices are in an one-to-one correspondence to an interval-system of the real line, and two vertices are connected if and only if the corresponding intervals have common points. A well-known property of interval-graphs is that they do not contain circuits without chords. Let's suppose that we have an one-to-one correspondence between the vertices of a complete graph G and the members of a system of pairwise intersecting 2-intervals. The edge between two vertices of G is coloured with red (with blue) if the corresponding 2-intervals meet each other in their first (second) components. Clearly G is an interval-graph if we consider only its edges of one colour. Now it is easy to see that Theorem 2. follows from the following

THEOREM 3. Let G be a complete graph, and its edges $E(G)$ coloured with red and blue. (An edge may be coloured with both colours.) Let us suppose that G contains neither blue nor red circuits of length 4 and 5 without chords. Then $V(G) = V(R) \cup V(B)$ where R and B complete red and blue subgraphs. (We denote the set of H's vertices by $V(H)$).

PROOF: by induction on the number of G's vertices. For $|V(G)| = 2$ the theorem holds. Supposing that it is true for $|V(G)| = n$, let G be a graph of the desired property with $|V(G)| = n+1$. If $p \in V(G)$ then $V(G-\{p\}) = V(R) \cup V(B)$. Let R_b and B_r be the subgraphs of R and B with which p is connected only with blue and red edges respectively. It is clear that $|V(R_b)|$ and $|V(B_r)| \neq 0$, otherwise we have nothing to prove. Let us choose the decomposition of $V(G-\{p\})$ so that $|V(B_r)| + |V(R_b)|$ should be minimal. Let us consider a $q \in V(B_r)$. Because of the minimality there exists a blue edge (q,r) between q and R. Let s be an element of $V(R_b)$. If $r \notin V(R_b)$ then (q,s) is not red, otherwise p,q,s,r,p would be

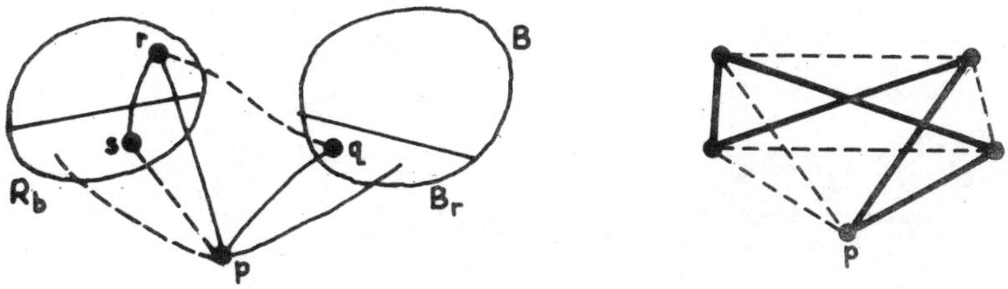

a red circuit without chords. Analogously we may suppose that in case of $u \in V(R_b)$ there exists a $v \in V(B_r)$ such that (u,v) is not blue. Now it follows that there is a circuit in G with points alternately to $V(R_b)$ and $V(B_r)$ and its edges are alternately blue and red. Let us consider the circuit C of minimal lenght of such type. It is easy to see that C is of length 4, so p and the points of C determine a circuit of length 5 without chords. This contradiction proves the theorem.

THEOREM 4. $\ell(3) = 4$

PROOF: the following example shows that $\ell(3) \geqq 4$:

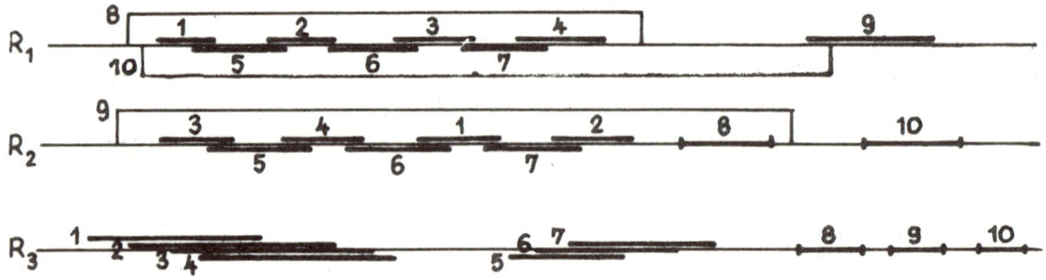

Let $\mathfrak{A} = \{A^\nu\}$ be a system of 3-intervals, any two of them having common points. $A^\nu = I_1^\nu \cup I_2^\nu \cup I_3^\nu$, $P \in R_1$ and $Q \in R_2$ are two points given arbitrarily.

Let us decompose $\mathfrak{A}$ as follows: $\mathfrak{A} = \mathfrak{A}_{11}(P,Q) \cup \mathfrak{A}_{12}(P,Q) \cup \cup \mathfrak{A}_{21}(P,Q) \cup \mathfrak{A}_{22}(P,Q) \cup B(P,Q)$ where

$$\mathfrak{A}_{11}(P,Q) = \{ A^\nu : I_1^\nu \subset (-\infty, P) \; I_2^\nu \subset (-\infty, Q) \}$$

$$\mathfrak{A}_{12}(P,Q) = \{ A^\nu : I_1^\nu \subset (-\infty, P) \; I_2^\nu \subset (Q, +\infty) \}$$

$$\mathfrak{A}_{21}(P,Q) = \{ A^\nu : I_1^\nu \subset (P, +\infty) \; I_2^\nu \subset (-\infty, Q) \}$$

$$\mathfrak{A}_{22}(P,Q) = \{ A^\nu : I_1^\nu \subset (P, +\infty) \; I_2^\nu \subset (Q, +\infty) \}$$

$$B(P,Q) = \{ A^\nu : (I_1^\nu \cup I_2^\nu) \cap \{P,Q\} \neq \emptyset \}$$

In the proof we often use the following simple proposition: let $\{X_i\}$ and $\{Y_i\}$ be two systems of intervals, $X_i \cap Y_j \neq \phi$ for all i and j, then $\cap X_i$ or $\cap Y_j$ is non-empty.

The third components of the 3-intervals from $\mathfrak{A}_{11}(P,Q)$ and $\mathfrak{A}_{22}(P,Q)$ satisfy the assumptions of this proposition, so we conclude that one of them has common points. The same reasoning can be applied to $\mathfrak{A}_{12}(P,Q)$ and $\mathfrak{A}_{21}(P,Q)$. Because of the symmetry we may suppose that the two systems are $\mathfrak{A}_{11}(P,Q)$ and $\mathfrak{A}_{12}(P,Q)$ that is $\bigcap_{A^\nu \in \mathfrak{A}_{11}(P,Q)} I_3^\nu \neq \phi$ and $\bigcap_{A^\nu \in \mathfrak{A}_{12}(P,Q)} I_3^\nu \neq \phi$.

We wish to emphasize that the two systems depend on P and Q which is denoted shortly by $\{P,Q\} \rightarrow \langle 11,12\rangle$. Now we want to define the points P^* and Q^* so that $\{P^*,Q^*\} \rightarrow \langle 11,12,21,22\rangle$ holds. Let $P^* = \max_{\{P,Q\}\rightarrow\langle 11,12\rangle} P$ (The lines are considered as sets ordered in the usual way.) P^* is the right endpoint of one and only one interval. Let Q be an arbitrary point of R_2 for which $\{P^*,Q\} \rightarrow \langle 11,12\rangle$. It is clear from the definition of P^* that if $(P^*,P']$ contains no endpoints of I_1^ν-s, then $\{P',Q\} \rightarrow \langle 11,12\rangle$ is false, so for example $\bigcap_{A^\nu \in \mathfrak{A}_{11}(P',Q)} I_3^\nu$ is empty. This means that if P^* is the right endpoint of the component $I_1^{\nu_1}$ then $A^{\nu_1} \in \mathfrak{A}_{11}(P',Q)$ and there exists an I_1^α where $A^\alpha \in \mathfrak{A}_{11}(P^*,Q)$ so that $I_3^{\nu_1} \cap I_3^\alpha = \phi$. Thus $\bigcap_{A^\nu \in \mathfrak{A}_{22}(P^*,Q)} I_3^\nu \neq \phi$ that is $\{P^*,Q\} \rightarrow \langle 11,12,22\rangle$.

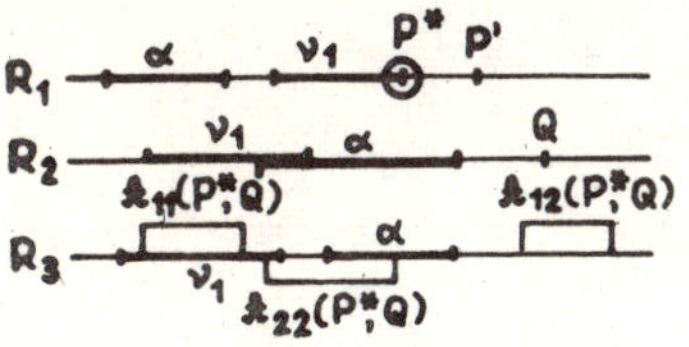

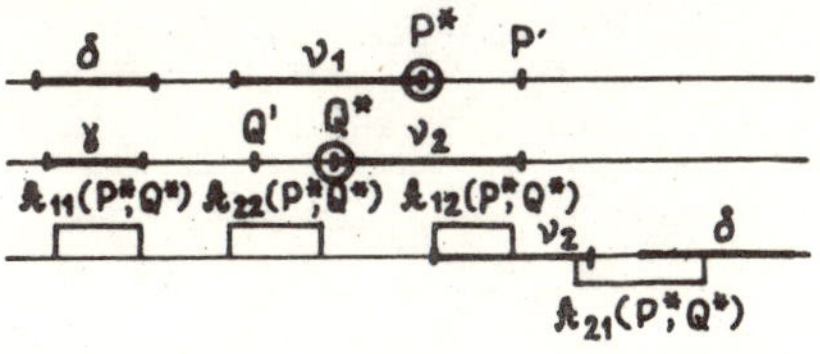

Q^* is defined similarly: $Q^* = \min_{\{P^*,Q\}\rightarrow\langle 11,12,22\rangle} Q$ (Q^* is the left endpoint of one and only one interval.)

If $A^{\nu_1} \in \mathfrak{A}_{12}(P', Q^*)$ then $\{P', Q^*\} \to \langle 11,12,22 \rangle$ is false, so there exists an $A^{\beta} \in \mathfrak{A}_{12}(P^*, Q^*)$ for which $I_3^{\nu_1} \cap I_3^{\beta} = \phi$, that is $\{P^*, Q^*\} \to \langle 11,12,22,21 \rangle$.

It follows that $A^{\nu_1} \in \mathfrak{A}_{11}(P', Q^*)$ since $Q^* \notin I_2^{\nu_1}$, so there is an $A^{\gamma} \in \mathfrak{A}_{11}(P', Q^*)$ so that $I_3^{\nu_1} \cap I_3^{\gamma} = \phi$. If $[Q', Q^*)$ does not contain endpoints then $\{P^*, Q'\} \to \langle 11,12,22 \rangle$ is false because of the minimality of Q^*. Let Q^* be the left endpoint of the interval $I_2^{\nu_2}$. In case of $A^{\nu_2} \in \mathfrak{A}_{22}(P^*, Q')$ every element of the set $\{I_3^{\nu}: A^{\nu} \in \mathfrak{A}_{22}(P^*, Q')\}$ meets I_3^{γ} and $I_3^{\nu_1}$, this also holds for $I_3^{\nu_2}$ which contradicts to the minimality of Q^*.

We conclude that $A^{\nu_2} \in \mathfrak{A}_{12}(P^*, Q')$ so we can find an $A^{\delta} \in \mathfrak{A}_{12}(P^*, Q^*)$ so that $I_3^{\delta} \cap I_3^{\nu_2} = \phi$. Thus $\{P^*, Q^*\} \to \langle 11,12,22,21 \rangle$. $\{I_3^{\nu}: A^{\nu} \in \mathfrak{A}_{11}(P^*, Q^*) \cup \mathfrak{A}_{22}(P^*, Q^*)\}$ and $\{I_3^{\nu}: A^{\nu} \in \mathfrak{A}_{12}(P^*, Q^*) \cup \mathfrak{A}_{21}(P^*, Q^*)\}$ are interval-systems of non-empty intersection, so we can choose the points R and S from these intersections. It is clear that for all A^{ν}, $\{P^*, Q^*, R, S\} \cap A^{\nu} \neq \phi$ which proves the theorem.

We close the first part with a note.

Let us suppose that $\mathfrak{A}$ is a system of c-intervals, any k of them having common points. $\ell^k(c)$ is defined in a similar manner as $\ell(c)$ above. It is easy to see that $\ell^k(c) = 1$ if $k \geq 2c$. The following scheme shows the values of $\ell^k(c)$ known to us.

k \ c	2	3	4	5	
2	2	4			
3	2	3			
4	1	2	2		
5	1	2	2		
6	1	1	2	2	

II.

Let A^{ν} be a subset of the real line R which is expressible as the union of c disjoint closed intervals. A^{ν} is also called shortly c-interval. Let

$\ell_t^*(c)$ be the least integer so that for every t-independent finite family $\mathcal{A} = \{A^\nu\}$ there exists a set $\{P_i\} \subset R$ with the property: $\{P_i\} \cap A^\nu \neq \phi$ if $A^\nu \in \mathcal{A}$ and $|\{P_i\}| = \ell_t^*(c)$. Obviously $\ell_t^*(1) = \ell_t(1) = t$.

The existence of $\ell_t^*(c)$ follows from

THEOREM 5. $\ell_t^*(c) \leqq [\ell_t^*(c-1)]^{c-1} \cdot \ell_t(c) + \sum_{i=1}^{c-1} [\ell_t^*(c-1)]^i$

This estimate is generally very far from being the best. We only sketch the proof:

If $A^\nu = I_1^\nu \cup I_2^\nu \cup \ldots \cup I_c^\nu$ (the components are indexed from left to the right) and $\mathcal{B}_1 = \{B_1^\nu\}$ is the system of (c-1)-intervals derived from A^ν by replacing I_1^ν and I_2^ν with their convex hull. There exists a set $\mathcal{P}_1 \subset R$ of $\ell_t^*(c-1)$ elements having the property:

$$P \cap B_1^\nu \neq \phi \quad \text{if} \quad B_1^\nu \in \mathcal{B}_1 \text{ and } P \in \mathcal{P}_1$$

We delete the sets A^ν for which $P \cap A^\nu \neq \phi$ and repeat the same reasoning by taking the system $\mathcal{B}_2 = \{B_2^\nu\}$ where B_2^ν is the $c-1$-interval of components I_1^ν, $\mathrm{conv}\{I_2^\nu \cup I_3^\nu\}\, I_4^\nu, \ldots, I_c^\nu$, which yields the set $\mathcal{P}_2$. We choose a point Q of $\mathcal{P}_2$, and delete from $\mathcal{B}_2$ the sets B_2^ν for which $Q \cap B_2^\nu \neq \phi$. Applying this method $c-1$ times, we see the remaining system $\mathcal{A}'$ "separated" by the points $P, Q, \ldots$ and as a consequence of Theorem 1. there is a set $\mathcal{P}$ (depending on the choice of $P, Q, \ldots$) of $\ell_t(c)$ elements enjoying the property: $\mathcal{P} \cap A^\nu \neq \phi$ if $A^\nu \in \mathcal{A}'$. This is repeated for every choice of $P \in \mathcal{P}_1$, $Q \in \mathcal{P}_2, \ldots$ and the theorem follows.

From this theorem $\ell^*(2) \leqq \ell^*(1) \cdot 2 + 1 = 3$. This is the best estimate of $\ell^*(2)$ as will be shown in Theorem 6.

Now we generalize the problems considered above in the following manner. We replace the real lines, intervals and c-intervals by trees, subtrees and forests of c components. (c-forests for the sake of shortness.)

All the theorems stated in part I. remain true in this generalized setting. We define some notions on the analogy of the ones having played an important role in the previous proofs.

Let G be a tree and F a subtree of G. The complement of F with respect to G being a forest, we denote by $F'(x)$ the component containing $x \in V(F)$. $x \in V(F)$ is said to be an extreme point of F if and only if $F'(x) \neq \{x\}$.

The assumption that two intervals have no common endpoints is replaced by the next one: if $F_1, F_2 \subset G$ then F_1 and F_2 do not contain common extreme points.

In the first part, intervals often appeared in extremal position. This is replaced by such an $F \subset G$ that $F'(x)$ is maximal.

At last we may clearly assume the $\ell_t(c)$ $(\ell_t^*(c))$ vertices to be extreme.

The existence of $\ell_t^*(c)$ in this general case also holds as a consequence of Theorem 1. The proof of the next theorem is much more difficult than in case of intervals.

THEOREM 6. $\ell^*(2) = 3$

PROOF: we assert that $\ell^*(2) \leqq 3$. Let $\mathcal{F} = \{F^\nu\}$ be a finite family of 2-forests, any two of them having non-empty intersection. The components of F^ν are denoted by H_1^ν and H_2^ν. It is clear that there is an (unique) S^ν path of G connecting an $x \in V(H_1^\nu)$ and a $y \in V(H_2^\nu)$, so that $V(S^\nu) \cap V(H_1^\nu) = \{x\}$ and $V(S^\nu) \cap V(H_2^\nu) = \{y\}$. The graphs $T^\nu = F^\nu \cup S^\nu$ are substrees of G, any two of them having common points, so we can choose a point S from $\bigcap_\nu T^\nu$.

Now we split $\mathcal{F}$ into two parts: $\mathcal{F}' = \{F^\nu : S \in F^\nu\}$ $\mathcal{F}'' = \mathcal{F} - \mathcal{F}'$. We may assume (by changing the indices if nessecery) that if $F^\nu \in \mathcal{F}'$ then $S \in H_1^\nu$. The edges of G incident to S divide G into subtrees $G_1, G_2, \dots, G_k$. $(G_i \cap G_j = S)$ It is obvious that in case of $F^\nu \in \mathcal{F}''$ H_1^ν and H_2^ν are subgraphs of different G_i-s. It may be supposed that every G_i contains some component of F^ν. ($F^\nu \in \mathcal{F}''$). We define a graph M in order to show the distribution of F^ν'-s components in the G_i-s. $V(M) = \{p_1, p_2, \dots, p_k\}$ and $(p_i, p_j) \in E(M)$ if and only if there exists an F^ν, the components of which are in G_i and G_j.

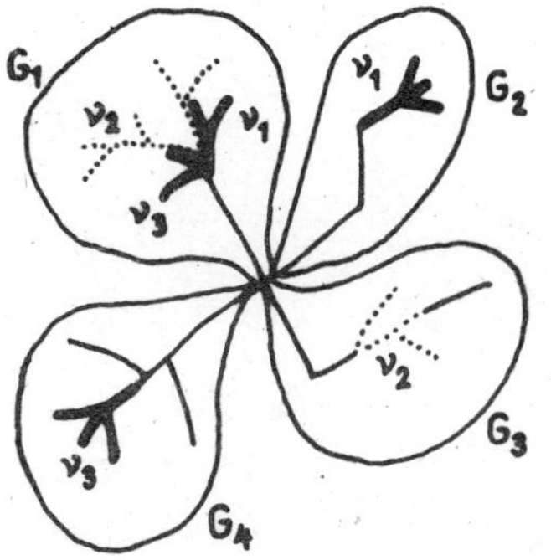

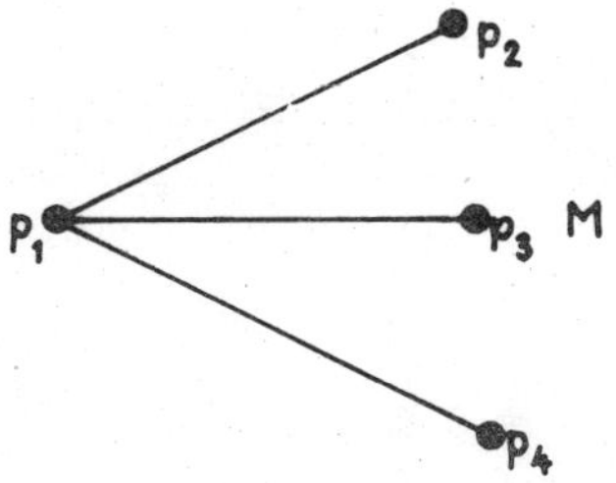

It is clear that any two edges of M have a common vertex and the degree of every point is at least one. Consequently M is a triangle or a star. If M is a star as shown on the figure, we define G' deleting from G_1 the point S with the edge incident to S and $G'' = \bigcup_{i=2}^{k} G_i$. Now for every $F^\nu \in \mathcal{F}''$ the two components of F^ν are in G' and G'' respectively so the existence of an $X \in V(G')$ and of an $Y \in V(G'')$ such that $\{X,Y\} \cap V(F^\nu) \neq \emptyset$ follows from Theorem 2. Thereby $\{X, Y, S\} \cap V(F^\nu) \neq \emptyset$ if $F^\nu \in \mathcal{F}$ which proves our assertion.

In the sequel we assume M to be a triangle, that is we have $G_1 \cup G_2 \cup G_3 = G$. We refine the splitting $\mathcal{F} = \mathcal{F}' \cup \mathcal{F}''$ in the following manner: $\mathcal{F}' = \bigcup_{i=1}^{3} \mathcal{F}_i$ and $\mathcal{F}'' = \mathcal{F}_{12} \cup \mathcal{F}_{23} \cup \mathcal{F}_{31}$ where

$\mathfrak{F}_i = \{ F^\nu : F^\nu \in \mathfrak{F}', H_2^\nu \subset G_i \}$ $\quad (i = 1,2,3)$

$\mathfrak{F}_{12} = \{ F^\nu : F^\nu \in \mathfrak{F}''$, the components of F^ν are in G_1 and $G_2 \}$

$\mathfrak{F}_{23} = \{ F^\nu : F^\nu \in \mathfrak{F}''$, the components of F^ν are in G_2 and $G_3 \}$

$\mathfrak{F}_{31} = \{ F^\nu : F^\nu \in \mathfrak{F}''$, the components of F^ν are in G_3 and $G_1 \}$

It is easy to prove that we can chose P,Q,R from $V(G_1), V(G_2)$ and $V(G_3)$ resp. with the property: $\{P,Q,R\} \cap V(F^\nu) \neq \emptyset$ if $F^\nu \in \mathfrak{F}''$. The details are left to the reader.

If P,Q,R are vertices of G_1, G_2 and G_3 respectively, then $G(S,P,Q,R)$ will denote the union of the (unique) paths $\overline{SP}$, $\overline{SQ}$, $\overline{SR}$. (This is the smallest subtree of G containing the vertices S, P, Q, R).

Let $G(S,P,Q,R)$ be minimal for set-theoretic inclusion with the property: $\{P,Q,R\} \cap V(F^\nu) \neq \emptyset$ if $F^\nu \in \mathfrak{F}''$.

We prove that $\{P,Q,R\} \cap V(F^\nu) \neq \emptyset$ if $F^\nu \in \mathfrak{F}''$ and $F^\nu \in \mathfrak{F}_i \cup \mathfrak{F}_j$ for suitable i and j $(i \neq j)$. If this does not hold, then there is, for example, an $F^\alpha \in \mathfrak{F}_1$ and an $F^\beta \in \mathfrak{F}_2$ so that $\{P,Q,R\} \cap V(F^\alpha \cup F^\beta) = \emptyset$. Let $H_1^{\nu_1}$, $H_1^{\nu_2}$ and $H_1^{\nu_3}$ be the (unique) subtrees with extreme points P, Q, R.

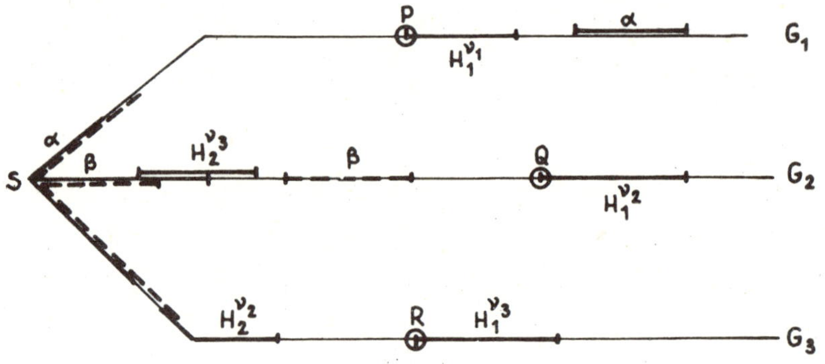

If $F^{\nu_3} \in \mathfrak{F}_{23}$ then $H_2^{\nu_3}$ meets H_1^α (in the interior of path $\overline{SQ}$ by minimality) and so $H_2^{\nu_3} \cap H_1^{\nu_2} = \emptyset$. It follows that $F^{\nu_2} \in \mathfrak{F}_{23}$, and the same argument shows $H_2^{\nu_2}$ to be disjoint from $H_1^{\nu_3}$; this leads to contradiction. Replacing F^α by F^β, the same argument can be repeated and the assertion follows.

We may assume (by symmetry): $\{P,Q,R\} \cap V(F^\nu) \neq \phi$ when $F^\nu \in \mathcal{F}'' \cup \mathcal{F}_2 \cup \mathcal{F}_3$. This is the property for which we choose a minimal tree $G(S,Q^*,R^*)$.

If $\{Q^*,R^*\} \cap V(F^\nu) \neq \phi$ for all $F^\nu \in \mathcal{F}_1$ we have nothing to prove. Let F^γ be such that $H_1^\gamma \cap \{Q^*,R^*\} = \phi$ $F^\gamma \in \mathcal{F}_1$, $H_1^{\nu_2}$ and $H_1^{\nu_3}$ will denote the two components belonging to the extreme points Q^* and R^*. Obviously $F^{\nu_2} \in \mathcal{F}_{12}$ and $F^{\nu_3} \in \mathcal{F}_{31}$ otherwise we have a contradiction. (Using F^γ as above.) Clearly $H_2^{\nu_2} \cap H_2^{\nu_3} \neq \phi$. We choose P^* as close as possible to the tree $H_2^{\nu_2} \cup H_2^{\nu_3}$ with the propety: $\{P^*,Q^*,R^*\} \cap F^\nu \neq \phi$ for all $F^\nu \in \mathcal{F}'' \cup \mathcal{F}_2 \cup \mathcal{F}_3$.

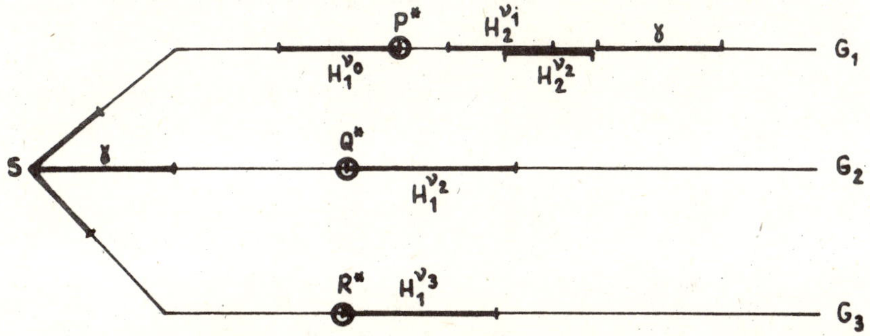

By the minimum-property of $G(S,Q^*,R^*)$ $P^* \notin H_2^{\nu_2} \cup H_2^{\nu_3}$. P^* is the extreme point of a tree $H_1^{\nu_0}$, but the assumptions $H_1^{\nu_0} \in \mathcal{F}_2$, $H_1^{\nu_0} \in \mathcal{F}_3$, $H_1^{\nu_0} \in \mathcal{F}_{12}$ lead to contradiction.

So $\{P^*,Q^*,R^*\} \cap F^\nu \neq \phi$ for all $F^\nu \in \mathcal{F}$. We conclude that $\ell^*(2) \leqq 3$. The following example proves that $\ell^*(2) \geqq 3$:

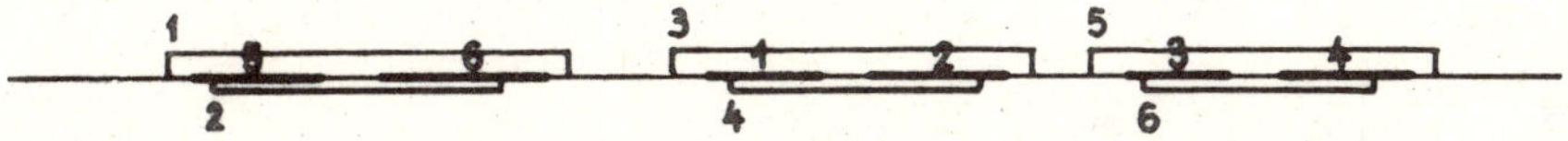

Finally we present a conjecture: if G is a tree and $\mathcal{F} = \{F^\nu\}$ is a finite family of its c-forests, any $c+1$ of them having non-empty intersection, then there exists a set $P \subset V(G)$ so that $|P| = c$ and $P \cap F^{\nu} \neq \phi$ for all $F^{\nu} \in \mathcal{F}$. We can prove this only in case of $c \leqq 3$, and if G is a path. (c is arbitrarily given)

L. Surányi's result is worth mentioning (oral communication); it gives a necessary and sufficient condition for a graph G to be representable as a system of subtrees of a suitable tree (between two vertices of G, there is an edge iff the corresponding subtrees have common vertices): A graph G is representable if and only if it does not contain any circuit without chords.

Using this theorem we can easily obtain the following two corollaries of our first and fourth theorem:

COROLLARY 1. Let G be a complete graph, and its edges coloured with c different colours. (An edge may be coloured with more than one colour.) Let us suppose that G does not contain any circuit of one colour without chords. Then there exists an integer $f(c)$ (which does not depend on $|V(G)|$) so that $V(G) = \bigcup_{i=1}^{f(c)} V(G_i)$ where G_i is a complete graph of one colour

COROLLARY 2. If $c=3$ then the least value of $f(3)$ is 4.

REFERENCE

[1] A. HAJNAL-J. SURÁNYI: Über die Auflösung von Graphen in vollständige Teilgraphen Ann. Univ. Sci. Budapest, 1. 958, p. 113.

Selecting the tth largest using binary errorless comparisons

by

A. Hadian and M. Sobel
Minnesota, USA

1. INTRODUCTION

There are given n numbers, or items with scalar attributes, $x_1, x_2, \ldots, x_n$ which are unkown and assumed to be pairwise unequal. For given t, we wish to find the tth largest of these numbers using only binary errorless comparisons; each comparison between two x's tells us only which is larger and which is smaller.

Two criteria are considered for evaluating and comparing procedures that accomplish our goal. A minimax (or M-optimal) procedura minimizes the maximum numoer of comparisons needed to find the tth largest. Assuming a random ordering at the outset (with equal probability for each of the $n!$ arrangements), an E-optimal procedure is one that minimizes the expected number of comparisons required. Since the same procedure in general does not satisfy both criteria, we may refer to these two criteria as separate problems, the M-problem and the E-problem.

Kislicyn [2] considers the M-problem and gives an upper bound

$P_t(n)$ for the maximum number of comparisons $M_t(n)$ required by an M-optimal procedure, namely

$$(1.1) \qquad M_t(n) \le P_t(n) = n-1+\sum_{i=1}^{t-1}[\log(n-i)] ,$$

where $[x]$ is the usual notation for the integer part of x and all logarithms are to the base 2. It is well known that $M_1(n) = n-1$ and many procedures, including the knock-out tournament, achieve this lower bound, Kislicyn's work is also discussed by Moon [3 - page 48].

Two problems related to ours are the problem of ordering the t largest and the problem of selecting the t largest (unordered) considered by Sobel in [6] and [7], respectively. For ordering the 2 largest items the best possible M-value $M_2^{(0)}(n)$ was already found by Schreier [4] and Slupecki [5], namely

$$(1.2) \qquad M_2^{(0)}(n) = n-1+[\log(n-1)] = P_2(n) .$$

Since we cannot find the 2nd largest item without also finding the largest item, it follows that $M_2(n)$ is also given by (1.2). Sobel gives several procedures for ordering the 2 largest items in [6], two of which are M-optimal, and it follows that all these procedures can also be used in our problem for $t=2$; the ordering of the t best is also considered in [1]. The problem of selecting the 2 best (unordered) in [7] is related to our problem for both $t=2$ and $t=3$ since in either case we know which are the two largest items at termination. Hence the $\operatorname{Min}\{M_2(n), M_3(n)\}$ is an upper bound for the M-value for the $t=2$ selection problem. Conversely the best possible M-value given by

$$(1.3) \qquad M_2^{(S)}(n) = n-1+[\log(n-2)]$$

for the $t=2$ selection problem in [7] is a lower bound for both $M_2(n)$ and $M_3(n)$ in our problem. The relationship of our problem to other research problems in considered in Section 6.

In this paper we improve on $P_t(n)$ in (1.1) by introducing three new procedures R_1, R_2 and R_3 which yield new sharper upper bounds $M_t(n|R_i)$ $(i = 1, 2, 3)$ of $M_t(n)$ such that for all t and n

$$(1.4) \qquad M_t(n) \le M_t(n|R_3) \le M_t(n|R_2) \le \mathrm{Min}(P_t(n), P_{n-t+1}(n))$$

and for $t < [\frac{n+1}{2}]$ we have $M_t(n|R_1) \le P_t(n)$.

One example for $N = 8$, $t = 4$ (which also affects the result for $n = 9$, $t = 5$) requires at most 12 comparisons $< M_4(8|R_3) = 13$ and shows that even our new bounds are not the best possible. However this counter-example is an isolated phenomenon and we have found that $M_t(n|R_3)$ is difficult to improve upon for most values of n and t. In fact, for an infinite sequence of n values and $t = (n+1)/2$, procedures R_2 and R_3 are identical and are conjectured to be both M-optimal and E-optimal.

In addition, expressions are derived for the expected number of comparisons $A_t(n|R_1)$ under procedure R_1.

2. PROCEDURE R_1.

The basic ideas of the procedure are the use of a binary expansion, the use of recursion methods, and the simple fact that the largest among any $n-t+2$ items has at most $t-2$ items above it and hence can be eliminated as a competitor for the position of tth largest in the entire set of n (which has exactly $t-1$ items above it).

Let the binary decomposition of $n-t+2$ be given by

$$(2.1) \qquad n-t+2 = 2^{r_1} + 2^{r_2} + \cdots + 2^{r_s},$$

where the integers $r_1 > r_2 > \cdots > r_s \ge 0$ and $s \ge 1$ are defined by (2.1). We note that (2.1) can also be written as

(2.2) $$r_i = [\log(n-t+2-\sum_{j=1}^{i-1} 2^{r_j})] \qquad (i = 1,2,\dots,s)$$

The procedure R_1 is described in four steps as follows.

1. Choose $n-t+2$ items randomly and divide them into s groups, the group size for the ith group being 2^{r_i} as indicated by (2.1). Find the largest in each group, using the well-known knockout tournament method. (The remaining $j-2$ items are temporarily held in reserve.)

2. Find the largest among these $n-t+2$ by comparisons among these s winners. In doing this we start with the smallest group (of size 2^{r_s}) and compare it's winner with that of the second smallest, then compare the largest of both groups with the largest in the third smallest group, etc. This gives a connected graph (or a tree) with $n-t+2$ vertices containing a unique maximal element, which we denote by $x^{(i)}$ if it comes from the ith group.

3. Since $x^{(i)}$ cannot be the tth largest of the entire set, we remove it from our graph along with all the line segments connecting it to other vertices; this gives us a number of unconnected subsets. Using exactly one of the $t-2$ items in reserve, we rebuild another monolithic connected structure (or tree) of size $n-t+2$. The largest as again removed and this step is repeated until all $t-2$ reserve units are used up.

4. We then conduct a simple play-the-winner tournament (eliminating all losers) among all competitors for 2nd place in the final structure with $n-t+2$ items. The winner is the tth largest in the original set of n items.

DERIVATION OF $M_t(n|R_1)$.

The maximum number of line segments (or connections) are lost in step 3 when $i=1$, i.e., the worst case is when $x^{(i)} = x^{(1)}$ comes from the first group of size r_1. This results in a loss of r_1+1 connections if $s>1$

or r_1 if $s = 1$ (which we write as $r_1 + 1 - \delta_{1s}$ for general s). Hence we need at most $r_1 + 1 - \delta_{1s}$ comparisons to rebuild another tree. Using the notation $M(t,n)$ for the number of comparisons required in step 3 above, we obtain the recursion

$$M(t,n) = r_1 + 1 - \delta_{1s} + M(t-1, n-1) \qquad \text{for} \quad t-2 > 0 \tag{2.3}$$

with the boundary condition given by $M(2,n) = 0$. We note that for $M(t-1,n-1)$ on the right side of (2.3) the value of $n-1-(t-1)+2$ is the same as for $M(t,n)$ and hence the r's defined in (2.1) do not change. From (2.3) by iteration we easily obtain

$$M(t,n) = (t-2)(r_1 + 1 - \delta_{1s}). \tag{2.4}$$

To obtain $M_t(n|R_1)$ from (2.4) we have to add on exactly $n-t+1$ comparisons for steps 1 and 2 and at most $r_1 - \delta_{1s}$ for step 4. This gives

$$\begin{aligned} M_t(n|R_1) &= n-t+1+(t-2)(r_1+1-\delta_{1s}) + r_1 - \delta_{1s} = \\ &= n-1+(t-1)(r_1 - \delta_{1s}). \end{aligned} \tag{2.5}$$

Using (2.2) with $i = 1$ it is readily observed that for all integers $s \geq 1$

$$r_1 - \delta_{1s} = [\log(n-t+1)] = \{\log(n-t+2)\} - 1 \tag{2.6}$$

where $\{x\}$ is the smallest integer not less than x. Hence we obtain from (2.5)

$$M_t(n|R_1) = n-1+(t-1)[\log(n-t+1)] = n-t+(t-1)\{\log(n-t+2)\}. \tag{2.7}$$

REMARK. It should be noted that the total number of comparisons $M_t(n|R_1)$ also satisfies the recursion (2.3) if we take $t \geq 1$ and change the boundary condition so that $M_1(n|R_1) = n-1$. We make use of this observation in writing the recursion formulas (4.1) for a new procedure R_3 in section 4.

The upper bound for $M_t(n)$ found by Kislicyn [2] is

(2.8) $$P_t(n) = n-1+\sum_{i=1}^{t-1}[\log(n-i)] .$$

Comparing this with the middle expression in (2.7), we note that our expression is obtained from (2.8) by replacing each term in the summation by the smallest summand. Hence

(2.9) $$P_t(n) - M_t(n|R_1) = \sum_{i=1}^{t-2}([\log(n-i)] - [\log(n-t+1)]) \geq 0 .$$

For certain subsequences of n-values the difference in (2.9) approaches zero and for others (as in (2.12) below) it approaches infinity.

We now consider a special sequence of n values and t values

(2.10) $$n = 2^{r+1} - 3 , \qquad t = 2^r - 1$$

which is indexed by r $(r = 1,2,\cdots)$. By (2.7) we find that

(2.11) $$M_t(n|R_1) = (r+1)(2^r - 2)$$

and the difference in (2.9) is

(2.12) $$P_t(n) - M_t(n|R_1) = 2^r - 3 \to \infty$$

as $r \to \infty$. It is interesting to note that for each r-value in (2.10), the procedure R_1 has no variation in the number of comparisons required, i.e., the minimum number of comparisons is also given by (2.11) (see also remark after (2.15)).

DERIVATION OF $A_t(n|R_1)$.

To derive the expected number of comparisons under procedure R_1 we again consider steps 1 and 2, step 3, and step 4 separately. For steps 1 and 2 we again use $n-t+1$ comparisons. In step 3 we use the fact that the probability that the ith group yields the largest of $n-t+2$ is $p_i = 2^{r_i}/(n-t+2)$ $(i=1,2,\ldots,s)$. The number of line connections broken by removing $x^{(i)}$ is $r_i+i-\delta_{is}$, since we have r_i connections to $x^{(i)}$ within the ith group, we pick up one connection for each of the $i-1$ larger groups, and we pick up $1-\delta_{is}$ connections from the smaller groups. In step 4 we have to find the best one of $r_i+i-\delta_{is}$ competitors and hence we use $r_i+i-1-\delta_{is}$ comparisons. At each iteration the items are associated with the group they came from and the new item is associated with the depleted i^{th} group. The marginal probability that the ith group yields the largest remains the same at each iteration and we obtain

$$A_t(n|R_1) = n-t+1+(t-2)\sum_{i=1}^{s} p_i(r_i+i-\delta_{is}) + \sum_{i=1}^{s} p_i(r_i+i-1-\delta_{is}) = \tag{2.13}$$

$$= n-t+(t-1)\sum_{i=1}^{s} p_i(i+r_i-\delta_{is})$$

Since

$$r_i-\delta_{is} = [\log(n-t+1-\sum_{j=1}^{i-1} 2^{r_j})] = \{\log(n-t+2-\sum_{j=1}^{i-1} 2^{r_j})\}-1, \tag{2.14}$$

it follows from (2.11) that

$$A_t(n|R_1) = n-1+\left(\frac{t-1}{n-t+2}\right)\sum_{i=1}^{s} 2^{r_i}(i+\{\log\sum_{j=i}^{s} 2^{r_j}\}). \tag{2.15}$$

For the subsequence (2.10) we find that $A_t(n|R_1)$ reduces to (2.11); this also follows from the above observation that there is no variation when procedure R_1 is used for any r-value in (2.10).

3. SYMMETRIZED VERSION OF PROCEDURE R_1

It should be noted that procedure R_1 is asymmetric with respect to the median and is generally better for $t \le [\frac{n+1}{2}]$ than for $t > [\frac{n+1}{2}]$;

for example, $M_2(4|R_1) = 4 < M_3(4|R_1) = 5$. In exceptional cases the reverse holds; for example $M_3(6|R_1) = 9 > M_4(6|R_1) = 8$. In this section we define a symmetrized version R_2 of procedure R_1 which takes advantage of these inequalities.

A procedure R_1' dual to R_1, is first defined by interchanging t and $n-t+1$ everywhere. Corresponding to equation (2.1) we write

$$t+1 = 2^{r_1'} + 2^{r_2'} + \dots + 2^{r_{s'}'} \tag{3.1}$$

and put $n-t-1$ items in reserve. We find the smallest in each group and, starting with the smallest group, find the smallest of all $t+1$ items. Then we eliminate items with t or more superiors.

We define the symmetrized procedure R_2 by using either R_1 or R_1', whichever gives a smaller maximum, This clearly gives the result

$$\begin{aligned} M_t(n|R_2) &= \mathrm{Min}\,(M_t(n|R_1), M_t(n|R_1')) = \\ &= \mathrm{Min}\,(M_t(n|R_1), M_{n-t+1}(n|R_1)) = \\ &= \mathrm{Min}\,(n-1+(t-1)[\log(n-t+1)], n-1+(n-t)[\log t]) \end{aligned} \tag{3.2}$$

We wish to show that for r large the rows of table 1 for n close to but not greater than 2^{r+1} form an arithmetic progression (AP) with common difference r for all values of t except at most for a few values close to $n/2$. Under procedure R_1 for any fixed $n = 2^r + x$ with $0 < x \le 2^r$ it is easy to show (proof omitted) that for t running from 1 to x the values of $M_t(n|R_1)$ form an AP with common difference r. In particular, for $n = 2^{r+1}$ the AP extends up to $t = 2^r$.

Under procedures R_2 this property does not hold exactly but we wish to show that the same property holds asymptotically as $r \to \infty$. Let $n = \Theta 2^r$ for fixed Θ with $1 < \Theta \le 2$ and let $t = \varepsilon 2^r$ for fixed ε with $0 < \varepsilon \le 1$. In order that $M_t(n|R_1)$ be not greater than $M_t(n|R_1')$, we can use the last expressions in (3.2) to obtain for asymptotically large r

$$(3.3) \qquad (\varepsilon 2^r - 1)[\log 2^r(\Theta - \varepsilon)] \le 2^r(\Theta - \varepsilon)[\log(\varepsilon 2^r)]$$

We note from (3.3) that the inequality holds for $r \to \infty$ both for $\varepsilon < \Theta/2$ and for $\varepsilon = \Theta/2$. In particular, for $\Theta = 2$ it holds for $1 \le t \le 2^r$. Hence for $r \to \infty$ the values of $M_t(n|R_2)$ are asymptotically the same as $M_t(n|R_1)$ for all $t < \Theta 2^{r-1} = n/2$. In particular, for $\Theta = 2$ the length of the AP in the row for $n = 2^{r+1}$ divided by the length 2^r of the whole row tends to 1 as $r \to \infty$.

We note from Table 1 that procedure R_3 defined in the next section affects only a small portion of the entries and, based on numerical studies, it is conjectured that the same properties proved above also hold for procedure R_3.

4. AN IMPROVED PROCEDURE, R_3.

The basic idea is to give us the flexibility whenever possible of deciding at each stage of the algorithm whether we wish to eliminate the next item off the top (because it has $n-t+1$ or more inferiors) or off the bottom (because it has t or more superiors). For $n-t+2 \le 4$ the break-up in step 3 of procedure R_1 gives us structures with at most 2 items connected; hence we are free to eliminate the next item from the top or the bottom. Conversely, for $n-t+2 \ge 7$ the break-up may contain structures of size 3, 4 or larger which are oriented to finding the largest and we then have to continue eliminating off the top. In between, for $n-t+2 = 5$ or 6, we have a similar flexibility if and only if the removal of $x^{(i)}$ breaks up the structure of size 4; our formulas below reflect this complication for $n-t+2$ equal to 5 or 6. A similar discussion holds for changes of strategy in the reverse direction if we replace $n-t+2$ above by $t+1$.

Let $f_{t,n}$ (and $f'_{t,n}$) denote the maximum number of comparisons under the new procedure R_3 if we eliminate the first item off the top (off the bottom). Then $M_t(n|R_3) = \min(f_{t,n}, f'_{t,n})$. The recursion formulas that define procedure R_3 can now be written as six equations but, since the second three equations are dual to the first three, we write only the latter as follows:

$$f_{t,n} = 2 + M_{t-1}(n-1|R_3) \qquad \text{for } n-t = 1 \text{ or } 2$$

$$f_{t,n} = \mathrm{Max}(3 + M_{t-1}(n-1|R_3),\ n-t-2+f_{t-1,n-1}) \qquad \text{for } n-t = 3 \text{ or } 4 \tag{4.1}$$

$$f_{t,n} = \{\log(n-t+2)\} + f_{t-1,n-1} \qquad \text{for } n-t \geq 5$$

The three dual equations are obtained by interchanging $f_{t,n}$ and $f'_{t,n}$ and replacing any t which is not a subscript by $n-t+1$. Since $f_{t,n}$ now represents the total number of comparisons, rather than the middle part as in (2.3), the boundary conditions are

$$f_{1,n} = n-1 = f'_{1,n}; \quad f_{t,t} = t-1 \tag{4.2}$$

for all n and t.

The results of this iteration are given in Table 1 for $n \leq 20$. In Table 1 the cases for which procedure R_3 gives an improvement over procedure R_2 are shown by a star on the appropriate entry.

It is interesting that for the special subsequence defined in (2.10) we have $n-t+2 = t+1 = 2^r$ and the equations (4.1) reduce to (2.3) so that

$$M_t(n|R_1) = M_t(n|R_2) = M_t(n|R_3) = (r+1)(2^r - 2) \tag{4.3}$$

and since there is 'no variation' in the results for these cases the common value in (4.3) is also the expectation for all 3 procedures.

5. A COUNTEREXAMPLE

To show that the results of procedure R_3 can be further improved in isolated cases we now give a procedure due to D.G. Doren (see acknowledgement below) for the special case $n=8$ and $t=4$; the only other case that this appears to affect is that of $n=9$ and $t=5$. The idea behind this procedure is to find a strategy that allows us to simultaneously eliminate items 'off the top' and 'off the bottom'; this appears to be possible only at one stage and only for $n=8$ and $t=4$.

We separate the 8 items into 2 groups of 4 and find the best item (by knock-out) in each group. Let the 2 groups be denoted by

$$x_1 < x_2 < x_4, \quad x_3 < x_4;$$

(5.1)

$$y_1 < y_2 < y_4, \quad y_3 < y_4.$$

We compare x_2 and y_2 and suppose (without loss of generality) that $x_2 < y_2$. This simultaneously eliminates x_1 and y_4 from contention and we need the 3rd largest of the remaining six. Compare x_2 and y_3; in the worst case $x_2 < y_3$ and we eliminate x_2 so that we now need the 3rd largest of 5 items with the relations $y_1 < y_2$ and $x_3 < x_4$. Since $M_3(5) = 6$ under any of our procedures and we use both relations, we need exactly 4 more comparisons. Thus we have a total of 6 (from (5.1)) $+ 2 + 4 = 12$ comparisons. All of our procedures R_1, R_1', R_2 and R_3 require 13 comparisons for this case.

Using the recursion (2.3), this result also gives a reduction of one for the case $n=9$, $t=5$, but does not affect any more cases.

6. SOME RELATED PROBLEMS AND ASSOCIATED FUTURE RESEARCH

To show that our problem of finding the tth largest is central to several other related problems, we briefly mention some of these.

I. Selecting the t largest of n items

Here we consider the procedure of putting 1 unit aside (call it x), find the tth largest (say y) of the remaining $n-1$ and finally compare x and y (let z denote the larger). Then the union of z and the $t-1$ units found to be larger than y constitute the t largest (unordered). If we use the basic procedure R_1 with $M_t(n)$ given by (2.7) then the maximum value $M_t^{(S)}(n)$ for the selection problem is

$$M_t^{(S)}(n) = n-1+(t-1)[\log(n-t)]. \tag{6.1}$$

It is conjectured that if we use an M-optimal procedure for finding the tth largest of $n-1$ then the resulting procedure (using the above) will be an M-optimal selection procedure.

II. For a fixed number of comparisons C $(C > t)$ find a subset of the n items such that (i) it contains the t largest (t given), and (ii) its maximum (or expected) size is as small as possible.

Using the expression for $M_t(n|R_1)$ in (2.7), we find the largest integer x such that

$$x-1+(t-1)[\log(x-t+1)] \le C; \tag{6.2}$$

for $C \ge t-1$ the solution $x \ge t$ is unique since the left side of (6.2) is (strictly) increasing in x. Then we can find the t largest in a randomly chosen subset of size x. Combining these t with the remaining $n-x$ units gives a subset which is proposed as a solution for II.

It is conjectured that this procedure will have some optimal properties, at least when equality holds in (6.2). For example, if $n = 2^{r+1}-3$, $t = 2^r-1$ and $C = (r+1)(2^r-2)$ then $x=n$ and the above procedure gives a subset of size exactly t, i.e., the minimum and maximum subset size are both t.

III. Find and order the t largest of a set of n items, for t, n both given and $1 \leq t \leq n/2$.

Among the possible procedures we can find the tth largest by starting with one of our procedures (say R_1) and then completely ordering the $t-1$ items which we know to be above it; the latter can be accomplished by one of the several procedures considered in [1] and [6]. This procedure gives good results for small values of t; in particular, it is optimal for $t = 2$ and appears to be optimal for $t = 3$, although the result for $t = 3$ is not proved.

ACKNOWLEDGEMENT

The authors wish to thank David G. Doren of Intech Inc., Minneapolis, Minnesota for contributing the example given in Section 5 and for some stimulating conversations on this research.

Table 1

The Maximum Number of Comparisons Under Procedure R_3

n \ t	1	2	3	4	5	6	7	8	9	10
2	1									
3	2	3								
4	3	4								
5	4	6	6§							
6	5	7	8							
7	6	8	10	11*						
8	7	9	11	13+						
9	8	11	12	14	16+					
10	9	12	14*	15	17					
11	10	13	16	18*	18	20				
12	11	14	17	20	23	21				
13	12	15	18	21	24	26	24§			
14	13	16	19	22	25	28	27			
15	14	17	20	23	26	29	30	35		
16	15	18	21	24	27	30	33	36		
17	16	20	22	25	28	31	34	37	40	
18	17	21	24*	26	29	32	35	38	41	
19	18	22	25*	30	30	33	36	39	42	45
20	19	23	27	30*	35	34	37	40	43	46

* These six entries are the ones for which procedure R_3 improve upon the result of procedure R_2; in each case there was a saving of exactly one.

\+ These two entries can be reduced by one if we use the counterexample explained in the text in section 5.

§ These two entries correspond to $r = 2$ and 3 in (2.10); in each of these cases the maximum and minimum number of comparisons are equal under any of our procedures.

REFERENCES

[1] HADIAN, A. (1969). Optimality properties of various procedures for ranking n different numbers using only binary comparisons. Technical Report No. 117, Department of Statistics, University of Minnesota.

[2] KISLICYN, S. S. (1964). On the selection of the kth element of an ordered set by pairwise comparison (Russian). Sibirsk. Mat. Ž. 5 557-564 (MR 29 No. 2198).

[3] MOON, J. W. (1968). Topics on Tournaments, Holt, Rinehart and Winston, New York.

[4] SCHREIER, J. (1932). On tournament elimination systems (Polish). Mathesis Polska 7 pp. 154-160.

[5] SLUPECKI, J. (1949-51). On the system S of tournaments. Colloq. Math. II pp. 286-290.

[6] SOBEL, M. (1968). On an optimal search for the t best using binary errorless comparisons: The ordering problem. Technical Report No. 113, Department of Statistics, University of Minnesota.

[7] SOBEL, M. (1968). On an optimal search for the t best using binary errorless comparions: The selection problem. Technical Report No. 114, Department of Statistics, University of Minnesota.

Proof of a conjecture of P. Erdős

by

A. Hajnal and E. Szemerédi
Budapest, Hungary

§ 1. INTRODUCTION

In this paper we are going to prove the following conjecture of P. ERDŐS. Let $\ell \geq 1$ be an integer and let $\mathcal{G}$ be a graph of m vertices, such that the valency of every point of $\mathcal{G}$ is less than ℓ. Let G denote the set of vertices of $\mathcal{G}$, and put $m = \ell n + r$, $0 \leq r < \ell$. Then there exist subsets $A_1, \dots, A_\ell$ of G such that A_i does not contain an edge for $i = 1, \dots, \ell$ and $|A_i| = n+1$ for $i = 1, \dots, r$, $|A_i| = n$ for $i = r+1, \dots, \ell$; $G = A_1 \cup \dots \cup A_\ell$.

To state the above statement in a shorter form, let us say that $G = A_1 \cup \dots \cup A_\ell$ is a uniform decomposition of length ℓ of G, if the A_i are disjoint and $||A_i| - |A_j|| \leq 1$ for every $i, j = 1, \dots, \ell$. The conjecture can be stated as follows.

If the valency of every vertex of $\mathcal{G}$ is less than ℓ, then G has a uniform decomposition into the union of ℓ independent sets.

This statement follows trivially from the following one.

(o) Let $\ell, n \geq 1$ be integers, and let $\mathcal{G}$ be a graph of ℓn vertices such that the valency of every vertex is less than ℓ. Then G is the union of ℓ independent sets of power n.

Several partial results had been obtained up to now. (o) is trivial if either $\ell = 1$ or $n = 1$.

For $\ell \geq 1$, $n = 2$ (o) was proved by G. Dirac before the general conjecture was stated. See [1].

For $\ell \geq 1$, $n = 3$ (o) was proved by K. Corrádi and A. Hajnal [2].

For $\ell = 2,3$ and $n \geq 1$ (o) was proved by B. Zelinka [3].

For $\ell = 4$ and $n \geq 1$ (o) was proved by B. Grünbaum [4].

For the general case a partial result had been obtained by K. Corrádi, saying that under the conditions of (o) there is a decomposition $G = A_1 \cup \dots \cup A_\ell$ into the union of ℓ, n element subsets such that $A_1, \dots, A_\mu$ are independent sets, and $A_{\mu+1}, \dots, A_\ell$ contain only a relatively small number of edges, where $\mu = \ell - \left[\frac{n-1}{2}\right]$ (hence $\mu \leq \ell - 1$ even for $n \geq 3$. [5]

Note that in some of the above-mentioned papers ([1], [2], [5]) a statement equivalent to (o) formulated for the complement of $\mathcal{G}$ is considered.

§ 2. DEFINITIONS. NOTATION

A graph $\mathcal{G}$ is considered to be an ordered pair $\langle G, g \rangle$ where G is the set of vertices, denoted by $x, y, \dots$ etc. and g is a set of unordered pairs $\{x, y\}$, $x \neq y$ $x, y \in G$.

The elements of g are called the edges of $\mathcal{G}$.

Thus we consider graphs without loops and multiple edges.

The graph $\mathcal{H} = \langle H, h \rangle$ is a subgraph of $\mathcal{G}$ if $H \subset G$ and $h \subset g$.

If H is a set $[H]^2 = \{\{x, y\} : x \neq y \text{ and } x, y \in H\}$.

Let $\mathcal{G}$ be a graph, $H \subset G$. We put

$$\mathcal{G}(H) = \langle H, g \cap [H]^2 \rangle$$

$\mathcal{G}(H)$ is the subgraph of $\mathcal{G}$, spanned by the vertices of H. If $\mathcal{G}(H)$ has no edges, we say that H is an independent subset of $\mathcal{G}$.

$|H|$ denotes the number of elements of H.

Let now $\mathcal{G}$ be a graph. We put $\alpha(\mathcal{G}) = |G|$. For every $x \in G$ we put $V(x,\mathcal{G}) = \{y \in G : \{x,y\} \in g\}$ and $v(x,\mathcal{G}) = |V(x,\mathcal{G})|$. $v(x,g)$ is the valency of x in $\mathcal{G}$. $v(\mathcal{G}) = \sup\{v(x,\mathcal{G}) : x \in G\}$ is the valency of $\mathcal{G}$. Let $H \subset G$, $x \in G$. Put $V(x,H,\mathcal{G}) = \{y \in H : \{x,y\} \in g\}$; $v(x,H,\mathcal{G}) = |V(x,H,\mathcal{G})|$. $v(x,H,\mathcal{G})$ is the valency of x in $\mathcal{G}$ relative to H.

Let $H_0, H_1 \subset G$. Put $V(H_0, H_1, \mathcal{G}) = \{\{x,y\} \in g$ and $y \in H_1\}$; $|V(H_0, H_1, \mathcal{G})| = v(H_0, H_1, \mathcal{G})$.

In what follows, if $\mathcal{G}$ is fixed and there is no danger of confusion, the reference to $\mathcal{G}$ will be omitted. We briefly speak about independent subsets, the valeny of x, the valency of x relative to H, and we write $v(x)$, $v(x,h)\ldots$ etc. instead of $v(x,\mathcal{G})$, $v(x,H,\mathcal{G})\ldots$ respectively.

§ 3. PROOF OF THE THEOREM

THEOREM 1.

Let $\ell \geq 1$, and $v(\mathcal{G}) < \ell$. Then G has a uniform decomposition of length ℓ into the union of ℓ independent sets.

THEOREM 2.

Let $\ell, n \geq 1$, and $v(\mathcal{G}) < \ell$, $\alpha(\mathcal{G}) = n\ell$. Then there are independent sets $A_1, \ldots, A_\ell$; $|A_i| = n$ for $i = 1, \ldots, \ell$ such that $G = A_1 \cup \ldots \cup A_\ell$.

First we prove that Theorem 2 implies Theorem 1.

Let $\mathcal{G}$ be a graph satisfying the requirements of Theorem 1.

Put $m = \alpha(\mathcal{G})$, $m = \ell n_0 + r$, $0 \le r < \ell$.

We define a graph $\mathcal{G}'$ as follows. Let $|H| = \ell - r$, $H \cap G = 0$, $n = n_0 + 1$. Put $G' = G \cup H$, $g' = g \cup [H]^2$. Then $\mathcal{G}' = \langle G', g' \rangle$ is defined and, by $|H| \le \ell$, $v(\mathcal{G}') < \ell$; $\alpha(\mathcal{G}') = n\ell$. $\mathcal{G}'$ satisfies the conditions of Theorem 2.

Let $G' = A'_1 \cup \ldots \cup A'_\ell$ be a decomposition of $\mathcal{G}'$ satisfying the requirements of Theorem 2. Considering that $\mathcal{G}'(H)$ is a complete graph, $|H \cap A'_i| \le 1$ for $i = 1, \ldots, \ell$.

Put $A_i = A'_i - H$ for $i = 1, \ldots, \ell$. Then

$$G = A_1 \cup \ldots \cup A_\ell$$

is obviously a uniform decomposition of G into the union of ℓ independent sets.

PROOF OF THEOREM 2.

3.1. Preliminaries

Let $n \ge 1$ be fixed. We prove the statement by induction on ℓ. For $\ell = 1$ the statement is trivial. Assume that

(1) $\ell > 1$, and the theorem is true for every graph $\mathcal{G}'$ with $\alpha(\mathcal{G}') = \ell' n$, $v(\mathcal{G}') < \ell'$ for every $1 \le \ell' \le \ell - 1$. It will be convenient to have the following brief notation:

def. (2) Let $\mathcal{G}$ be a graph, and let $n_1, \ldots, n_s$ be a sequence of integers, $H \subset G$. We put

$$[n_1, \ldots, n_s, \mathcal{G}, H] = \{(A_1, \ldots, A_s) : \bigcup_{i=1}^{s} A_i = H,$$

A_i is independent in $\mathcal{G}$, $|A_i| = n_i$ for $i = 1, \ldots, s$ and $A_i \cap A_j = 0$ for $i \ne j$, $i, j = 1, \ldots, s\}$.

If $\mathcal{G}$ is a fixed graph, we omit the reference to $\mathcal{G}$. If $G = H$ we omit the reference to H as well. We will use further self-explanatory abbreviations e.g.

$$[n \times \ell, \mathcal{G}, H] \quad \text{denotes} \quad [n_1, \ldots, n_\ell, \mathcal{G}, H]$$

where $n_i = n$ for $i = 1, \ldots, \ell$

$[n-1, n\times(\ell-2), n+1, \mathcal{G}, H]$ denotes $[n_1, \dots, n_\ell, \mathcal{G}, H]$

where $n_1 = n-1$, $n_i = n$ for $i = 2,\dots,\ell-1$, $n_\ell = n+1$ etc.

def. (3) Let $H \subset G$, $|H| = nj-1$, $j \geq 1$. We put

$$C(H) = \{x \in G : \text{There is } (A_1,\dots,A_j) \in [n\times j, \mathcal{G}, H \cup \{x\}]\}.$$

We now assume that the theorem is false for ℓ i.e. there is a $\mathcal{G}'$ with $\alpha(\mathcal{G}') = n\ell$, $v(\mathcal{G}') < \ell$ and $[n\times\ell, \mathcal{G}'] = 0$ and in what follows in the proof

def. (4) $\mathcal{G}$ is a graph with minimal number of edges* satisfying the conditions

$$\alpha(\mathcal{G}) = n\ell; \quad v(\mathcal{G}) < \ell, \quad [n\times\ell, \mathcal{G}] = 0.$$

From now on, if not indicated differently, all the concepts defined relative to a graph, refer to this graph $\mathcal{G}$.

We now prove

(5) $$[n-1, n\times(\ell-2), n+1] \neq 0$$

PROOF: By $[n\times\ell] = 0$, $\mathcal{G}$ has an edge.

Let $\{x_0, x_1\} = e \in g$. Let $\mathcal{G}' = \langle G, g'\rangle$ where $g' = g - \{e\}$. Then $\alpha(\mathcal{G}') = n\ell$; $v(\mathcal{G}') < \ell$.

By the minimality of the number of the edges of $\mathcal{G}$, there is $(A'_1,\dots,A'_\ell) \in [n\times\ell, \mathcal{G}']$. By $(A'_1,\dots,A'_\ell) \notin [n\times\ell]$ there is i, $1 \leq i \leq \ell$ with $e \subset A'_i$. By $v(\mathcal{G}) < \ell$, $e \in g$ we have $v(e, G-e) \leq 2(\ell-1) - 2 < 2(\ell-1)$.

Hence there are $j \neq i$, $1 \leq j \leq \ell$ and $k < 2$ with $v(x_k, A'_j) = 0$. Put $A_1 = A'_i - \{x_k\}$, $A_\ell = A'_j \cup \{x_k\}$ and let $(A_2,\dots,A_{\ell-1})$ by any permutation of

*Note that the minimality of the number of edges will be used only in the proof of (5)

$$\{A'_\sigma : 1 \le \sigma \le \ell ; \sigma \ne i, \sigma \ne j\}.$$

Then $(A_1, \dots, A_\ell) \in [n-1, n \times (\ell-2), n+1]$ and this proves (5).

Before outlining the idea of our proof, we prove

(6) Let $(A_1, \dots, A_\ell) \in [n-1, n \times (\ell-2), n+1]$. Then there are $2 \le k \le \ell-1$, $x \in A_k$ with $v(x, A_1) = 0$

PROOF: Put $H = A_2 \cup \dots \cup A_\ell$. We have $v(A_1, H) \le$ $\le (\ell-1)|A_1| = (\ell-1)(n-1) < n(\ell-1)$ since $\ell > 1$.

Hence there is $2 \le k \le \ell$, with $v(A_1, A_k) < n$. Considering $|A_k| \ge n$, then there is $x \in A_k$ with $v(x, A_1) = 0$. If $k = \ell$, then

$$(A_1 \cup \{x\}, A_\ell - \{x\}, A_2, \dots, A_{\ell-1}) \in [n \times \ell]$$

in contradiction to the indirect assumption (4). Hence $k \le \ell - 1$.

3.2. The definition of a chain. Outline of the proof

def. (7) Let $(A_1, \dots, A_\ell) \in [n-1, n \times (\ell-2), n+1]$; $k \ge 1$ $(A_1, \dots, A_\ell)$ is said to be a k-chain if it satisfies the following condition.

For every $2 \le j \le k+1$, there are $1 \le s < j$, $x \in A_j$ with $v(x, A_s) = 0$. By (5) and (6) there is a 1-chain.

The basic idea of our proof is to show by induction on k, using the indirect assumption, that there are k-chains for $1 \le k \le \ell - 1$. If this will be established for $k = \ell-1$ then, as it will be seen, it follows trivially that $[n \times \ell] \ne 0$ and this is a contradiction.

The technical execution of the proof is, however, not quite so simple. Our detailed plan is the following.

def. (8) Let i be the maximal number with the following property. For every $(A_1, \dots, A_\ell) \in [n-1, n \times (\ell-2), n+1]$ there is a permutation

$(k_1, \dots, k_\ell)$ of the integers $1, \dots, \ell$ with $k_1 = 1$, $k_\ell = \ell$ such that $(A_{k_1}, \dots, A_{k_\ell})$ is an i-chain.

By (6) we have $i \geq 1$

def. (9) In what follows in the proof, $(A_1^o, \dots, A_\ell^o)$ will denote a fixed i-chain which shows that i is maximal i.e. $(A_{k_1}^o, \dots, A_{k_\ell}^o)$ with $k_1 = 1$, $k_\ell = \ell$ is never an $i+1$ -chain.

By (5) and (6) such an $(A_1^o, \dots, A_\ell^o)$ exists.

We put

def. (10) $$q = \ell - 1 - i$$

$$A^o = A_1^o \cup \dots \cup A_{i+1}^o \ ; \qquad |A^o| = (i+1)n - 1$$

$$B^o = A_{i+2}^o \cup \dots \cup A_\ell^o \ ; \qquad |B^o| = qn + 1$$

$$A^o \cup B^o = G \ ; \qquad A^o \cap B^o = 0$$

def. (11) Let $(A_1, \dots, A_\ell)$ be an i-chain $(A_1, \dots, A_\ell)$ will be said an i, A^o-chain, if

$$A_1 \cup \dots \cup A_{i+1} = A^o ; \ (A_{i+2} \cup \dots \cup A_\ell = B^o).$$

In 3.3 we will prove several lemmas describing the structure of k-chains in general.

In 3.4 we will prove the first main lemma saying that in every i, A^o-chain there are at least $2q+1$ sets A_j which "can stand at the end" so that the i, A^o -chain property is maintained.

In 3.5 we will consider k-chains of maximal length having the above permutation property and obtain a contradiction from $k < \ell - 1$.

It is hard to say anything about the idea of the proof of the details. They are rather technical, and the authors sincerely hope that simpler proofs will be found in the future. However, all the proofs we give have the following common feature. They are all based on easy computations using that, by $v(\mathcal{G})<\ell$, $v(A,B)$ is small for certain subsets $A \subset A^o$, $B \subset B^o$. In the first part of the proof in 3.3, and 3.4 we mainly use estimations of the type $v(A^o, B^o) \leq (\ell-1)|A^o|$, while in the second part 3.5 $v(A^o, B^o) \leq (\ell-1)|B^o|$ is used more heavily.

3.3. PRELIMINARIES. LEMMAS

(12) Let $(A_1, \ldots, A_j) \in [n-1, n\times(j-1), H]$; $j \geq 1$, $x \in G-H$ and assume $1 \leq k \leq j$, $x \in C(A_1 \cup \ldots \cup A_k)$.
Then $x \in C(A_1 \cup \ldots \cup A_j)$

PROOF: By the assumption there is $(A'_1, \ldots, A'_k) \in$ $\in [n\times k, A_1 \cup \ldots \cup A_k \cup \{x\}]$.

Then $(A'_1, \ldots, A'_k, A_{k+1}, \ldots, A_j) \in [n\times j, A_1 \cup \ldots \cup A_j \cup \{x\}]$ hence $x \in C(A_1, \ldots, A_j)$.

(13) Let $(A_1, \ldots, A_j) \in [n-1, n\times(j-1), H]$; $j \geq 2$, $x \in G-H$.

Assume there is $y \in A_j$, $y \in C(A_1 \cup \ldots \cup A_{j-1})$ and either $v(x, A_j) = 0$ or $V(x, A_j) = \{y\}$ for a $y \in C(A_1 \cup \ldots \cup A_{j-1}$. Then $x \in C(A_1 \cup \ldots \cup A_j)$

PROOF: If $v(x, A_j) = 0$ then let $y \in C(A_1 \cup \ldots \cup A_{j-1})$ if $v(x, A_j) \neq 0$, let y be the unique element of $V(x, A_j)$. By the assumption, there is $(A'_1, \ldots, A_{j-1}) \in [n\times(j-1), A_1 \cup \ldots \cup A_{j-1} \cup \{y\}]$.
Put $A'_j = (A_j - \{y\}) \cup \{x\}$. Then $|A'_j| = n$, A'_j is independent. Hence $(A'_1, \ldots, A'_j) \in [n\times j, A_1 \cup \ldots \cup A_j \cup \{x\}]$ i.e. $x \in C(A_1 \cup \ldots \cup A_j)$.

(14) Let $(A_1, \ldots, A_\ell)$ be a k-chain.

Assume that $2 \leq j \leq k+1$. We have

(i) If $x \in A_j$, and there is $1 \leq s \leq j$ with $v(x, A_s) = 0$ then $x \in C(A_1 \cup \ldots \cup A_{j-1})$

and

(ii) There is $x \in A_j$

with $x \in C(A_1 \cup \dots \cup A_{j-1})$

PROOF: We prove (i) and (ii) by induction on j.

For $j = 2$ both statements are true, by definition (7).

Assume $2 < j$, and both statements are true for $2 \leq s < j$.

Assume $x \in A_j$, $v(x, A_s) = 0$ for some $1 \leq s < j$.

If $s = 1$, then $x \in C(A_1)$ and, by (12), (i) holds.

If $2 \leq s$, then (ii) holds for s, and thus, by (13), $x \in C(A_1 \cup \dots \cup A_s)$, hence, by (12), $x \in C(A_1 \cup \dots \cup A_{j-1})$

Thus (i) holds for j. Then, by definition (7), (ii) holds for j too.

As a corollary of the indirect assumption we have

(15) Let $(A_1, \dots, A_\ell)$ be a k-chain for $k \geq 1$, $x \in A_\ell$.

Then

(i) $x \notin C(A_1 \cup \dots \cup A_s)$ for $1 \leq s \leq \ell - 1$.

(ii) $v(x, A_j) \neq 0$ for $1 \leq j \leq k+1$.

(iii) There is no $\ell - 1$ chain.

(iv) $q \geq 1$ (see def. (10))

PROOF: (i) If $x \in C(A_1 \cup \dots \cup A_s)$. Then, by (12), $x \in C(A_1 \cup \dots \cup A_{\ell-1})$. There is $(A'_1, \dots, A'_{\ell-1}) \in [n \times (\ell - 1), A_1 \cup \dots \cup A_{\ell-1} \cup \{x\}]$.

Put $A'_\ell = A_\ell - \{x\}$. Then $(A'_1, \dots, A'_\ell) \in [n \times \ell]$ is a contradiction to (4). (ii) follows from (i), (13) and (14). (iii) follows from (i). (iv) follows from (iii).

We now state an important lemma which uses the induction hypothesis (1) heavily.

(16) Let $(A_1, \dots, A_\ell)$ be a k-chain such that there is no permutation $n_{k+2}, \dots, n_\ell$ of the integers $k+2, \dots, \ell$ such that $(A_1, \dots, A_{k+1}, A_{n_{k+2}}, \dots, A_{n_\ell})$ is a k+1-chain. Put $B = A_{k+2} \cup \dots \cup A_\ell$.

Then the following statements are true.

(i) $v(x, A_j) \neq 0$ for $1 \leq j \leq k+1$; $x \in B$.

(ii) Let $x \in B$, $p = \ell - k - 1$. Then there is

$$(B_1, \dots, B_p) \in [n \times p, B - \{x\}]$$

(iii) $x \notin C(A_1 \cup \dots \cup A_{k+1})$ for every $x \in B$

PROOF: Ad (i) Let $x \in B$. If $x \in A_\ell$ then (i) follows from (15).

Let $x \in A_s$, $k+2 \leq s \leq \ell - 1$. Then $v(x, A_j) \neq 0$ for $1 \leq j \leq k+1$ otherwise $(A_1, \dots, A_{k+1}, A_s, A_{k+2}, \dots, A_\ell)$ is a k+1-chain.

Ad (ii). Put $A = G - B$. By (i), $v(y, A) \geq k+1$ for every $y \in B$. Put $B' = B - \{x\}$, $\mathcal{G}' = \mathcal{G}(B')$.

Then $v(\mathcal{G}') < \ell - (k+1) = p$ and $|B'| = \alpha(\mathcal{G}') = np$.

Hence, by (1), there is $(B_1, \dots, B_p) \in [n \times p, B - \{x\}]$. This proves (ii).

Ad (iii). Assume $x \in C(A) \cap B$. By $x \in C(A)$, there is $(A'_1, \dots, A'_{k+1}) \in [n \times (k+1), A \cup \{x\}]$ and by (ii) there is $(B_1, \dots, B_p) \in [n \times p, B - \{x\}]$. But then

$(A'_1, \dots, A'_{k+1}, B_1, \dots, B_p) \in [n \times \ell]$ is a contradiction to (4).

(17) Let $(A_1, \dots, A_\ell)$ be a k-chain for $k \geq 1$.

Let $1 \leq s < j$, $2 \leq j \leq k+1$, $x \in A_j$, $v(x, A_s) = 0$.

Then $(A_1, \dots, A_s, A_j, A_{s+1}, \dots, A_{j-1}, A_{j+1}, \dots, A_\ell)$ is a k-chain too.

PROOF: By definition (7).

(18) Let $(A_1, \dots, A_\ell)$ be a k-chain, $k \geq 1$, $A = A_1 \cup \dots \cup A_{k+1}$.
Put $\bar{A}_j = \{ y \in A_j : v(y, A_s) \neq 0$ for $1 \leq s < j \}$ $1 \leq j \leq k+1$.

Then

(i) $v(y, A_1 \cup \dots \cup A_{j-1}) \geq j-1$ for every $y \in \bar{A}_j$, $1 \leq j \leq k+1$.

(ii) Let $x \in G-A$, $x \notin C(A)$. Then either $v(x, A_j) \geq 2$ or $v(x, \bar{A}_j) = 1$ for $1 \leq j \leq k+1$.

PROOF: (i). If $v(y, A \cup \dots \cup A_{j-1}) < j-1$ then there must be $1 \leq s < j$, with $v(y, A_s) = 0$.

(ii) Let $x \in G-A$, $x \notin C(A)$. Assume (ii) is false for a j, $1 \leq j \leq k+1$. Then $v(x, A_j) \leq 1$, $V(x, A_j) \subset A_j - \bar{A}_j \neq 0$. If $y \in A_j - \bar{A}_j$, then by (14), $y \in C(A_1 \cup \dots \cup A_{j-1})$. Thus, by (13), $x \in C(A_1 \cup \dots \cup A_j)$, hence, by (12), $x \in C(A)$ is a contradiction.

3.4. i, A° -chains

(19) Let $(A_1, \dots, A_\ell)$ be in i, A°-chain. Then the requirements (i)...(iii) of (16) hold for $(A_1, \dots, A_\ell)$ with $k = i$

PROOF. Assume $(A_1, \dots, A_{i+1}, A_{n_{i+2}}, \dots, A_{n_\ell})$ is an $i+1$ chain for a permutation $n_{i+2}, \dots, n_\ell$ of the integers $i+2, \dots, \ell$. Then, by (13) and (14), there is $x \in A_{n_{i+2}}$ with $x \in C(A^\circ)$.

By definition (9), $(A_1^\circ, \dots, A_\ell^\circ)$ is an i-chain satisfying the conditions of (16) with $i = k$. Thus $x \in C(A)$ contradicts to (16) (iii). It follows that $(A_1, \dots, A_\ell)$ satisfies the conditions of (16) with $k = i$. Hence (19) follows from (16).

(20) There is no $(A_1, \dots, A_\ell) \in [n-1, n \times (\ell-2), n+1]$ satisfying the following condition. There is $x \in A_\ell$ with

$$|\{ 1 \leq j < \ell : v(x, A_j) = 0 \}| \geq q.$$

PROOF: Let $(A_1,\dots,A_\ell) \in [n-1, n\times(\ell-2), n+1]$; $x \in A_\ell$.

Put $N = \{1 \le j < \ell : v(x, A_j) = 0\}$ and assume that $|N| \ge q$.
By the definition (8) there is a permutation $k_1,\dots,k_\ell$ of the integers $1,\dots,\ell$ with $k_1 = 1$, $k_\ell = \ell$ such that $(A_{k_1},\dots,A_{k_\ell})$ is an i-chain. By $q+i+1 \ge \ell > \ell-1$ there is $k_j \in N$; with $1 \le j \le i+1$. This contradicts to (15).

(21) Let $(A_1,\dots,A_\ell)$ be an i, A°-chain.

Put $X_0 = \{y \in A_{i+1} : v(y, A^\circ) \ge i\}$.

Assume $x \in B$, $v(x, X_0) = 0$ and there is $y \in V(x, A_{i+1})$ with $v(y, A^\circ) < i-q$.

Then $v(x, A_{i+1}) \ge 3$

PROOF: Assume this is false, and $v(x, A_{i+1}) \le 2$.

By (19) we know that $x \notin C(A^\circ)$. By (18) (i) and (ii) we know that then $v(x, A_{i+1}) = 2$. Put $V(x, A_{i+1}) = \{y, z\}$. By the assumptions $v(y, A^\circ) < i-q$, and $v(z, A^\circ) < i$. There is $1 \le s \le i$ with $v(z, A_s) = 0$, hence, by (14), $z \in C(A_1 \cup \dots \cup A_i)$ i.e.

There is $(A'_1,\dots,A'_i) \in [n \times i, A_1 \cup \dots \cup A_i \cup \{z\}]$.

There is $1 \le r \le i$ with $v(y, A'_r) = 0$. Put $A'_{i+1} = (A_{i+1} - \{y,z\}) \cup \{x\}$. $|A'_{i+1}| = n-1$; and A_{i+1} is independent. By (16), there is $(B_1,\dots,B_q) \in [n\times q, B^\circ - \{x\}]$. Put $A''_\ell = A'_r \cup \{y\}$; $A''_1 = A'_{i+1}$; let $A''_2,\dots,A''_i$ be the sets A'_j , $1 \le j \le i+1$, $j \ne r$, $j \ne i+1$ in an arbitrary order, and let $A''_{i+1},\dots,A''_{\ell-1}$ be the sets $B_1,\dots,B_q$ in an arbitrary order.

Then $(A''_1,\dots,A''_\ell) \in [n-1, n\times(\ell-2), n+1]$

$y \in A''_\ell$, and $v(y, A^\circ) < i-q$ and $A''_2,\dots,A''_i$ are $i-1$ sets contained in A°.

Thus $|\{2 \le j \le i : v(y, A''_j) = 0\}| \ge i-1-(i-q-1) = q$. Hence $|\{2 \le j < \ell : v(y, A''_j) = 0\}| \ge q$. This contradicts to (20), and proves (21).

(22) Let $(A_1,\dots,A_\ell)$ be an i, A°-chain.

Then there is $y \in A_{i+1}$ with $v(y, A^o) < i-2q$.

(And as a corollary we have $i \geq 2q+1$)

PROOF: Let $X_0 = \{ y \in A_{i+1} : v(y, A^o) \geq i \}$

$X_1 = \{ y \in A_{i+1} : i-q \leq v(y, A^o) < i \}$;

$X_2 = \{ y \in A_{i+1} : i-2q \leq v(y, A^o) < i-q \}$.

Put $|X_r| = t_r$ for $r = 0,1,2$. Assume (22) is false

Then $A_{i+1} = X_0 \cup X_1 \cup X_2$, $t_0 + t_1 + t_2 = n$.

We also have $v(X_r, A^o) \geq (i-rq)t_r$ for $r = 0,1,2$.

Put now $Y_0 = \{ x \in B^o : v(x, X_0) \neq 0 \}$

$Y_1 = \{ x \in B^o - Y_0 : v(x, X_1) \geq 2 \}$; $\quad Y_2 = \{ x \in B^o - (Y_0 \cup Y_1) : v(x, X_1 \cup X_2) \geq 3 \}$.

First of all we prove $B^o = Y_0 \cup Y_1 \cup Y_2$.

Let $x \in B^o$. By (16) and (18) we know that either $x \in Y_0$ or $v(x, A_{i+1}) \geq 2$. Assume $x \notin Y_0$. Then, by (21), either $x \in Y_1$ or $x \in Y_2$.

Considering that $v(X_0, A^o) \geq t_0 i$, we have $v(X_0, B^o) \leq qt_0$ hence $|Y_0| \leq qt_0$, and thus, by (10),

(i) $\quad |Y_1 \cup Y_2| \geq (n-t_0)q+1 = (t_1+t_2)q+1$.

By the definition of Y_1

$$t_1(\ell-1) \geq v(X_1) \geq v(X_1, A^o) + v(X_1, Y_1) \geq (i-q)t_1 + 2|Y_1|$$

i.e. $2qt_1 \geq 2|Y_1|$ hence

(ii) $\quad |Y_1| \leq qt_1$

From (i) and (ii) we get $|Y_2| \geq qt_2 + 1 + qt_1 - |Y_1|$; $2|Y_1| + 3|Y_2| \geq$ $\geq 2qt_1 + 3qt_2 + 3$; hence by the definition of Y_2 we have $(t_1+t_2)(\ell-1) \geq$ $\geq v(X_1 \cup X_2) \geq v(X_1, A^o) + v(X_2, A^o) + v(X_1 \cup X_2, Y_1 \cup Y_2) \geq (i-q)t_1 + (i-2q)t_2 +$ $+ 2|Y_1| + 3|Y_2| \geq (i-q+2q)t_1 + (i-2q+3q)t_2 + 3$.

Considering that $i+q = \ell-1$ this is a contradiction which proves (22).

We now conclude this § with the lemma already indicated in the introduction.

(23) Let $(A_1,\dots,A_\ell)$ be an i, A°-chain.

Let $M = \{2 \le j \le i+1 :$ There is a permutation $n_1,\dots,n_{i+1}$ of integers $1,\dots,i+1$ with $n_1 = 1$ and $n_{i+1} = j$ such that $(A_{n_1},\dots,A_{n_{i+1}}, A_{i+2},\dots,A_\ell)$ is an i, A°-chain $\}$

Then $$|M| \ge 2q+1$$

PROOF: (By (22) we know that $i \ge 2q+1$. Then $|A_{i+1-t}| = n$ for $t = 0,\dots,2q$. We prove, by induction on t, $0 \le t \le 2q$ that there is a permutation $k_1^t,\dots,k_{i+1}^t$ with $k_1^t = 1$; $k_j^t = j-t$ for $i+1-2q+t \le j \le i+1$, such that $(A_{k_i^t},\dots,A_{k_{i+1}^t},\dots,A_\ell)$ is an i, A°-chain. This obviously holds for $t = 0$. Assume $0 \le t < 2q$ and the statement is true with $k_1^t,\dots,k_{i+1}^t$.

By (22), there is $y \in A_{k_{i+1}^t}$ with $v(y, A^\circ) \le i-(2q+1)$.

Thus if $N = \{1 \le j < i+1 : v(y, A_j) = 0\}$ then $|N| \ge 2q+1$.

Let τ be the minimal number for which $k_\tau^t \in N$. Then $\tau \le i-2q$. Then, by (17), $(A_{k_1^t},\dots,A_{k_\tau^t}, A_{k_{i+1}^t}, A_{k_{\tau+1}^t},\dots,A_{k_i^t}, A_{i+2},\dots,A_\ell)$ is an i, A°-chain.

If we put $$(k_1^{t+1},\dots,k_{i+1}^{t+1}) = (k_1^t,\dots,k_\tau^t, k_{i+1}^t, k_{\tau+1}^t,\dots,k_i^t)$$

we see that $k_j^{t+1} = j-(t+1)$ holds for $i+1-2q+t+1 \le j \le i+1$ hence, by $t < 2q$, this holds for $j = i+1$.
This proves (23).

3.5. Chains of maximal length

def. (24) Let k be the maximal number for which there is a k-chain $(A_1,\dots,A_\ell)$ satisfying the following condition (□).

(□) Let $M = \{2 \le t \le k+1 :$ There is a permutation $(n_1,\dots,n_{k+1})$ of the integers $1,\dots,k+1$ such that $n_1 = 1$, $n_{k+1} = t$ and $(A_{n_1},\dots,A_{n_{k+1}},A_{k+2},\dots,A_\ell)$ is a k-chain $\}$.

Then $|M| \ge 2q+1$

By (23) $i=k$ is a number having property (□), hence k is defined. and $k \ge i$.

def. (25) In what follows $(A_1,\dots,A_\ell)$ will denote a fixed k-chain having property (□), and M the corresponding set defined in (□). We put

$$p = \ell-1-k \; ; \qquad p \le q$$

$$A = A_1 \cup \dots \cup A_{k+1} \; ; \qquad |A| = n(k+1)-1$$

$$B = A_{k+2} \cup \dots \cup A_\ell \; ; \qquad |B| = np+1$$

and by (15) we have $p \ge 1$

$(n_1,\dots,n_s)$ will always denote a permutation of the integers $1,\dots,s$.

Let $t \in M$, and let $(n_1^t,\dots,n_{k+1}^t)$ be such that $n_1^t = 1$, $n_{k+1}^t = t$ and $(A_{n_1^t},\dots,A_{n_{k+1}^t},A_{k+2},\dots,A_\ell)$ is a $k+1$-chain. To have a brief notation, we put

$$A_{n_j^t} = A_j^t \quad \text{for} \quad 1 \le j \le k+1 \quad \text{and}$$

$$\bar{A}_j^t = \{y \in A_j^t : v(y,A_s^t) \ne 0 \quad \text{for} \quad 1 \le s < j\} \quad \text{for} \quad 1 \le j \le k+1 ; \; t \in M$$

def. (26) Put $C_t = \{y \in B : v(y,A_t) = 1\}$; $D_t = B - C_t$ for $1 \le t \le k+1$.

We have

(27) Let $x \in B$. Then

(i) $v(x, A_s) \neq 0$ for $1 \leq s \leq k+1$

(ii) $x \notin C(A)$

(iii) $v(x, A_j^t) \leq 1$ implies $v(x, \bar{A}_j^t) = 1$ for $t \in M$, $1 \leq j \leq k+1$

PROOF: First we prove that there is no permutation $n_{k+2}, \dots, n_\ell$ of the integers $k+2, \dots, \ell$ such that $(A_1, \dots, A_{k+1}, A_{n_{k+2}}, \dots, A_{n_\ell})$ is a $k+1$-chain. Assume this is false with $n_{k+2}, \dots, n_\ell$ and put

$$A'_j = A_j \; ; \quad A'^t_j = A^t_j \qquad \text{for} \quad 1 \leq j \leq k+1 \; ; \qquad A'_j = A_{n_j}$$

for $k+2 \leq j \leq \ell$. Then $(A'_1, \dots, A'_\ell)$ is a $k+1$-chain, hence there are $1 \leq s \leq k+1$, $y \in A'_{k+2}$ with $v(y, A'_s) = 0$. Let now $t \in M$. Then there is r_t $r_t < k+1$ if $t \neq s$ such that $A'_s = A'^t_{r_t}$. Then, by (17), $(A'^t_1, \dots, A'^t_{r_t}, A'_{k+2}, A'^t_{r_t+1}, \dots, A'^t_{k+1}, A'^t_{k+3}, \dots, A'_\ell)$ is a $k+1$-chain too. Let now

$M' = \{ 1 \leq h \leq k+2$: There is $(n_1, \dots, n_{k+2})$ such that

$n_1 = 1$, $n_{k+2} = h$, $(A'_{n_1}, \dots, A'_{n_{k+2}}, A'_{n_{k+3}}, \dots, A'_\ell)$ is a $k+1$-chain $\}$. It follows from the above consideration that $M' \supset M$ if $s \notin M$ $M' \supset (M - \{s\}) \cup \{k+2\}$ if $s \in M$, hence $|M'| \geq 2q+1$.
This contradicts to the maximality of k. Thus our first statement is proved. Then (i) and (ii) are true by (16), and (iii) follows from (ii) and (18).

We now prove lemmas describing the structure of $\mathcal{G}(B)$.

(28) Let $t \in M$, $y \in A_t = A^t_{k+1}$ and put $V(y, C_t) = V$ $V_1 = V(y, B)$. Let $z_1 \neq z_2$; $z_1 \in V$;

$z_2 \in V_1$; $v(z_2, B) \geq p-1$.

Then $\{z_1, z_2\} \in g$

PROOF: Assume $\{z_1,z_2\} \notin g$. First of all let us remark that, by (27) (i), $v(z_2,B) \geq p-1$ implies $v(z_2,A) = k+1$, $v(z_2,B) = p-1$, hence by (27) (iii) we also have

(i) $z_2 \in V$ and $v(z_2, A_j^s) = v(z_2, \bar{A}_j^s) = 1$

for every $s \in M$, $1 \leq j \leq k+1$.

Put now $A'_j = A_j$ for $1 \leq j \leq k+1$, $j \neq t$; $A'_t = A_t \cup \{z_2\} - \{y\}$.

By (i), A'_t is independent and $|A'_t| = n$. By (27) and (17), there is $(B_1,\ldots,B_p) \in [n \times p, B-\{z_2\}]$. By $V \neq 0$ and by (27), $y \in \bar{A}^t{}_{k+1}$, hence $v(y,A) \geq k$, $v(y,B) \leq p$, $z_2 \in V$; $v(y,B-\{z_2\}) \leq p-1$.
Thus there is $1 \leq s \leq p$ with $v(y,B_s) = 0$.

Put $A'_\ell = B_s \cup \{y\}$ and let $A'_{k+2},\ldots,A'_{\ell-1}$ be any enumeration of B_r, $1 \leq r \leq p$; $r \neq s$.

Then $(A'_1,\ldots,A'_\ell) \in [n-1, n\times(\ell-2), n+1]$.

Let now $s \in M$, $1 \leq j \leq k+1$, and put $A_j'^s = A'_\sigma$ for the unique σ for which $A_j^s = A_\sigma$. Let j_s be the number for which $A_{j_s}^s = A_t$. $j_s = k+1$ if and only if $s = t$.

We now prove that

(ii) $(A_1'^s,\ldots,A_{k+1}'^s, A'_{k+2},\ldots,A'_\ell)$ is a k-chain too.

We have to verify that for every $2 \leq j \leq k+1$ there are $u \in A_j'^s$ and $1 \leq \sigma < j$ with $v(u, A_\sigma'^s) = 0$.

We distinguish the following cases a. $2 \leq j \leq j_s - 1$ b. $j = j_s$ c. $j_s+1 \leq j \leq k+1$.

We always use that $(A_1^s,\ldots,A_{k+1}^s, A_{k+2},\ldots,A_\ell)$ is a k-chain.

a) $A_j'^s = A_j^s$ for every $1 \leq j \leq j_s - 1$.

b) There are $u \in A_{j_s}^s = A_t$ and $1 \leq \sigma < j_s$ with $v(u, A_\sigma^s) = 0$.

$A^s_\sigma = A'^s_\sigma$; $A'^s_{j_s} = A'_t = A_t \cup \{z_2\} - \{y\}$; $u \notin \bar{A}^s_{j_s}$.

By $\{z_2, y\} \in g$ and by (i), $y \in \bar{A}^s_{j_s}$ hence $u \neq y$. Thus $u \in A'^s_{j_s}$ and $v(u, A'^s_\sigma) = 0$.

c) There are $u \in A^s_j$ and $1 \le \sigma < j$ with $v(u, A^s_\sigma) = 0$. $A'^s_j = A^s_j$. If $\sigma \neq j_s$ then $A'^s_\sigma = A^s_\sigma$ and the statement is obvious. Assume $\sigma = j_s$. Then $A'^s_{j_s} = A'_t$, $A'_t = A_t \cup \{z_2\} - \{y\}$. We have $v(u, A'^s_{j_s}) = 0$ provided that $\{u, z_2\} \notin g$. But $u \notin \bar{A}^s_j$ hence this follows from (i).

Thus (ii) is proved.

By the construction, $z_1 \in A'_{k+2} \cup \dots \cup A'_\ell$ and by $\{z_1, z_2\} \notin g$ $v(z_1, A'_t) = 0$. Thus, by (15) and by (ii) $z_1 \notin A'_\ell$.

Thus we can assume $z_1 \in A'_{k+2}$.

Then, by (ii) and by (17), $(A'^s_1, \dots, A'^s_{k+1}, A'_{k+2}, \dots, A'_\ell)$

and hence $(A'^s_1, \dots, A'^s_{j_s}, A'_{k+2}, A'^s_{j_s+1}, \dots, A'^s_{k+1}, A'_{k+3}, \dots, A'_\ell)$

is a $k+1$-chain for every $s \in M$.

That means that $(A'_1, \dots, A'_\ell)$ is a $k+1$-chain and if $M' = \{2 \le j \le k+2:$ there is a permutation $(n_1, \dots, n_{k+2})$ with $n_1 = 1$, $n_{k+2} = j$ such that $(A'_{n_1}, \dots, A'_{n_{k+2}}, A'_{k+3}, \dots, A'_\ell)$ is a $k+1$-chain $\}$ then $M' \supset (M - \{t\}) \cup \{k+2\}$, hence $|M'| \ge 2q+1$.

This contradicts the maximality (24) of k.

Thus the indirect assumption $\{z_1, z_2\} \notin g$ is false and (28) is proved.

We now prove the main lemma describing $\mathcal{G}(B)$.

(29) Let $t \in M$, $y \in A_t = A^t_{k+1}$, put $V = V(y, C_t)$.

Then $[V]^2 \subset g$ i.e. $\mathcal{G}(V)$ is a complete subgraph of $\mathcal{G}$.

PROOF: Put $V_1 = V(y, B)$. Let $u_1 \neq u_2 \in V$.

Assume (29) is false and $\{u_1, u_2\} \notin g$.

By $V \neq 0$ and by (27), $y \in \bar{A}^t_{k+1}$, hence there is $z \neq y$ $z \in A^t_{k+1} - \bar{A}^t_{k+1}$. By (14) then $z \in C(A_1 \cup \dots \cup A_k)$ hence there is $(A'_1, \dots, A'_k) \in [n \times k, A_1 \cup \dots \cup A_k \cup \{z\}]$.

Put $A'_{k+1} = A^t_{k+1} \cup \{u_1, u_2\} - \{y, z\}$. By $\{u_1, u_2\} \subset V \subset C_t$ $|A'_{k+1}| = n$ and A'_{k+1} is independent.

Put $B' = (B - \{u_1, u_2\}) \cup \{y\}$; $\mathcal{G}(B') = \mathcal{G}'$. Then, by (25), $\alpha(\mathcal{G}') = np$. We prove $v(\mathcal{G}') < p$.

Because of $y \in \bar{A}^t_{k+1}$; $v(y, A) \geq k$; $v(y, B) \leq p$; $v(y, B') \leq p-2$.

Let $x \in B$. Then, by (27), $v(x, A) \geq k+1$, hence $v(x, B) \leq p-1$.

Assume $x \in B' - V_1$, $x \neq y$. Then $v(x, B') \leq v(x, B) \leq p-1$.

Assume $x \in B' \cap V_1$. If $v(x, B) \leq p-2$ then $v(x, B') \leq p-2+1 = p-1$. Assume $v(x, B) = p-1$.

Then, by (28), e.g. $\{x, u_1\} \in g$, hence $v(x, B') \leq p-1-1+1 = p-1$. Thus $v(\mathcal{G}') < p$.

By the induction hypothesis (1), then there is

$(B_1, \dots, B_p) \in [n \times p, \mathcal{G}', B']$ and that means $(B_1, \dots, B_p) \in [n \times p, B']$.

Considering $G = A'_1 \cup \dots \cup A'_{k+1} \cup B'$ then

$(A'_1, \dots, A'_{k+1}, B_1, \dots, B_p) \in [n \times \ell]$ is a contradiction to (4).

Thus $\{u_1, u_2\} \in g$ for an arbitrary pair $u_1 \neq u_2 \in V$.

Hence $[V]^2 \subset g$ and (29) is proved.

From (29) we are going to obtain a contradiction with an easy computation in a few steps.

First of all let us remark

(30) Let $t \in M$. Then $v(B, A_t) \geq np+1+|D_t|$.

PROOF: By def. (26) $v(x, A_t) \neq 1$ for $x \in D_t$.

Thus, by (27) (iii) $v(x, A_t) \geq 2$ for $x \in D_t$.

Hence $v(B, A_t) \geq 2|D_t| + |C_t| = np+1+|D_t|$.

We also have

(31) $v(B, A_t) \geq np+1$ for $1 \leq t \leq k+1$.

PROOF: By (27) (i), $v(x, A_t) \neq 0$ for every $1 \leq t \leq k+1$, $x \in B$ hence $v(B, A_t) \geq |B| = np+1$.

(32) $$v(B,A) \leq \sum_{x \in B} (p-1-v(x,B)) + (k+1)(np+1)$$

PROOF: By $B \cap A = 0$, $v(B,A) = \sum_{x \in B} v(x,A)$.

By $v(\mathcal{G}) < \ell$; $v(x,A) + v(x,B) \leq \ell - 1$ for every $x \in B$, hence $v(x,A) \leq \ell - 1 - v(x,B) = k+1+p-1-v(x,B)$

thus (32) follows.

(33) There is $t \in M$ with $\sum_{x \in B} (p-1-v(x,B)) \geq (2q+1)|D_t|$.

PROOF: By (30) and (31) we have

$$v(B,A) = \sum_{t=1}^{k+1} v(B, A_t) \geq \sum_{t \in M} |D_t| + (k+1)(np+1)$$

Hence, by (32)

$$\sum_{t \in M} |D_t| \leq \sum_{x \in B} (p-1-v(x,B))$$

Let $|D_t| = \min\{|D_{t'}| : \text{ for } t' \in M\}$. Then, by (24)

$$(2q+1)|D_t| \le |M||D_t| \le \sum_{x \in B} (p-1-v(x,B))$$

and (33) follows.

We now obtain a contradiction by proving

(34) For every $t \in M$, $\sum_{x \in B} (p-1-v(x,B)) < (2q+1)|D_t|$.

PROOF: Let $y \in A_t$. Put $v_y = v(y, C_t)$.

If $v_y > 0$, then, by (27), $y \in \bar{A}^t_{k+1}$, hence $v(y,A) \ge k$ and thus $v_y \le p$. Put $X_r = \{ y \in A_t : v_y = r \}$, $s_r = |X_r|$ for $r = 0, \dots, p$.

Then $A_t = X_0 \cup \dots \cup X_p$, $n = s_0, \dots, s_p$, the X_r are disjoint.

Put now $Y_r = \bigcup_{y \in X_r} V(y, C_t)$ for $r = 1, \dots, p$.

Considering the definition of C_t, $V(y_1, C_t) \cap V(y_2, C_t) = 0$ for $y_1 \neq y_2 \in A_t$. Thus we have

(35) $|Y_r| = s_r r$ for $r = 1, \dots, p$.

On the other hand, obviously

(36) $C_t = \bigcup_{r=1}^{p} Y_r$ where the summands are disjoint, hence

$$|C_t| = \sum_{r=1}^{p} rs_r = \sum_{r=0}^{p} rs_r .$$

Considering (25) and (26) it follows from (36) that

$$|D_t| = np+1-\sum_{r=0}^{p} rs_r \qquad \text{hence, by} \qquad n = \sum_{r=0}^{p} s_r ,$$

(37) $$|D_t| = \sum_{r=0}^{p}(p-r)s_r+1 .$$

On the other hand it follows from the main lemma (29) that

(38) $v(x,B) \geq r-1$ for every $x \in Y_r$, $r = 1,\dots,p$ and trivially $v(x,B) \geq 0$ for $x \in D_t$.

Thus by (35), (36) and (38) we have

(39) $$\sum_{x\in B}(p-1-v(x,B)) \leq \sum_{r=1}^{p}(p-r)|Y_r|+(p-1)|D_t| = \sum_{r=1}^{p}(p-r)rs_r+(p-1)|D_t| .$$

Thus by (37) and (39)

$$(2q+1)|D_t| - \sum_{x\in B}(p-1-v(x,B)) \geq (2q-p+2)\Big(\sum_{r=0}^{p}(p-r)\, s_r+1\Big) - \sum_{r=1}^{p}(p-r)rs_r$$

Considering that by (25), $p \leq q$, $2q-p+2-r > 0$ since $r \leq p$ we have $(2q-p+2-r)(p-r)s_r \geq 0$ for $r = 0,\dots,p$ and $2q-p+2 > 0$, the right-hand side is positive, hence $(2q+1)|D_t| > \sum_{x\in B}(p-1-v(x, B))$ and this proves (34).

(34) contradicts to (33), hence our indirect assumption (4) leads to a contradiction.

Q. E. D.

REFERENCES

[1] G.A. DIRAC, Some theorems on abstract graphs, Proc. London Math. Soc. (3) 2 (1952) pp. 69-81.

[2] K. CORRÁDI and A. HAJNAL, On the maximal number of independent circuits in a graph, Acta Math. Acad. Sci. Hung. 14 (1963) pp. 423-439.

[3] B. ZELINKA, On the number of independent complete subgraphs, Publ. Math. Debrecen 13 (1966), pp. 95-97.

[4] B. GRÜNBAUM, A result on graph-coloring, Michigan Math. Journal. 15 (1968) pp. 381-383.

[5] K. CORRÁDI, On a problem concerning finite graphs, Annales Univ. Sci. Budapest 9 (1966) 157-165.

Problems involving graphical numbers

by

F. Harary

Ann. Arbor, USA

There is a wonderful tradition in Hungarian mathematical circles of presenting conjectures and unsolved problems in public speeches. There is an even nicer tradition in Hungary of solving some of them! The material presented here represents the beginning of a problem area and it is hoped that this will stimulate research on the topic of graphical numbers.

Several authors* have independently discovered a method for

---

*As a way of indicating independent discovery of this coding system, we list these four names in alphabetical order:

(1) H. J. W. Duijvesteijm of the Technological University Twente in Enschede, The Netherlands, discovered this coding for a graph in his doctoral dissertation dealing with the dissection of triangles into squares.

(2) B. R. Heap of the National Physical Laboratory in Teddington, Middlesex, England, has developed computer cards for each of the 1,044 graphs with seven points and is in process of developing the same information for the 12,344 graphs with eight points.

(3) D. Lefkovitz of The Moore School of Electrical Engineering, University of Pennyslvania, Philadelphia, developed this coding system in connection with the classification of chemical compounds.

(4) R. C. Read and his colleagues in the Computing Centre of the University of the West Indies, Kingston, Jamaica, developed the same code for the same purpose.

associating a unique number with a given graph G, which characterizes G up to isomorphism. This was motivated by the need for developing a method for coding large and complicated chemical compounds. The device which accomplishes the coding is the assignment of a unique labeling to the points of G so that the resulting adjacency matrix A is in a certain canonical form.

We will associate with each graph G two different coding numbers, one maximum and one minimum, and will see that these are related to each other by way of the complement of a graph. In addition we tabulate both of these coding numbers for all graphs with four or five points. We will follow the notation and terminology of the books [1] for graphs and [2] for digraphs.

THE MAXIMUM CODE FOR A GRAPH

Consider a graph at random, say the graph G_1 of Figure 1.

G_1:

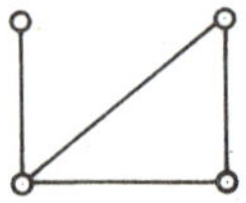

Figure 1.

By a labeling of a graph G with p points is meant an assignment of the numbers $1, 2, \ldots, p$ to its points. For each labeling of G, there is an adjacency matrix A which is a $p \times p$ symmetric binary matrix in which i, j entry is 1 whenever the points labeled i and j are adjacent, and this entry is zero otherwise. To illustrate we show in Figure 2 one labeling of the graph of Figure 1 and the corresponding adjacency matrix, A_1

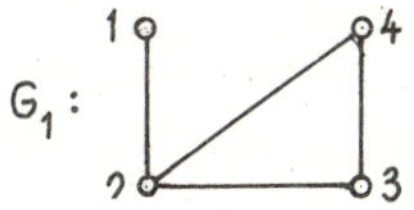

$$A_1 = \begin{bmatrix} 0 & 1 & 0 & 0 \\ 1 & 0 & 1 & 1 \\ 0 & 1 & 0 & 1 \\ 0 & 1 & 1 & 1 \end{bmatrix} = \begin{bmatrix} \diagdown & 1 & 0 & 0 \\ & \diagdown & 1 & 1 \\ & & \diagdown & 1 \\ & & & \diagdown \end{bmatrix}$$

Figure 2.

A binary sequence can be extracted from this matrix A which determines A uniquely and hence also determines the graph G. Since A is symmetric and has diagonal zero, we may ignore the diagonal and also all entries below it. There are six remaining entries, which if written consecutively by rows as 100, then 11, and finally 1 result in the binary sequence 100111.

It was shown in [3] that for any graph G with p points in which the order of the automorphism group is s (standing for the number of symmetries), the number L of different labelings of G is given $L = p!/s$.

For each of these L labelings, there results a different binary sequence as just illustrated with the labeling in Fig. 2. Thus we may immediately define the (maximum) code of a given graph G as the maximum number M(G) which can be obtained by taking the binary sequences resulting from all the labelings of G, and regarding it as a binary number. (Later we shall denote the minimum such number by m(G).) For example, the graph G_1 of Fig. 1 does not have the labeling of Fig. 2, but rather that of Fig. 3 to obtain its code M(G).

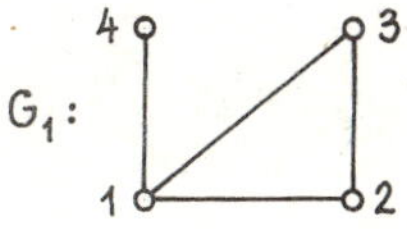

$$A = \begin{bmatrix} \diagdown & 1 & 1 & 1 \\ & \diagdown & 1 & 0 \\ & & \diagdown & 0 \\ & & & \diagdown \end{bmatrix}$$

$$M(G_1) = 111100 = 60$$

Figure 3.

We are now ready to show in Table 1 all of the graphs with four points and their codes given in both binary and decimal forms.

Table 1. The maximum code of all graphs with four points.

Graph G	Binary M(G)	Decimal M(G)
	000000	0
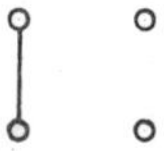	1000000	32
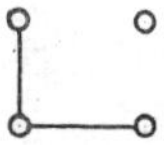	110000	48
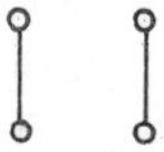	100001	33
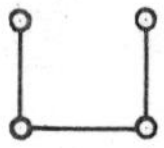	110010	50
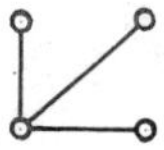	111000	56
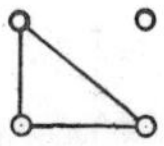	110100	52
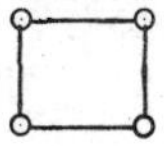	110101	51
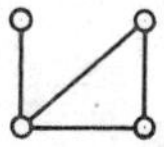	111100	60
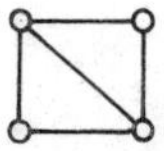	111110	62
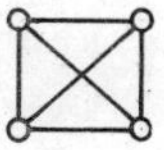	111111	63

For the sake of completeness we note that the code numbers $M(G)$ for the graphs with three points are 0, 4, 6, and 7. This data together with the numbers contained in Table 1 suggest many questions. Let us say that a graphical number is the code of some graph.

1. Which numbers are graphical? For example, the number 53 is not graphical, as can be seen from Table 1.

2. If a number is graphical, when does its uniquely determined graph have certain specified properties? For example, which graphical numbers correspond to connected graphs, trees, blocks, planar graphs, eulerian graphs, hamiltonian graphs, etc.? Also, which have given values for its invariants, such as connectivity, chromatic number, the various covering and independence numbers, etc.?

3. Develop an efficient algorithm for a given graph G to obtain that one of its L different labelings which will result in the code number.

4. When does one graph have a bigger code than another?

5. Which graphs have odd codes? Which have prime codes?

Several of the participants at the Symposium in Balatonfüred made certain observations concerning these and related problems, as follows.

G. BARON observed that (1) the number of disconnected graphs with p points whose code number is less than the minimum of the code numbers of all connected gr--hs with p points is exactly $\pi(p)-1$, where $\pi(p)$ stands for the number of partitions of p, (2) the number of connected graphs with p points whose code number is greater than the maximum of all the code numbers of the disconnected graphs with p points is exactly g_{p-1}, where g_n denotes the number of graphs with n points. The proof of (1) is that such disconnected graps G are precisely the linear forests, which are those forests in which every component is a path. The proof of (2) follows since the graphs G in question are all those whose maximum degree is as large as possible, namely

$p-1$. From this one can obtain the number of disconnected graphs G whose code is greater than the minimum of the codes of all connected graphs by subtraction from the known number of disconnected graphs with p points.

A. RECSKI experimented with the data to appear below (Table 2) which gives the codes for all the graphs with five points. From these data, he was able to observe the following phenomena.

(1) In general, the graph G with p points and at least one line for which $M(G)$ is minimum is of course that graph with exactly one line for which $M(G) = 2^{\binom{p-1}{2}}$.

Similarly map $M(G)$ is obtained naturally for $G = K_p$, and $M(K_p) = 2^{\binom{p}{2}} - 1$.

(2) This interval $[2^{\binom{p-1}{2}}, 2^{\binom{p}{2}} - 1]$ for nonzero graphical numbers can be narrowed further if one can specify a sequence of closed subintervals $[\alpha_i, \beta_i]$ within which all values of $M(G)$ occur for these p point graphs G which are not totally disconnected. Only the numbers α_i can be immediately determined since they are the binary numbers of the form 11...100...0.

The actual determination of the code of a given graph G depends on finding that labeling of G which results in the maximum binary number obtained from the corresponding adjacency matrix A. As a beginning, it can be stated unequivocally that the point to be labeled v_1 must be one of the points of G having maximum degree $\Delta = \Delta(G)$, and that the points to be labeled $v_2, v_3, \ldots, v_{\Delta+1}$ must be, in some order, the Δ points adjacent to v_1. For only by such an initial labeling can the first $\Delta+1$ digits in the binary form of $M(G)$ be all 1's. The process of choosing which of the points of maximum degree to label v_1 and how to order the points in the neighborhood of v_1 is a bit more subtle.

We now define the mincode $m(G)$ as the minimum number whose binary form is obtained from the L different labelings of G in the same manner as in the definition of the code $M(G)$. As usual we denote the complement of G

by $\bar{G}$. The next observation links the code, mincode, and the complement. It is so easily verified that its proof is omitted.

OBSERVATION

For any graph G with p points,

$$M(G) + m(\bar{G}) = 2^{\binom{p}{2}} - 1. \tag{1}$$

In Appendix I of the book [1], each graph G with $p = 5$ is designate by an ordered triple $(5, q, i)$ where q is the number of lines of G. The index i is assigned to G more or less arbitrarily with the proviso that when G is the $(5, q, i)$ graph, then its complement $\bar{G}$ is the $(5, 10-q, i)$ graph, except for the graphs with $q = 5$, two of which are self-complementary. For the sake of completeness, we need to display in Figure 4 these ordered triple designations for the graphs with $p = 5$, for $q = 0, 1, 2, 3, 4, 5$ only. We note that for $q = 5$, the graphs $(5,5,2)$ and $(5,5,6)$ are self-complementary, while the remaining four 5, 5 graphs occur in the two complementary pairs: $(5,5,1)$ and $(5,5,5)$; $(5,5,3)$ and $(5,5,4)$.

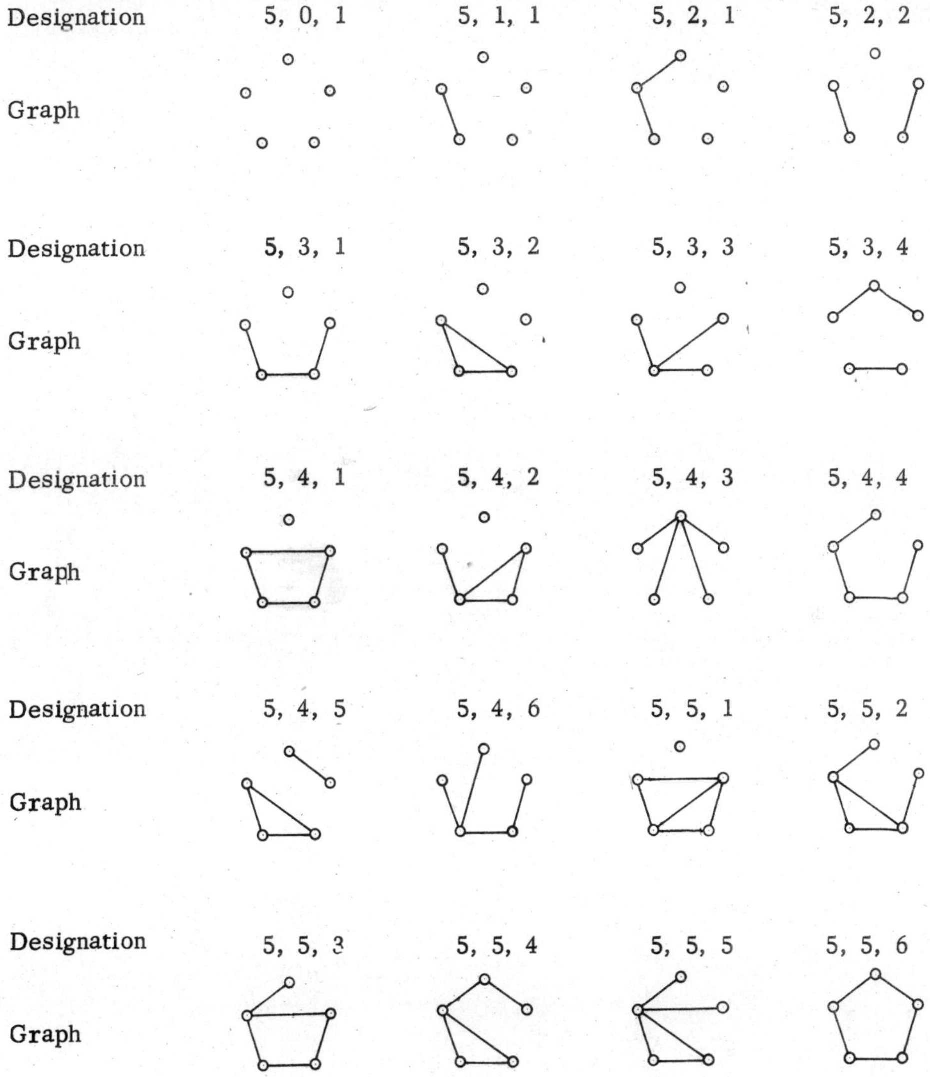
Designation
5, 0, 1
5, 1, 1
5, 2, 1
5, 2, 2
Graph
Designation
5, 3, 1
5, 3, 2
5, 3, 3
5, 3, 4
Graph
Designation
5, 4, 1
5, 4, 2
5, 4, 3
5, 4, 4
Graph
Designation
5, 4, 5
5, 4, 6
5, 5, 1
5, 5, 2
Graph
Designation
5, 5, 3
5, 5, 4
5, 5, 5
5, 5, 6
Graph

Figure 4.

We conclude with Table 2, which presents both the code and mincode of all 5 point graphs.

Table 2. Codes and mincodes of 5 point graphs

G	M(G)	m(G)	G	M(G)	m(G)
5, 0, 1	0	0	5, 10, 1	1023	1023
5, 1, 1	512	1	5, 9, 1	1022	511
5, 2, 1	768	3	5, 8, 1	1020	255
5, 2, 2	516	12	5, 8, 2	1011	507
5, 3, 1	784	13	5, 7, 1	1010	239
5, 3, 2	800	7	5, 7, 2	1016	223
5, 3, 3	896	11	5, 7, 3	1012	127
5, 3, 4	769	76	5, 7, 4	947	254
5, 4, 1	788	30	5, 6, 1	993	235
5, 4, 2	928	15	5, 6, 2	1008	95
5, 4, 3	960	75	5, 6, 3	948	63
5, 4, 4	786	86	5, 6, 4	937	237
5, 4, 5	801	116	5, 6, 5	907	222
5, 4, 6	904	77	5, 6, 6	946	119
5, 5, 1	944	31	5, 5, 5	992	79
5, 5, 2	936	87	5, 5, 2	936	87
5, 5, 3	906	94	5, 5, 4	929	117
5, 5, 6	787	236	5, 5, 6	787	236

REFERENCES

[1] F. HARARY, Graph Theory. Addison-Wesley, Reading, Mass., 1969.

[2] F. HARARY, R. Norman, and D. Cartwright, Structural Models: An introduction to the theory of directed graphs. Wiley, New York, 1965.

[3] F. HARARY, E. M. Palmer, and R. C. Read, The number of ways to label a structure. Psychometrika 32 (1967) 155-156.

A composition theorem for the enumeration of certain subsets of symmetric semigroups and the corresponding sets of labeled functional digraphs

by

B. Harris and L. Schoenfeld
Wisconsin, USA Buffalo, USA

1. INTRODUCTION

Let X be a non-empty set of distinct elements and let T be the set of all mappings of X into X. Then T is a semigroup with identity under composition; that is, if $x \in X$ and $\alpha, \beta \in T$, then $(\alpha\beta)(x) = \alpha(\beta(x))$. This semigroup is called the symmetric semigroup on X. There is no loss of generality in considering $X_n = \{1, 2, \ldots, n\}$ and letting T_n be the symmetric semigroup in X_n, since any two finite sets X, Y with the same cardinality have isomorphic symmetric semigroups.

To every $\alpha \in T_n$, there corresponds a unique labeled functional diagraph $G(\alpha)$ obtained as follows. Let X_n be the set of n vertices of the graph. In each $x \in X_n$ draw the directed edge from x to $\alpha(x)$; if $x = \alpha(x)$ draw a loop at x. Such a graph has the property that from each vertex, exactly one directed edge emanates. Conversely, every such graph with this property uniquely determines an $\alpha \in T_n$. Because of this correspondence, it is convenient to employ the same terminology and notation for both the elements $\alpha \in T_n$ and the corresponding graphs $G(\alpha)$.

The components of the graph are the equivalence classes of X_n with respect to α determined by the equivalence relation $x \sim y$ if and only if there are non-negative integers r, s such that $\alpha^r(x) = \alpha^s(y)$. We denote the component containing x by $K_\alpha(x)$.

It can easily be seen that each component has exactly one cycle. If x is cyclic, we denote the cycle containing x by $C_\alpha(x)$. For any $x \in X_n$, we denote the cycle in $K_\alpha(x)$ by $D_\alpha(x)$. C_α will denote the set of cyclic points in X_n, that is, $C_\alpha = \bigcup_{x \in X_n} D_\alpha(x)$. In terms of the mapping $\alpha \in T_n$, x is cyclic, if there exists an integer $m > 0$ with $\alpha^m(x) = x$. The least such positive integer is called the length of the cycle $C_\alpha(x)$. Finally, we call the length of the longest directed path from any acyclic vertex to the cycle the height of the graph and we let T_{j_n}, $0 \leq j \leq n-1$ be the set of $\alpha \in T_n$ of height not exceeding j. We designate the restriction of α to the elements of C_α by α^*. Thus α^* is an element of S_k, the symmetric group on k elements, for some k, $1 \leq k \leq n$.

In what now follows, we present a brief summary of certain enumerative results for some classes of subsets of symmetric semigroups. The complete details of the proofs of these results as well as a number of applications of the main theorems will be subsequently published. Here we shall only state the results, provide a brief outline of some of the proofs, giving shorter arguments in detail, and list some classes of subsets of symmetric semigroups which satisfy the required hypotheses. This listing will serve the purpose of exhibiting the generality of the results to be described as well as exhibiting both familiar and new combinatorial results in enumeration which are consequences of these theorems. In particular, with appropriate specialization of these results, we obtain earlier results of the authors (Harris [7], Harris and Schoenfeld [8], [9], [10], [11], [12]), as well as results due to Riordan ([18], [20]), Rényi and Szekeres [16] and Tainiter [21].

2. ENUMERATION OF VARIOUS SUBSETS OF SYMMETRIC SEMIGROUPS AND THE COMPOSITION THEOREM

A subset M_k of S_k is said to be self-conjugate, if for every $\lambda \in S_k$, $\lambda M_k \lambda^{-1} = M_k$. To every element $\gamma \in S_k$, we can associate a partition $(r_1, r_2, \ldots, r_k)$, where r_i is the number of cycles of length i. Clearly, $\sum_{i=1}^{k} i r_i = k$. The k-tuple $(r_1, r_2, \ldots, r_k)$ is called the class of γ. A subset of S_k is self-conjugate if and only if it is the set of all permutations in a subset of the possible classes. Thus, there is a correspondence between the self-conjugate subsets of S_k and the subsets of non-negative k-tuples $(r_1, r_2, \ldots, r_k)$ with $\sum_{i=1}^{k} i r_i = k$. In $k = 1, 2, \ldots$ let W_k be given self-conjugate subsets of S_k. We will denote the corresponding sets of k-tuples by $R(W_k)$. Let $w_k = |W_k|$, $w_0 = 1$ and let w denote the sequence $\{w_n\}_{n=0}^{\infty}$. In addition, let

$$(1) \qquad \phi_w(z) = \sum_{k=0}^{\infty} \frac{w_k z^k}{k!} .$$

Now define W_{kn}, $k \le n$, as the set of all $\beta = \gamma \tau \gamma^{-1}$, where γ is a one-to-one mapping of X_k into X_n and $\tau \in W_k$. Each such β permutes k of the integers of X_n and is obtained from $\tau \in S_k$ by relabeling the integers in X_k by integers in X_n without changing the class.

We now consider the problem of enumerating the set of $\alpha \in T_n$ satisfying

(a) $\alpha^* \in W_{kn}$ for some $k \le n$

(b) $\alpha \in T_{jn}$.

Note in particular that $R(W_{kn}) = R(W_k)$.

We let $V_{w,j,n}$ be the number of such $\alpha \in T_n$ and let $V_{w,j,0} = 1$

for all j and w_k. Clearly $V_{w,j,n} \le n^n$. In addition, we denote $V_{w,n-1,n}$ by $V_{w,n}$. This is the number of all labeled functional digraphs on n points satisfying (a). We now have the following theorem.

THEOREM 1. If $n \ge 0$ and $0 \le j \le n$, then

$$V_{w,j,n} = \sum_{\substack{k_0+k_1+\dots k_j = n \\ k_0, k_1, \dots k_j \ge 0}} \frac{n!}{k_0! k_1! \dots k_j!} w_{k_0} k_0^{k_1} k_1^{k_2} \dots k_{j-1}^{k_j} . \tag{2}$$

Proof. The demonstration of (2) may be accomplished by elementary combinatorial reasoning and the details will not be given here.

The following corollary is of interest, since it provides an explicit closed form expression for a number of problems in enumerative combinatorics.

Corollary. Let $V_{w_k,n}$ be the number of $\alpha \in T_n$ with $\alpha^* \in W_{kn}$, where k is fixed and $1 \le k \le n$ and let $V_{w,n}$ be the number of $\alpha \in T_n$ satisfying (a). Then

$$V_{w,n} = \sum_{k=1}^{n} V_{w_k,n} \tag{3}$$

and

$$V_{w_k,n} = \binom{n-1}{k-1} w_k n^{n-k} . \tag{4}$$

Proof. Clearly

$$V_{w_{k_0},n} = \sum_{\substack{k_1+k_2+\dots+k_{n-1} = n-k_0 \\ k_1, k_2, \dots, k_{n-1} \ge 0}} \frac{n!}{k_0! k_1! \dots k_{n-1}!} w_{k_0} k_0^{k_1} k_1^{k_2} \dots k_{n-2}^{k_{n-1}} .$$

From a combinatorial lemma due to Katz [14], we have

$$\sum \frac{n!}{k_0!\,k_1!\dots k_{n-1}!}\, w_{k_0} k_0^{k_1} k_1^{k_2} \dots k_{n-2}^{k_{n-1}} = w_{k_0} k_0 \binom{n}{k_0} n^{n-k_0-1},$$

and the corollary is established.

Now consider the function $V_{w,j,n}(t_0, t_1, \dots, t_j)$ defined by

$$(5)\quad V_{w,j,n}(t_0,t_1,\dots,t_j) = \sum_{\substack{k_0+k_1+\dots+k_j=n \\ k_0,k_1,\dots,k_j \ge 0}} \frac{n!}{k_0!\,k_1!\dots k_j!}\, w_{k_0} k_0^{k_1} k_1^{k_2} \dots k_{j-1}^{k_j} t_0^{k_0} t_1^{k_1} \dots t_j^{k_j}$$

If in (5), we replace t_m, $0 \le m \le j$, by zt_m and let $\zeta = \max\{|zt_0|, |zt_1|, \dots, |zt_j|\}$, then

$$|z^n V_{w,j,n}(t_0,t_1,\dots,t_j)| = |V_{w,j,n}(t_0 z, t_1 z, \dots, t_j z)| \le$$

$$\le V_{w,j,n}(\zeta, \zeta, \dots, \zeta) =$$

$$= \zeta^n V_{w,j,n}(1, 1, \dots, 1) =$$

$$= \zeta^n V_{w,j,n} \le \zeta^n n^n .$$

Thus if we now define

$$(6)\quad \Psi_{w,j}(z; t_0, t_1, \dots, t_j) = \sum_{n=0}^{\infty} V_{w,j,n}(t_0, t_1, \dots, t_j)\, z^n/n! \quad ,$$

then the series converges absolutely for $\zeta < \frac{1}{e}$; that is, we have absolute convergence whenever $|zt_0|, |zt_1|, \dots, |zt_j| < \frac{1}{e}$. We now have the following theorem.

THEOREM 2. If $\zeta < \frac{1}{e}$ and $|zt_{j-1}e^{zt_j}| < \frac{1}{e}$, then

(6a) $$\Psi_{w,0}(z;t_0) = \phi_w(zt_0)$$

and for $j \geq 1$.

(7) $$\Psi_{n,j}(z;t_0,t_1,\dots,t_j) = \Psi_{w,j-1}(z;t_0,t_1,\dots,t_{j-2},t_{j-1}e^{zt_j}).$$

Proof. As $V_{w;0,n}(t_0) = w_n t_0^n$, we clearly get (6a), since $|zt_0| < \frac{1}{e}$. To establish (7), observe that

$$\Psi_{w,j}(z,t_0,t_1,\dots,t_j) = \sum_{n=0}^{\infty} \frac{z^n}{n!} \sum_{q=0}^{n} \sum_{\substack{k_0+k_1+\dots+k_{j-1}=q \\ k_0,k_1,\dots,k_{j-1}\geq 0}} \frac{n!}{k_0!k_1!\dots k_{j-1}!(n-q)!} \cdot$$

$$\cdot w_{k_0} t_0^{k_0}(k_0t_1)^{k_1}\dots(k_{j-2}t_{j-1})^{k_{j-1}}(k_{j-1}t_j)^{n-q} =$$

$$= \sum_{q=0}^{\infty} \frac{z^q}{q!} \sum_{\substack{k_0+k_1+\dots+k_{j-1}=q \\ k_0,k_1,\dots,k_{j-1}\geq 0}} w_{k_0} t_0^{k_0}(k_0t_1)^{k_1}\dots(k_{j-2}t_{j-1})^{k_{j-1}} \sum_{n=q}^{\infty} \frac{(k_{j-1}t_jz)^{n-q}}{(n-q)!} .$$

Thus

$$\Psi_{w,j}(z;t_0,t_1,\dots,t_j) = \sum_{q=0}^{\infty} \frac{z^q}{q!} V_{w,j-1,q}(t_0,t_1,\dots,t_{j-2},t_{j-1}e^{zt_j}).$$

The conclusion follows since $|zt_{j-1}e^{zt_j}| < \frac{1}{e}$.

We now define $\Lambda_0(z_0) = z_0$, $\Lambda_1(z_0,z_1) = z_0e^{z_1}$, and for $j \geq 2$,

(8) $$\Lambda_j(z_0,z_1,\dots,z_j) = \Lambda_{j-1}(z_0,z_1,\dots,z_{j-2},z_{j-1}e^{z_j}).$$

From the definition, we have the following theorem.

THEOREM 3. In $j \geq 1$,

(9) $$\Lambda_j(z_0,z_1,\dots,z_j) = z_0 e^{\Lambda_{j-1}(z_1,z_2,\dots,z_j)}.$$

Proof. The conclusion is obvious for $j=1$. Therefore assume that the result holds for $j-1$ where $j \geq 2$. Then

$$\Lambda_j(z_0,z_1,\dots,z_j) = \Lambda_{j-1}(z_0,z_1,\dots,z_{j-2},z_{j-1}e^{z_j}) =$$

$$= z_0 \exp\{\Lambda_{j-2}(z_0,\dots,z_{j-2},z_{j-1}e^{z_j})\} =$$

$$= z_0 \exp\{\Lambda_{j-1}(z_1,z_2,\dots,z_{j-1},z_j)\},$$

completing the induction.

Corollary. If $j \geq 0$ and $|z_m| \leq \frac{1}{4}$ for $m = 0,1,\dots,j$ then

$$|\Lambda_j(z_0,z_1,\dots,z_j)| < \frac{1}{e}.$$

Proof. The conclusion is obvious for $j=0$. Assume that it holds for $j-1$ where $j \geq 2$. Then

$$|\Lambda_j(z_0,z_1,\dots,z_j)| \leq |z_0| \exp\{|\Lambda_{j-1}(z_1,z_2,\dots,z_j)|\} \leq$$

$$\leq \frac{1}{4}\exp\left(\frac{1}{e}\right) < \frac{1}{e}.$$

We now obtain the principal theorem of this paper.

THEOREM 4 (Composition theorem). Given $j \geq 0$, $|zt_m| < \frac{1}{e}$ for $0 \leq m \leq j$ and $|\Lambda_m(zt_{j-m}, \dots, zt_j)| < \frac{1}{e}$ for $1 \leq m \leq j$; then

(10) $$\Psi_{w,j}(z; t_0, t_1, \dots, t_j) = \phi_w(\Lambda_j(zt_0, zt_1, \dots, zt_j)).$$

Proof. For $j = 0$, this follows from theorem 2. Now suppose the result holds for $j-1$, $j \geq 1$. From the hypotheses,

$$|zt_{j-1} e^{zt_j}| = |\Lambda(zt_{j-1}, zt_j)| < \frac{1}{e}.$$

Hence from Theorem 2,

$$\Psi_{w,j}(z; t_0, t_1, \dots, t_j) = \Psi_{w,j-1}(z; t_0, t_1, \dots, t_{j-2} t_{j-1} e^{zt_j}).$$

Now for $1 \leq m \leq j-1$,

$$|\Lambda_m(zt_{j-1-m}, \dots, zt_{j-2}, zt_{j-1} e^{zt_j})| =$$

$$= |\Lambda_{m+1}(zt_{j-1-m}, \dots, zt_{j-2}, zt_{j-1}, zt_j)| < \frac{1}{e}.$$

Then, from the induction hypothesis, we conclude that

$$\Psi_{w;j-1}(z; t_0, t_1, \dots, t_{j-2}, t_{j-1} e^{zt_j}) =$$

$$= \phi_w(\Lambda_{j-1}(zt_0, \dots, zt_{j-2}, zt_{j-1} e^{zt_j})) =$$

$$= \phi_w(\Lambda_j(zt_0, \dots, zt_{j-2}, zt_{j-1}, zt_j)).$$

Corollary 1. Let $j \geq 0$ and $|zt_m| \leq \frac{1}{4}$ for $m = 0,1,\ldots,j$, then

$$\Psi_{w;j}(z;t_0,t_1,\ldots,t_j) = \phi_w(\Lambda_j(zt_0,zt_1,\ldots,zt_j)) . \tag{11}$$

Proof. This follows immediately from the corollary to theorem 3.

Theorem 4. provides an enumerating formula for labeled functional digraphs satisfying the hypotheses of theorem 1. by number of points, by number of points on cycles, and by number of points at distance m from the cycle, $1 \leq m \leq j$. An important specialization is given by the following corollary.

Corollary 2. Let $\Lambda_j(z) = \Lambda_j(z,\ldots,z)$, then for $|z| \leq \frac{1}{4}$ the exponential generating function of $V_{w,j,n}$ is

$$\Psi_{w,j}(z) = \phi_w(\Lambda_j(z)) = \sum_{k=0}^{\infty} V_{w,j,n} z^n/n! , \tag{12}$$

where $\Lambda_0(z) = z$ and for $j \geq 1$,

$$\Lambda_j(z) = z \exp\{\Lambda_{j-1}(z)\} . \tag{13}$$

Now we have

Corollary 3. Let $V_{w,j,n,k}$ be the number of $\alpha \in T_n$ satisfying (a) and (b) and with exactly k cyclic points, then

$$\sum_{n=0}^{\infty} \sum_{k=0}^{n} V_{w,j,n,k} \frac{t^k z^n}{n!} = \phi_w(\Lambda_j(zt,z,\ldots,z)) . \tag{14}$$

The same methods used to obtain theorem 4 may also be employed to enumerate by components as well as number of points. To see this observe that the number of components of α is the number of cycles of α^* (J. Dénes [3], [4], [5] calls α^* the main permutation of α). Let W_{kr} be the subset of W_k with r cycles. Clearly if W_k is self-conjugate, so is W_{kr}. Then defining $\bar{V}_{w,j,n,r}$ as the number of $\alpha \in T_{jn}$ with r components satisfying (a), $\bar{V}_{w_{kr},n}$ as the number of $\alpha \in T_n$ with r components and $\alpha^* \in W_{kr}$, $k \le n$ then

(15) $$\Psi_{w_r,j}(z) = \Phi_{w_r}(\Lambda_j(z))$$

and

(16) $$V_{w_{kr},n} = \binom{n-1}{k-1} w_{kr}\, n^{n-k}$$

To enumerate by number of components, the coefficient of $u^r z^n/n$ in

(17) $$\ell_w(r,z) = \sum_{r=1}^{n} \Phi_{w_r}(\Lambda_j(z))\, u^r$$

is the number of $\alpha \in T_{jn}$ with $\alpha^* \in W_{kr}$ some $k \le n$ and with exactly r components.

A different technique for enumeration by components has also been studied, but this will be reported on in the more complete version if this work.

We now list a number of illustrations of the types of problems that these results may be applied to.

4. EXAMPLES

To illustrate some of the many possible applications of the preceding theorems, we list a number of examples.

Example 1. Connected functional digraphs. Here each α has exactly one cycle, so that W_k is the set of k cycles and $w_k=(k-1)!$, $w_0 = 1$. The corollary to theorem 1. gives the formula for the number of connected functional digraphs on n points (Katz [14], Riordan [19], Harris [6]).

Example 2. Enumerating all of T_{jn}. Here $W_k = S_k$ and $|w_k| = k!$. Here the corollary to theorem 1 enumerates functional digraphs by number of cyclic points, a result previous given in Harris [6], Riordan [19].

Example 3. Let W_k be the alternating group on k elements. Then $w_0 = 1$, $w_1 = 1$, $w_k = \frac{k!}{2}$, $k \geq 2$.

Example 4. Enumeration of all $\alpha \in T_n$ with $\alpha^{j+1} = \alpha^j$. This is equivalent to the enumeration of forests of rooted labeled trees by height (Riordan [20]). Here $W_k = \{e_k\}$, the identity in S_k. In particular, $j=1$ gives the enumeration of the number of idempotent elements of T_n (see Harris and Schoenfeld [8], [9], [10], Harris [7], and Tainiter [21]). The results of this paper also yield the well-known formula for forests of rooted labeled trees on n vertices (Riordan [20]). With a slight extension of this example, we obtain the enumeration of rooted labeled trees on n vertices (Rényi and Szekeres [16], Riordan [18]), Clarke's theorem [2] and Cayley's formula for the enumeration of trees, both rooted and unrooted (see Riordan [17]). The method for enumeration by components gives a result in Harris [7].

Example 5. The solutions of $\alpha^{j+k} = \alpha^j$ in T_n. If k is a prime, Theorem 4 gives the result in Harris and Schoenfeld [10]. For k a prime $\phi_w(z)$ has been given by Jacobsthal [13], for general k by Moser and Wyman [15] and Chowla, Herstein and Scott [1].

Example 6. The enumeration of the solutions of $\alpha^{j+2} = \alpha^{j}$ which are not solutions of $\alpha^{j+1} = \alpha^{j}$ in T_n. Here

$$\phi_w(z) = e^{z^2/2}$$

W_k is a self-conjugate set and the results in this paper apply to this case.

Example 7. The enumeration of $\alpha \in T_{jn}$ such that α has no fixed points. Here

$$\phi_w(z) = \frac{e^{-z}}{1-z} .$$

The purpose of this list is to exhibit that many well-known combinatorial results are contained in these theorems as well as new ones. Hopefully, this will exhibit the generality of these results.

REFERENCES

[1] S. CHOWLA, I.N. HERSTEIN and W.R.SCOTT, The solutions of $x^d = 1$ in symmetric groups, Norske Vid. Selsk. Forhandlingen, 25 (1952), 29-31.

[2] L.E. CLARKE, On Cayley's formula for counting trees, J. London Math. Soc. 33 (1958), 471-475.

[3] J. DÉNES, On transformations, transformation-semigroups, and graphs, Theory of Graphs, Proceedings of the Colloquium held at Tihany, Hungary, 1966, 65-75.

[4] J. DÉNES, On some properties of commutator semigroups, Publicationes Math. Debrecen, 15 (1965) 283-285.

[5] J. DÉNES, On graph representations of semigroups, Proceedings of the Calgary International Conference on Combinatorial Structures and their Applications, to appear.

[6] B. HARRIS, Probability distributions related to random mappings, Ann. Math. Stat. 31 (1960), 1045-1062.

[7] B. HARRIS, A note on the number of idempotent elements in symmetric semigroups, Ann. Math. Monthly, 79 (1967) 1234-1235.

[8] B. HARRIS and L. SCHOENFELD, The number of idempotent elements in symmetric semigroups, J. Combinatorial Theory 3 (1961) 121-135.

[9] B. HARRIS and L. SCHOENFELD, Asymptotic expansions for coefficients of analytic functions, Ill. Journal of Math. 12 (1968), 264-277.

[10] B. HARRIS and L. SCHOENFELD, The number of solutions of in symmetric semigroups, Proceedings of the Calgary International Conference on Combinatorial Structures and their Applications, to appear.

[11] B. HARRIS and L. SCHOENFELD, The number of generalized idempotent elements in symmetric semigroups, MRC Technical Summary Report 998 (1969).

[12] B. HARRIS and L. SCHOENFELD, The number of solutions of in symmetric semigroups, MRC Technical Summary 1000 (1969).

[13] E. JACOBSTHAL, Sur le nombre d'éléments du group symmétrique dont l'ordre est un nombre premier, Norske Wid. Selsk. Forhandlingen, 21 (1949) 49-51.

[14] L. KATZ, Probability of indecomposability of a random mapping function, Ann. Math. Statist., 26 (1955), 512-517.

[15] L. MOSER and M. WYMAN, On solutions of $x^d=1$ in symmetric groups, Canadian J. Math. 7 (1955), 159-168.

[16] A. RÉNYI and G. SZEKERES, On the height of trees, J. Australian Math. Soc. 1 (1967) 491-501.

[17] J. RIORDAN, An introduction to combinatorial analysis, John Wiley and Sons, New York, 1958.

[18] J. RIORDAN, The enumeration of trees by height and diameter, IBM Journal of Research and Development, 4 (1960), 413-418.

[19] J. RIORDAN, Enumeration of linear graphs for mappings of finite sets, Ann. Math. Statist., 33 (1962), 178-185.

[20] J. RIORDAN, Forest of labeled trees, J. Combinatorial Theory, 5 (1968), 90-103.

[21] M. TAINITER, A characterization of idempotens in semigroups, J. Combinatorial Theory, 5 (1968), 370-373.

Graphs with transitive Abelian automorphism group

by

W. Imrich
Vienna, Austria

We will consider finite undirected graphs without multiple edges or loops. If X is a graph let $V(X)$ denote its set of vertices and $E(X)$ its set of edges, which we will consider as unordered pairs $[a,b]$ of vertices a,b. A permutation φ on $V(X)$ is called an automorphism of X, if $[\varphi a, \varphi b] \in E(X)$ if and only if $[a,b] \in E(X)$. The automorphisms of X form a group $G(X)$. It is known that the automorphism group of a graph is the direct product of cyclic groups of order two if it is transitve and Abelian. This has been shown by CHAO [1] and SABIDUSSI [5].

MCANDREW [3] has reported that there exist finite graphs X with transitve Abelien $G(X)$ if and only if $G(X) \cong C(2)^n$, with $n \neq 2,3,4$, where $C(2)$ is the cyclic group of order two. A proof of the existence of such graphs beginning with $n = 8$ has appeared in [2], where the infinite case is also treated.

It is easily seen that every transitive Abelian permutation group G on V is regular, i.e. transitive and for $\alpha \in G$, $a \in V$ the relation $\alpha a = a$ implies that α is the identity ι of G.

If H is a subset of G not containing the identity, the color graph $X_{G,H}$ is defined by $V(X_{G,H}) = G$ and $E(X_{G,H}) = \{[g, gh] : g \in G, h \in H\}$. It has been shown by SABIDUSSI [4] that every connected graph X which admits a regular group G of automorphisms is isomorphic to a color graph $X_{G,H}$, where H is a set of generators of G. H and an isomorphism of X onto $X_{G,H}$ can be found as follows: One chooses a vertex v of X, maps every vertex x of $V(X)$ into the uniquely defined automorphism $\varphi_x \in G$ with $\varphi_x v = x$, and defines $H = \{\varphi_x : x \in V(X), [v,x] \in E(X)\}$.

THEOREM. For every natural number n different from 2,3 or 4 there exists a graph X with transitive Abelian $G(X)$ isomorphic to the direct product $C(2)^n$.

PROOF. For $n = 1,2$ the theorem is trivial. For $n = 3,4$ we use a result from [2], namely that every graph X whose automorphism group is a $C(2)^n$, $n > 1$, contains a spanning n-dimensional cube. If X is a graph with $G(X) \cong C(2)^n$, we also have $G(X') \cong C(2)^n$ for the complementary graph X' of X defined by $V(X') = V(X)$ and $E(X') = \{[a,b] : a,b \in V(X), [a,b] \notin E(X)\}$. This means we can restrict our attention in the case $n = 3$ to graphs with at most 14 edges and in the case $n = 4$ to graphs with at most 60 edges.

Let X be a graph with $G(X) \cong C(2)^3$ and at most 14 edges. X contains a spanning three-dimensional cube K_3 with 12 edges, but $X \neq K_3$ because $G(K_3) \ncong C(2)^3$. Therefore we have to introduce additional edges in K_3 to obtain X. Suppose we have to introduce k edges at a vertex v of K_3. By the transitivity of $G(X)$ it follows that we have to introduce k edges at every vertex of K_3, giving altogether a total of $12 + 4k > 14$ edges in X, which is impossible.

Let X be a graph with $G(X) \cong C(2)^4$ and at most 60 edges. To obtain X we have to introduce no more than 28 edges into the four-dimensional cube K_4. Consequently at most $[28/8] = 3$ edges have to be introduced at every vertex of the cube. Let the neighbours of a selected vertex v of K_4 be

a,b,c,d. By SABIDUSSI's theorem there exists a one-to-one correspondence between $x \in V(X)$ and $\varphi_x \in G(X)$ defined by $\varphi_x v = x$. This permits us to identify henceforth φ_x and x. Thus we can say that we have to introduce at most three new edges of the type $[v, z_i]$, $1 \le i \le 3$, in K_4, where the z_i are products of two or more elements from $\{a,b,c,d\}$. By the above, a,b,c,d and the z_i are a set H of generators of G. It is easily seen that every automorphism φ of G mapping H onto itself can be considered as an automorphism of X leaving v invariant. This means that $G(X)$ is not regular if φ is nontrivial. Thus we will have to show the existence of a nontrivial φ leaving H invariant, no matter how we introduce new neighbours of v. We will do this by indicating the permutation φ_H induced by φ in H.

We note that every element in G can be uniquely expressed as a product of elements in $\{a,b,c,d\}$, such that every element occurs at most once. In the following we will mean by a factor of an element g in G an element occuring in this unique representation of g.

If only one edge is introduced, we can assume that it connects v with ab, abc or $abcd$. In any case the transposition (a,b) generates the desired nontrivial automorphism φ of G.

The two-edge case is easily reduced to the one-edge case if $abcd$ is one of the z_i. If neither z_1 nor z_2 are of the form $abcd$, we consider several cases:

1. z_1 and z_2 are of the form xy. If z_1 and z_2 have a factor in common, we can assume $z_1 = ab$ and $z_2 = bc$. Then a nontrivial φ is generated by (a,c). Otherwise we have without loss of generality $z_1 = ab$ and $z_2 = cd$, and we can take $\varphi_H = (a,b)$.

2. z_1 has two factors and z_2 has three. If the two factors of z_1 are factors of z_2 we can suppose $z_1 = ab$, $z_2 = abc$ and take $\varphi_H = (a,b)$. Should only one factor of z_1 be a factor of z_2 we can set $z_1 = ab$, $z_2 = acd$ and $\varphi_H = (c,d)$.

3. $z_1 = abc$, $z_2 = abd$; $\varphi_H = (a,b)$.

We observe that this also settles the three-edge case if $z_3 = abcd$, because then z_1 and z_2 have to be as above, and all φ_H considered up to now leave z_3 invariant. This leaves the following possibilities to be considered:

4. All three z_i are of type xy. Should d not be a factor of any z_i we set $\varphi_H = (a,b)$. If on the other hand every element of $\{a,b,c,d\}$ occurs as a factor of at least one z_i, we have essentially two possibilities, namely $z_1 = ab$, $z_2 = bc$, $z_3 = cd$ or $z_1 = ab$, $z_2 = ac$, $z_3 = ad$. In the first case let φ_H be $(a,d)(b,c)$, and in the second (c,d).

5. The vertex $z_1 = abc$ and z_2, z_3 are of type xy. We list in tabellary form the possibilities for z_2, z_3 and indicate a φ_H:

ab, bc; (a,c)	ab, ad; $(a)(b)(c)(d,ad)$
ab, cd; (a,b)	ad, bd; (a,b)

6. z_1 and z_2 are of type xyz, and z_3 is of type xy. We can assume again $z_1 = abc$, and list z_2, z_3 and φ_H as above:

abd, ab; (a,b)	abd, ac; $(a)(b)(c)(d,abd)$
abd, cd; (c,d)	

7. All z_i are of type xyz. We have $z_1 = abc$, $z_2 = abd$, $z_3 = acd$ and can take $\varphi_H = (b,c,d)$.

To show the existence of graphs X with $G(X) \cong C(2)^n$ for $n \geq 5$, we consider the product G of n cyclic groups of order 2 with the generators a_i, $1 \leq i \leq n$, and a subset H of G defined as follows:

$$H = \{a_i, a_k a_{k+1}, a_1 a_2 a_{n-2} a_{n-1}, a_1 a_2 a_{n-1} a_n : 1 \leq i \leq n, 1 \leq k < n\}$$

For any $a \in G$ the mapping $\alpha : x \to ax$, $x \in V(X_{G,H}) = G$ is an automorphism of $X_{G,H}$. Identifying α and a, we can say that G is a group of automorphisms of $X_{G,H}$. If we can show that every automorphism of $X_{G,H}$ is in G we have $G(X_{G,H}) \cong C(2)^n$. It is easily seen that G acts regularly on $V(X_{G,H})$. Should there be automorphisms of $X_{G,H}$ not in G, then there would exist a nontrivial automorphism which maps at least one vertex into itself. Since $G(X_{G,H})$ is transitive, there would exist a $\varphi \in G(X_{G,H})$, $\varphi \notin G$, which maps the vertex v corresponding to the unit element of G into itself. We thus have to show that every $\varphi \in G(X_{G,H})$ with $\varphi v = v$ is the trivial automorphism. In fact it suffices to prove that φ maps every element of H into itself. For this purpose we consider the subgraph X of $X_{G,H}$ spanned by H. X can be described as follows: Any two consecutive vertices a_i, a_{i+1} are joined by an edge and connected with the vertex $a_i a_{i+1}$ for $1 \le i < n$. In addition, $a_1 a_2$ is connected with $a_{n-2} a_{n-1}$ and $a_{n-1} a_n$. The vertices $a = a_1 a_2 a_{n-2} a_{n-1}$ and $b = a_1 a_2 a_{n-1} a_n$ are isolated vertices in X. The only nontrivial automorphism of X is a transposition of a and b. Thus φ maps the vertices a_i, $1 \le i \le n$, and $a_i a_{i+1}$, $1 \le i < n$, into themselves. But if $a_1 a_2$ is fixed, so are a and b because they are vertices of the nontrivial component of the graph spanned by the neighbours of $a_1 a_2$. This proves the theorem.

REFERENCES

[1] CHAO, CHONG-YUN: On a theorem of Sabidussi, Proc. Amer. Math. Soc. 15 (1964) 291-292.

[2] IMRICH, W.: Graphen mit transitiver Automorphismengruppe, Mh. Math., to be published.

[3] MCANDREW, M.H.: On Graphs with Transitive Automorphism Groups, Amer. Math. Soc. Notices 12 (1965) 575.

[4] SABIDUSSI, G.: On a class of fixed point-free graphs, Proc. Am. Math. Soc. 9 (1958) 800-804.

[5] SABIDUSSI, G.: Vertex-transitive Graphs, Mh. Math. 68 (1964) 426-438.

Constellations et graphes topologiques

par

A. Jacques
Paris, France

Nous présentons ici une partie de la Thèse de Doctorat 3° cycle (Faculté des Sciences de Paris, 27 Janvier 1969) où nous avons développé la théorie des constellations qui fait apparaitre les propriétés algébriques et combinatoires des graphes topologiques.

1. NOTATIONS ET DEFINITIONS

Nous appelons graphe orienté (multigraphe orienté chez BERGE [1]) un quadruplet $G = (A, S, s, b)$ où A est un ensemble d'arcs, S un ensemble de sommets et s et b deux applications de A dans S définissant l'une le sommet-source, l'autre le sommet-but de l'arc considéré.

Deux graphes orienté $G = (A, S, s, b)$ et $G' = (A', S', s', b')$ sont isomorphes s'il existe une bijection λ de A sur A' et une bijection μ de S sur S' telles que:

$$\begin{cases} s'_o \lambda = \mu_o s \\ b'_o \lambda = \mu_o b . \end{cases}$$

Nous appelons graphe non orienté ou encore graphe orienté symétrique un quadruplet $G = (A, S, \alpha, s)$ où α est une involution de l'ensemble A des arcs et s une application de A dans S. En posant $b = s.\alpha$ nous obtenons alors un graphe orienté muni d'une "symétrie" due à l'involution α.

Deux graphes non orientés $G = (A, S, \alpha, s)$ et $G' = (A', S', \alpha', s')$ sont isomorphes si il existe une bijection λ de A sur A' et une bijection μ de S sur S' telles que:

$$\begin{cases} \lambda . \alpha = \alpha' . \lambda \\ s' . \lambda = \mu . s \end{cases}$$

Ce qui implique en posant $b = s.\alpha$ et $b' = s'.\alpha'$:

$$b' . \lambda = \mu . b .$$

Etant donnée une surface Σ orientée et fermée (variété à deux dimensions), nous appelons graphe topologique (J.W.T. YOUNGS [13 a]) sur Σ une famille $(h_a)_{a \in A}$ d'homéomorphismes du segment $[0,1]$ dans Σ tels que les images des ouverts $]0,1[$ soient deux à deux disjointes.

Deux graphes topologiques $(h_a)_{a \in A}$ et $(h'_{a'})_{a' \in A'}$ sur Σ sont dits homéomorphes s'il existe un homéomorphisme $h : \Sigma \to \Sigma$ (homéomorphisme de surfaces orientées) et une bijection $\lambda : A \to A'$ tels que:

$$\forall a \in A \qquad h'_{\lambda a} = h_a . h .$$

R e m a r q u e 1: A un graphe topologique $(h_a)_{a \in A}$ on peut associer un graphe $G = (A, S, s, b)$ de la manière suivante:

- S est l'ensemble des images de 0 et de 1,

- les applications s et b sont définies par:

$$s(a) = h_a(0) \quad \text{et} \quad b(a) = h_a(1) .$$

Par abus de langage c'est ce graphe associé que désignera le terme de graphe topologique.

R e m a r q u e 2: Deux graphes topologiques homéomorphes sont aussi isomorphes en tant que graphes (voir remarque 1). La relation d'homéomorphisme entre deux graphes topologiques est une relation d'équivalence, et c'est en fait une classe de graphes topologiques homéomorphes que nous appellerons par un nouvel abus de langage un graphe topologique.

Etant donné un graphe $G = (A, S, s, b)$ nous appellerons représentation topologique de G sur la surface orientée Σ un graphe topologique sur Σ isomorphe à G en tant que graphe. Sur une surface Σ donnée, un graphe peut admettre aucune, une ou plusieurs représentations topologiques; le genre d'une représentation topologique sur Σ est le genre de la surface Σ. Le genre du graphe G est le genre minimum de ses représentations topologiques sur toutes les surfaces possibles. Une représentation de G de genre minimum $g(G)$ est dite minimale. Un graphe G admettant une représentation topologique de genre $g(G) = 0$ est dit planaire: les graphes planaires sont les graphes qui admettent une représentation topologique sur toutes les surfaces orientées.

Considérons l'union des arcs de Jordan fermés d'une représentation topologique: c'est un fermé K de la surface Σ; alors Σ / K est un ouvert qui se décompose en ouverts connexes disjoints appelés faces de la représentation topologique. Nous dirons que la représentation topologique est simple (2-cell imbedding dans la terminologie anglaise) si les faces sont toutes des ouverts simplement connexes (homéomorphes à un disque plan).

Il résulte d'un théoreme de J. W. T. YOUNGS [13] que toute représentation topologique minimale est simple.

2. REPRESENTATIONS TOPOLOGIQUES SIMPLES ET CONSTELLATIONS

2.1) Le théorème d'EDMONDS

Toute représentation topologique d'un graphe sur une surface orientée définit en chaque sommet une permutation circulaire des arcs incidents à ce sommet, en considérant la rotation dans le sens positif autour de la normale orientée à la surface.

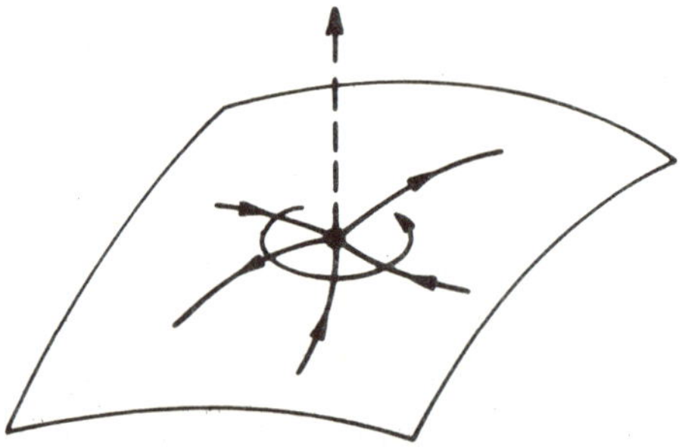

Cependant, il y a une infinité de représentations topologiques d'un graphe sur une surface orientée qui admettent la même famille de permutations circulaires des arcs incidents à chaque sommet: il suffit à partir d'une représentation topologique d'adjoindre des "anses" à la surface, sans modifier les arcs représentés. Une telle classe de représentations topologiques d'un graphe admet un représentant de genre minimum qui est unique: ce représentant a ses faces parfaitement définies par les contours "polygonaux" obtenus en suivant un arc (selon son sens ou non) et en tournant autour d'un sommet selon la permutation qui y est définie.

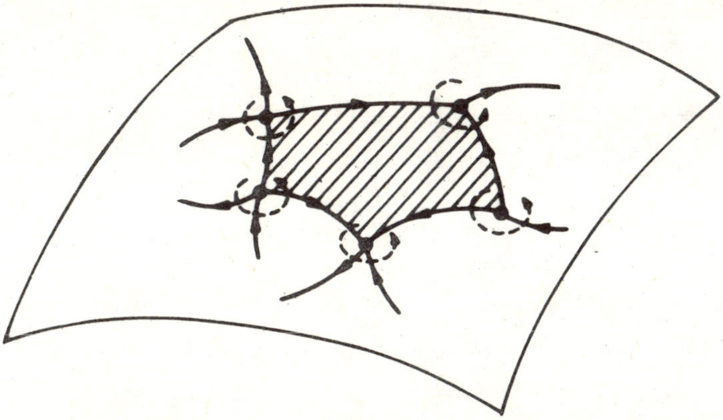

Ce représentant est de genre minimum puisque toutes ses faces sont simplement connexes et donc en nombre maximum.

Ainsi la classe des représentations topologiques d'un graphe qui admettent la même famille de permutations circulaires des arcs incidents à chaque sommet, est caractérisée par la représentation topologique simple qui appartient à cette classe.

Ce théorème très important dû à J. EDMONDS (cf. YOUNGS [13.a]) s'énonce:

> Toute représentation topologique simple d'un graphe sur une surface orientée est caractérisée par la donnée, en chaque sommet, d'une permutation circulaire des arcs incidents à ce sommet.

2.2) Les constellations

Si l'on considère un graphe orienté symétrique (cf. I.1.2) et si l'on convient de représenter les arcs opposés (ceux qui se correspondent dans l'involution) par des arcs de Jordan aussi voisins que l'on veut, alors on peut se restreindre à prendre en chaque sommet des permutations circulaires des arcs sortants (GUSTIN [3]). Ainsi pour une représentation topologique d'un graphe orienté symétrique, chaque arc appartient à une permutation circulaire et une seule: la représentation topologique définit une permutation de tous les arcs dont le restriction à chacun de ses cycles est circulaire sur les arcs sortant d'un sommet.

Soit $G = (A, S, \alpha, s)$ un graphe symétrique défini par l'involution α sur A et l'application "source" qui à chaque arc $a \in A$ fait correspondre le sommet source $s(a) \in S$; toute représentation topologique de G définit donc une permutation σ de A dont l'équivalence de transitivité (σ) est l'équivalence d'application de s. En vertu du théorème d'EDMONDS, toute représentation topologique simple de G est caractérisée par une permutation σ de A dont l'équivalence de transitivité (σ) est l'équivalence d'application de s. Mais on peut identifier S à l'ensemble des classes de transitivité de la permutation σ, et l'application s à la surjection canonique de A sur l'ensembl quotient $A/(\sigma) = S$.

Ainsi à toute représentation topologique simple d'un graphe symétrique $G = (A, S, \alpha, s)$ on peut faire correspondre le triplet (A, α, σ) que nous appelons constellation.

Réciproquement, à toute constellation $C = (A, \alpha, \sigma)$ où α est une involution et σ une permutation de A, on peut faire correspondre un graphe $G = (A, S, \alpha, s)$ où S est l'ensemble quotient $A/(\sigma)$ de A par l'équivalence de transitivité de σ et s la surjection canonique de A sur $A/(\sigma)$; de plus, en vertu du théorème d'EDMONDS, la constellation C définit une représentation topologique simple du graphe G sur une surface orientée.

REMARQUE 1.

Notre construction qui permet d'associer une constellation à une représentation topologique simple d'un graphe symétrique, suppose que deux arcs opposés soient contigus sur la surface:

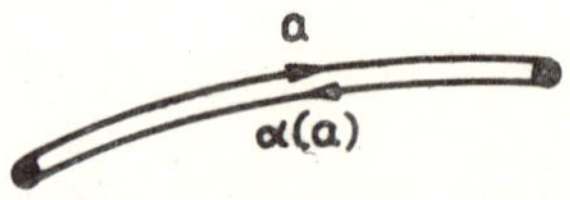

ce qui interdit, par exemple, la figure ci-dessous:

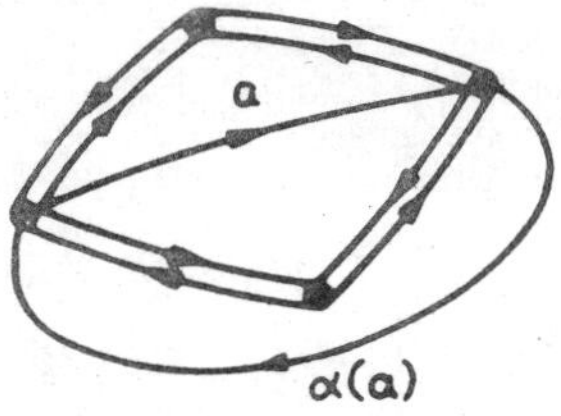

REMARQUE 2.

On convient généralement de représenter un graphe symétrique par un graphe non-orienté, c'est-à-dire de remplacer les arcs opposés par une arête.

Nous proposons, cependant, une autre façon de représenter un graphe symétrique: nous remplaçons un couple d'arcs opposés par deux demi-arêtes, que nous nommerons "brins".

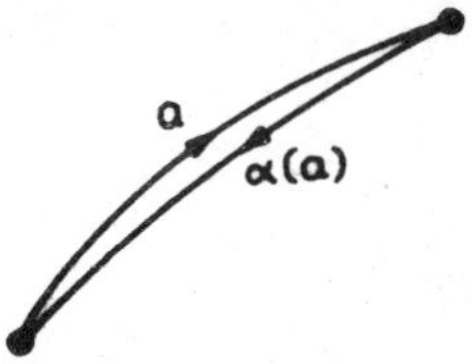

devient alors

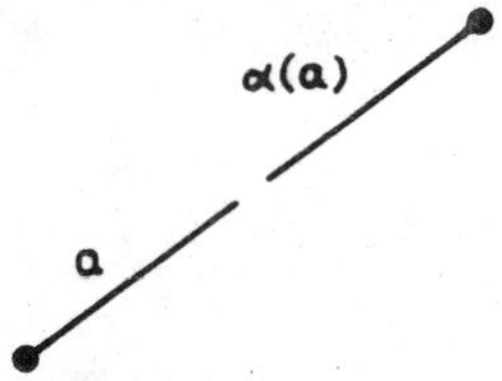

Les boucles qui sont leur propre opposé dans l'involution α, deviennent des "arêtes singulières" réduites à un seul brin:

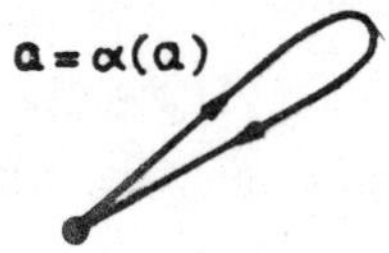

se transforme en

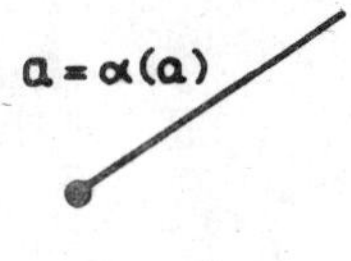

REMARQUE 3.

Il est clair que si le graphe est connexe, le groupe de permutations $[\sigma,\alpha]$ engendré par σ et α est transitif. De façon générale, les classes de transitivité du groupe $[\sigma,\alpha]$ définissent les composantes connexes du graphe associé à la constellation (A,α,σ) nous supposerons cependant, sauf mention explicite du contraire, que le graphe est connexe, c'est-à-dire que le groupe $[\sigma,\alpha]$ est transitif.

2.3) Constellation duale et genre

A toute représentation topologique d'un graphe sur une surface orientée on peut faire correspondre un graphe topologique dual (BERGE [1]):

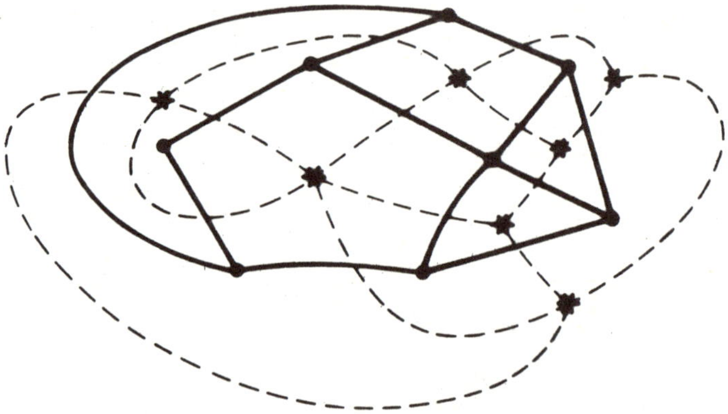

Mais les faces de la représentation topologique simple s'obtiennent aisément sur la constellation $C = (A,\alpha,\sigma)$: il suffit en effet de considérer les orbites de la permutation $\varphi = \sigma\alpha$ (GUSTIN [3], YOUNGS [13. a,b]). Il est clair alors que la constellation associée au graphe topologique dual est $C^* = (A,\alpha,\varphi)$ et que nous avons bien une dualité pusque:

$$\sigma = \varphi\alpha .$$

Nous appellerons donc constellation duale de $C = (A, \alpha, \sigma)$ la constellation $C^* = (A, \alpha, \varphi = \sigma\alpha)$.

Mais une représentation topologique simple d'un graphe étant une décomposition cellulaire d'une surface orientée, la caractéristique d'Euler-Poincaré de la surface (LEFSCHETZ [6], COXETER [2]) est:

$$\chi = N - M + F$$

où N est le nombre de sommets du graphe, M le nombre d'arêtes et F le nombre de faces. Dans le cas d'une surface orientée, le genre g est lié à la caractéristique χ par la relation:

$$\chi = 2 - 2g \qquad \text{ou} \qquad g = 1 - \frac{\chi}{2} .$$

D'où le genre de la surface pour une représentation topologique simple:

$$g = 1 + \frac{M - N - F}{2} .$$

Nous définissons donc le genre d'une constellation par l'entier positif ou nul:

$$g = z(\sigma, \alpha) + \frac{z(\varepsilon) - z(\alpha) - z(\sigma) - z(\sigma\alpha)}{2} ,$$

où $z(\sigma,\alpha)$ désigne le nombre de classes de transitivité du groupe $[\sigma,\alpha]$; $z(\alpha)$, $z(\sigma)$, $z(\sigma\alpha)$ et $z(\varepsilon)$ désignent les nombres de classes de transitivité de $\alpha,\sigma,\sigma\alpha$ et $\varepsilon = Id_A$. En effet, si l'on considère que le graphe associé à la constellation est connexe alors $z(\sigma,\alpha) = 1$; d'autre part, il est clair que:

$$N = z(\sigma), \quad M = z(\varepsilon) - z(\alpha) \quad \text{et} \quad F = z(\sigma\alpha).$$

Pour une démonstration algébrique du fait que le nombre g précédemment défini est un entier non négatif, nous renvoyons le lecteur à une publication que nous avons faite et dans laquelle nous généralisons cette inégalité: JACQUES [4].

Il est clair sur la formule qu'une constellation et sa duale ont même genre, car le groupe $[\sigma,\alpha]$ est identique au groupe $[\sigma\alpha,\alpha]$. Pour une constellation dont le graphe associé n'est pas connexe, le genre est la somme des genres des constellations définies par les classes de transitivité du groupe $[\sigma,\alpha]$.

3. PROPRIETES ALGEBRIQUES DES CONSTELLATIONS ET APPLICATIONS

3.1) Algèbre des constellations

Considérons la structure algébrique définie par les trois opérations unaires $\sigma, \bar{\sigma}$ et α linées par les identités:

$$\begin{cases} \alpha\alpha x = x \\ \sigma\bar{\sigma}x = \bar{\sigma}\sigma x = x \end{cases}$$

Nous appellerons constellations, les algèbres définies par cette structure algébrique (ce sont encore des automates dont l'alphabet d'entrée est l'ensemble $\{\sigma, \bar{\sigma}, \alpha\}$).

Soit C une constellation, il est clair que σ et α sont des permutations de C, et que la donnée de la structure de constellation sur C est équivalente à la donnée d'un homomorphisme du produit libre $Z * Z/2$ dans $\mathfrak{S}_C$.

Nous pouvons donc introduire de la façon habituelle les concepts traditionnels de l'algèbre générale: sous-algèbre, algèbre quotient, homomorphisme, etc ...

Remarques élémentaires:

1) La constellation C est monogène si et seulement si le groupe de

permutations $[\sigma,\alpha]$ engendré par σ et α est transitif sur C. Alors tout élément de C est générateur.

2) Toute constellation est union de constellations monogènes disjointes, ce qui détermine toutes les sous-constellations d'une constellation.

3) Une congruence d'une constellation monogène a pour classes un systeme d'imprimitivité du groupe $[\sigma,\alpha]$ et réciproquement. Les classes d'une congruence ont donc toutes même cardinal.

4) Le groupe d'automorphismes d'une constellations monogène est "fixed-point free" (ou semi-régulier, WIELANDT [12]). On peut aussi remarquer que ce groupe est le centralisateur de $[\sigma,\alpha]$ et utiliser un théorème de KUHN (WIELANDT [12] p. 9). Il s'en suit le résultat suivant qui généralise un théorème de TUTTE & HARARY [9]:

L'ordre du groupe d'automorphisme d'une constellation monogène finie divise le cardinal de la constellation.

Rappelons que le nombre de classes de transitivité du groupe $[\sigma,\alpha]$ est aussi le nombre de composantes connexes du graphe associé à la constellation. En particulier, une constellation est monogène si et seulement si son graphe associé est connexe (aussi parlerons-nous de constellation connexe au lieu de constellation monogène).

3.2) Groupes d'automorphismes des constellations et de leur graphe

Etant donnée une constellation $C = (A,\alpha,\sigma)$ nous avons vu (en 3.1) que le groupe des automorphismes de C est le centralisateur du groupe $[\sigma,\alpha]$ dans $\mathfrak{S}_A$:

$$\mathrm{Aut}\ (A,\alpha,\sigma) = Z[\sigma,\alpha] = Z(\sigma) \cap Z(\alpha).$$

Nous allons chercher quels liens lient ce groupe au groupe des automorphismes du graphe associé à la constellation. Si $G = (A,S,\alpha,s)$ est un graphe symétrique nous avons vu en 1.2 que ses automorphismes sont les couples $(\lambda,\mu) \in \mathfrak{S}_A \times \mathfrak{S}_S$ tels que

$$\begin{cases} \lambda\alpha = \alpha\lambda \\ \mu s = s\lambda \, . \end{cases}$$

Cependant si (λ,μ) et (λ,μ') sont deux automorphismes ayant même première coordonnée, on a alors;

$$\mu s = s\lambda = \mu' s \ ,$$

et comme μ et μ' sont inversibles cela entraine:

$$\mu = \mu' \, .$$

Ainsi le groupe $\mathrm{Aut}G$ des automorphismes d'un graphe $G = (A,S,\alpha,s)$ est isomorphe au groupe $\mathrm{Aut}_A G$ des permutations $\lambda \in \mathfrak{S}_A$ telles que:

$$\begin{cases} \lambda\alpha = \alpha\lambda \\ x,y \in A \, , \quad sx = sy \Longrightarrow s\lambda x = s\lambda y \, . \end{cases}$$

Nous considérons maintenant les automorphismes du graphe G comme opérant sur A, c'est-à-dire appartenant au groupe $\mathrm{Aut}_A G$.

Considérons alors la constellation $C = (A,\alpha,\sigma)$, les permutations $\lambda \in \mathrm{Aut}_A G$ sont telles que:

$$\begin{cases} \lambda\alpha = \alpha\lambda \\ x,y \in A : \ x \equiv y \pmod{\sigma} \Longrightarrow \lambda x = \lambda y \pmod{\sigma} \, ; \end{cases}$$

mais cette deuxième condition traduit que :

$$\lambda [\sigma] \lambda^{-1} = [\sigma]$$

où $[\sigma]$ note le groupe engendré par σ.

Le groupe $\mathrm{Aut}_A G$ des automorphismes du graphe G associé à la constellation $C = (A,\alpha,\sigma)$ est l'intersection du centralisateur de α et du normalisateur du groupe $[\sigma]$ dans $\mathfrak{S}_A$:

$$\mathrm{Aut}_A G = Z(\alpha) \cap N[\sigma] .$$

Le centralisateur d'un groupe étant un sous-groupe distingué de son normalisateur (cf. KUROSH [5] vol. I, p. 84), le groupe d'automorphismes d'une constellation C est un sousgroupe distingué du groupe d'automorphismes de son graphe associé G :

$$\mathrm{Aut}\, C = Z(\alpha) \cap Z(\sigma) \subset Z(\alpha) \cap N[\sigma] = \mathrm{Aut}_A G .$$

3.3) Théorème sur l'ordre du groupe d'automorphismes d'un graphe

Etant donné un graphe $G = (A,S,\alpha,s)$, toutes ses représentations topologiques simples sont obtenues par les constellations $C = (A,\alpha,\sigma)$ telles que σ ait pour équivalence de transitivité l'équivalence d'application de s.

Considérons une telle constellation C, ses conjuguées par les permutations $\omega \in \mathrm{Aut}_A G$ sont:

$$\omega^{-1} C \omega = (A,\alpha,\omega^{-1}\sigma\omega) .$$

Le nombre de ces conjuguées est égal à l'ordre du groupe-quotient $\mathrm{Aut}_A G/\mathrm{Aut}\, C$. D'autre part, toutes ces constellations conjuguées ont le graphe G pour graphe associé puisque $\omega \in \mathrm{Aut}_A G$, et de plus elles ont même genre (puisque conjuguées par une permutation). Par conséquent, l'ensemble des constellations de genre g associées à un même graphe G se décompose en classes de cardinal:

$$|\mathrm{Aut}_A G/\mathrm{Aut}\, C| \quad \text{où } C \text{ est un représentant}$$

de chaque classe.

Soit N_g le nombre de représentations topologiques simples du graphe G sur une surface orientée de genre g, et soit γ l'ordre du groupe $\mathrm{Aut}_A G$ des automorphismes du graphe G, alors:

$$N_g = \sum_{i \in I} |\mathrm{Aut}_A G / \mathrm{Aut}\, C_i| = \gamma \sum_{i \in I} \frac{1}{|\mathrm{Aut}\, C_i|} .$$

Mais nous avons établi en 3.1 que l'ordre du groupe d'automorphismes d'une constellation divise le cardinal de A:

$$|A| = k_i |\mathrm{Aut}\, C_i| .$$

D'où le nombre N_g :

$$N_g = \frac{\gamma}{|A|} \sum_{i \in I} k_i .$$

THEOREME:

L'ordre γ du groupe d'automorphismes d'un graphe divise le produit du cardinal de l'ensemble A de ses arcs par le nombre N_g de ses représentations topologiques simples sur une surface orientée de genre g:

$$\gamma \mid N_g \cdot \mathrm{Card}\, A$$

Corollaire: γ divise le produit du cardinal de A par le PGCD des nombres N_g, pour tout $g \geq 0$.

Application aux graphes planaires

Pour les graphes planaires, le nombre N_0 a été déterminé par Mc LANE [7] qui a donné son expression sous la forme d'un produit de factorielles entièrement déterminées par la décomposition du graphe en sous-graphes maximaux triplement connexes.

Pour les graphes planaires triplement connexes, il résulte d'un théorème classique de WHITNEY [11] qu'ils n'admettent que deux représentations planaires se déduisant l'une de l'autre par changement de l'orientation de la sphère (ou du plan).

Nous retrouvons alors le résultat de TUTTE & HARARY [9] (cf. aussi WEINBERG [10]):

L'ordre du groupe d'automorphismes d'un graphe planaire triplement connexe divise $4M$ (M est le nombre d'arêtes, ou encore $2M$ est le nombre d'arcs).

BIBLIOGRAPHIE

[1] BERGE, C. Théorie des graphes, Dunod, Paris, 1963.

[2] COXETER, H.S.M. Regular polytopes, Macmillan, New-York, 1963.

[3] GUSTIN, W. Orientable embeddings of Cayley graphs, Bull. A. M. S. v. 69 (1963) p. 272/5.

[4] JACQUES, A. Sur le genre d'une paire de substitutions, C.R. Acad. Sc. Paris, t. 267, p. 625/7, 28 Oct. 1968.

[5] KUROSH, A.G. Theory of groups, Chelsea, New-York, 1955.

[6] LEFSCHETZ, S. a) Introduction to Topology, Princeton Univ. Press, 1949.
b) Topology, Chelsea, New-York, 1956.

[7] MAC LANE, S. a)A combinatorial condition for planar graphs, Fund. Math. v. 28 (1937), p. 22/32.
b) A structural characterization of planar combinatorial graphs, Duke Math. J. v. 3 (1937), p. 340/472.

[8] RINGEL, G. a) Färbungsprobleme auf Flächen und Graphen, Veb. deut. Verlag der Wissenschaften, Berlin, 1959.
b) Über das Problem der Nachbargebiete auf orientierbaren Flächen, Abh. Math. Sem. Univ. Hamburg v. 25, p. 105/27, 1961.

[9] TUTTE, W.T. & HARARY, F. On the order of the group of a planar map, J. Combin. Th. v. 1, N° 3 (1966), p. 394/5.

[10] WEINBERG, L. On the maximum order of the automorphism group of a planar triply connected graph, J. Siam Appl. Math. v. 14 N° 4 (1966), p. 729/38.

[11] WHITNEY, H. Congruent graphs and the connectivity of graphs, Amer. J. Math. v. 54 (1932), p. 150/68.

[12] WIELANDT, H. Finite permutation groups, Acad. Press, New-York, 1964.

[13] YOUNGS, J.W.T. a) Minimal imbeddings and the genus of a graph, J. Math. Mech. v. 12 (1963), p. 303/15.

b) The Heawood map coloring conjecture, chap. 12 in Graph Theory and Theoretical Physics, Acad. Press, New-York, 1967.

On subgraphs without cycles in a tournament

by

H. A. Jung
Berlin - West

A tournament T_n is an oriented graph with n vertices in which any two different vertices are joined by exactly one arc. A set E of arcs in T_n is consistent if the subgraph of T_n defined by E contains no oriented cycle. $c(T_n)$ denotes the maximal cardinality of a consistent set in T_n.

P. ERDŐS and J.W.MOON* gave the following bounds for the function $f(n) = \min_{T_n} c(T_n)$:

$$1/4\,(n^2-1) \leqq f(n) \leqq 1/4\,(n^2-1) + (1/2 + o(1))\,(n^3 \log n)^{1/2} .$$

In the sequel the lower bound is improved.

Theorem: $f(n) \geqq 1/4\,(n^2-1) + 1/4\,([\log n]-1)\,2^{[\log n]}$,

hence $f(n) \geqq 1/4\,(n^2-1) + 1/8\,(\log n - 2)\,n$.

The idea of the argument is contained in the proof of the following lemma.

Lemma: Let $n \equiv o(2)$, $n = n_1 + n_2$ and $n_1 \leqq n_2$. Then

$$f(n) \geqq f(n_1) + f(n_2) + n_1 n_2/2 + n_1/2 .$$

Proof: Let V be the vertex set of the tournament T_n; for $v \in V$ let $d_i(v)$ and $d_o(v)$ denote the in- and outdegree of v in T_n respecitvely. From $d_i(v) + d_o(v) = n-1$ we deduce $d_i(v) \geqq n/2$ or $d_o(v) \geqq n/2$. Hence there exist $n/2$ vertices v with $d_i(v) \geqq n/2$, or there exist $n/2$ vertices v with $d_o(v) \geqq n/2$. Assuming the first case, we choose a set $V' \subseteq V$ with $|V'| = n_1$ and $d_i(v) \geqq n/2$ for all $v \in V'$. For the number t of edges issuing in $V-V'$ and terminating in V' we get

$$t = \sum_{v \in V'} d_i(v) - \binom{n_1}{2} \geqq n_1 \frac{n}{2} \binom{n_1}{2} .$$

The subtournaments of T_n defined by V' and $V-V'$ contain consistent sets of cardinality $f(n_1)$ and $f(n_2)$ respecitvely.

Hence $c(T_n) \geqq f(n_1) + f(n_2) + t$ which proves the lemma.

By induction on m using the lemma in the case $n_1 = n_2$ it is easily seen that

$$f(2^m) \geqq 2^{2m-2} + 1/4(m-1)2^m . \tag{1}$$

If more generally $n = \sum_{\lambda=1}^{\ell} 2^{m_\lambda}$ with $0 < m_1 < m_2 < \dots < m_\ell$ we get by induction on ℓ, using (1) and the lemma in the case $n_2 = 2^{m_\ell}$:

$$f(n) \geqq 1/4 n^2 + \sum_{\lambda=1}^{\ell} 1/4(2\ell - 2\lambda + m_\lambda - 1) 2^{m_\lambda} . \tag{2}$$

From (2) and $m_\ell = [\log n]$ we deduce in the case $n \equiv o(2)$ that $f(n) \geqq 1/4 n^2 + 1/4([\log n]-1) 2^{[\log n]}$. For odd n we use the inequality $f(n) \geqq f(n-1) + 1/2(n-1)$ which was proved in the cited paper of P.ERDŐS and

J. W. MOON. Since for odd n $[\log n] = [\log(n-1)]$ the assertion of the theorem follows from the last two inequalities.

Note that for $n = 2,4,6,8$ in (2) equality holds.

A covering problem

by

H. J. L. Kamps and J. H. van Lint
Eindhoven, The Netherlands

1. INTRODUCTION

Let R_k^n denote the set of vectors with n components chosen from the ring of integers mod k. We define the Hamming distance $d(\underline{x},\underline{y})$ of $\underline{x} \in R_k^n$, $\underline{y} \in R_k^n$ to be the number of places in which the two vectors differ. With this definition R_k^n is a metric space. If $\underline{x} \in R_k^n$ and e is a positive integer, we define the sphere $B(\underline{x},e)$ by

$$B(\underline{x},e) := \{\underline{y} \in R_k^n : d(\underline{x},\underline{y}) \le e\} .$$

There has been an increasing interest in the past few years in covering problems of the following type.

Let f be a function defined on R_k^n and $C = C(i,e)$ a condition on the values of f in the spheres $B(\underline{x},e)$ where $f(\underline{x}) = i$. One can ask for constructions of functions satisfying the condition C or for the maximal (minimal) number of points where such a function has certain prescribed values; for the sum of the values of f etc.

Examples

(Ex. 1) If $n = 2$, $k = 8$ we can consider R_k^n as a chessboard. The points attacked (or covered) by a rook placed at $\underline{x}$ form the sphere $B(\underline{x}, 1)$. For this reason such a sphere is called a rook domain. Let $f(\underline{x}) = 1$ if a rook is placed at $\underline{x}$ and $f(\underline{x}) = 0$ otherwise. The condition that every point is attacked (or covered) by at least one rook can be written as:

$\sum_{\underline{y} \in B(\underline{x},1)} f(\underline{y}) \geq 1$ for every $\underline{x} \in R_k^n$. The minimal number of rooks necessary for this covering is $\sum_{\underline{x} \in R_k^n} f(x)$.

This number is generally denoted by $\sigma(n,k)$. A special case of this problem is the main subject of this paper.

(Ex. 2) In [1] the problem is considered of coloring the points of R_k^n in such a way that in each point all the colors are "seen" where "$\underline{x}$ sees $\underline{y}$" means that $d(\underline{x},\underline{y}) \leq 1$, i.e., $\underline{y}$ is in the rook domain of $\underline{x}$. The maximal number of colors that can be used is called $w(n,k)$. In our formulation we number the colors and define $f(\underline{x}) = i$ if $\underline{x}$ has color number i. The condition C is that for every i and every $\underline{x} \in R_k^n$ there is a $\underline{y} \in R_k^n$ with $d(\underline{x},\underline{y}) \leq 1$ and $f(\underline{y}) = i$. The problem is to determine the maximum of f.

(Ex. 3) In (Ex. 2) one can simplify the problem by leaving certain points uncolored (and then defining $f(\underline{x}) = 0$) and by replacing the condition by: every uncolored point (= blank) must see every color, i.e., if $f(\underline{x}) = 0$ then for every i there is a $\underline{y} \in R_k^n$ with $d(\underline{x},\underline{y}) = 1$ and $f(\underline{y}) = i$. We shall use a construction of such a coloring in our treatment of (Ex. 1).

(Ex. 4) An interesting example of a covering problem of our type is the following conjecture (cf. [1]): If $f(\underline{x})$ is a non-negative integer for every $\underline{x} \in R_k^n$ and $f(\underline{x}) = 0$ implies $\sum_{\underline{y} \in B(\underline{x},1)} f(y) \geq k$ then

$\sum_{\underline{x} \in R_k^n} f(\underline{x}) \geq k^n/n$. It is known that this statement is true for $n = 2$ ([1]) and also if $f(\underline{x}) = 0$ or 1 for every $\underline{x}$ ([2]).

2. THE FUNCTION $\sigma(n,k)$

Determining the value of $\sigma(n,k)$ is generally a very hard combinatorial problem. A number of theorems have been proved concerning the cases $n=2$, $n=3$ and $k=p^{\nu}$ where p is a prime (cf. [6], [7], [1]). The results are

(2.1) $\sigma(2,k) = k.$

(2.2) $\sigma(3,k) = [\frac{1}{2}(k^2+1)].$

(2.3) $\sigma(n,k) = \dfrac{k^n}{1+n(k-1)}$ if this is an integer and k a power of a prime.

Other values of $\sigma(n,k)$ have been computed (cf. [3], [4], [5]). These can be found in the following table.

Known values of $\sigma(n,k)$

k \ n	2	3	4	5	6	7	8	9	10	11	12	13
2	2	2	4	7	12	16	32					
3	3	5	9	3^3								3^{10}
4	4	8	24	4^3								
5	5	13			5^4							
6	6	18	72									
7	7	25					7^6					

It is known that $\sigma(7,6) > 6^5$ (cf. [1]). In [4] the bound $\sigma(9,2) \geq 54$ is given. We have proved (unpublished) that $46 \leq \sigma(4,5) \leq 57$ but better bounds may be known. A case of special interest is the football-pool-problem. In a football-pool one wishes to forecast the outcome (win, lose or draw) of n matches.

The problem is to make a number of forecasts in such a way that, no matter what the outcome of the matches, at least one of the forecasts will have at most one mistake. This means that one has to determine $\sigma(n,3)$ and to construct the corresponding covering set. The proof of $\sigma(5,3) = 3^3$ in [5] took 9 pages. For larger values of n the problem looks hopeless. Especially interesting is of course to find $n < 13$ where $\sigma(n,3)$ is substantially smaller than the trivial bound 3^{n-2}. A small contribution in this direction is given by the theorem below.

3. THE CASE $n = 9$, $k = 3$

We shall prove

THEOREM 1: $\sigma(9,3) \leq 2.3^6$

Remark: This is $2/3$ of the trivial bound and although we cannot show this to be optimal, the method used may be of some interest.

Before going into the proof of theorem 1 we consider the following case of (Ex. 3). Let the points of R_3^5 be colored with nine colors, some being left blank. Let every blank see every color. Maximize the number of blank points. We do not solve this problem but give a construction of a coloring where $1/3$ of the points remains blank. (Maybe this is maximal.) Define

$$H := \begin{pmatrix} 0 & 0 & 1 & 1 & 1 \\ 0 & 1 & 0 & 1 & 2 \\ 1 & 1 & 1 & 1 & 1 \end{pmatrix}$$

and consider every $\underline{x} \in R_3^5$ as a column vector with elements in GF(3). Then if

$$H\underline{x} = \begin{pmatrix} \alpha_1 \\ \alpha_2 \\ \alpha_3 \end{pmatrix}$$ we define the coloring as follows:

(i) If $\alpha_3 = 0$ then $\underline{x}$ is blank ($f(\underline{x}) = 0$),

(ii) if $\alpha_3 \neq 0$ then there are 9 possible values of $\binom{\alpha_1}{\alpha_2}$, each determining a color (e.g. we could take $f(\underline{x}) = 1 + \alpha_1 + 3\alpha_2$ where α_1 and α_2 are considered as reals).

If $\underline{x}$ is blank, i.e., $x_1 + x_2 + x_3 + x_4 + x_5 = \alpha_3 = 0$, then the other 10 points in the rook domain of $\underline{x}$ are colored. Since, for every choice of $\binom{\beta_1}{\beta_2}$ there is a column of H starting with $\binom{\beta_1}{\beta_2}$ or $-\binom{\beta_1}{\beta_2}$; each color occurs in this rook domain and the color corresponding to $\binom{\alpha_1}{\alpha_2}$ occurs twice.

We now construct a covering set for R_3^9. First remark that $\sigma(4,3) = 9$ and that the corresponding covering set is perfect (it is the perfect simple-error-correcting code of word length 4 over $GF(3)$). We can describe the set as follows: $\underline{y} \in R_3^4$ is in the covering set if $\begin{pmatrix} 0 & 1 & 1 & 1 \\ 1 & 0 & 1 & 2 \end{pmatrix} \underline{y} = \underline{0}$. A translate of this covering set is obtained by adding a vector from the rook domain of $\underline{0}$ to the vectors of the covering set. Each of these 9 translates is a covering set and their union is R_3^4. We number the translated covering sets from 1 to 9. Next, consider R_3^9 as a direct sum of R_3^5 and R_3^4. If $\underline{x} \in R_3^5$ has color i and $\underline{y} \in R_3^4$ is in the translated covering set with number i, then $\underline{x} \oplus \underline{y}$ is taken as element of the covering set of R_3^9. Suppose $\underline{z} = \underline{x} \oplus \underline{y} \in R_3^9$ (with $\underline{x} \in R_3^5$, $\underline{y} \in R_3^4$). If $\underline{x}$ has color i, then a change of at most 1 component of $\underline{y}$ leads to a point in the covering set, i.e. $\underline{z}$ is covered. If, on the other hand, $\underline{x}$ is blank and $\underline{y}$ is in the covering set with number i, a change of one component leads to a point $\underline{x}' \oplus \underline{y}$ where $\underline{x}'$ has color i and therefore $\underline{x}' \oplus \underline{y}$ is in the covering set and covers $\underline{z}$. This proves theorem 1.

This proof has the advantage that it shows how the construction works and how to generalize the idea. We now give a simpler direct proof.

Proof of theorem 1: Define, over $GF(3)$,

$$A := \begin{pmatrix} 0 & 0 & 0 & 1 & 1 & 1 & 2 & 2 & 2 \\ 0 & 1 & 2 & 0 & 1 & 2 & 0 & 1 & 2 \end{pmatrix}$$

and $\underline{e}^T := (1\ 1\ 1\quad 1\ 1\ 1\quad 1\ 1\ 1)$.

Then $S := \{\underline{x} \in R_3^9 : A\underline{x} = \underline{0}\ \&\ (\underline{e},\underline{x}) \neq 0\}$ is a covering set for R_3^9 with $3^7 - 3^6 = 2\cdot 3^6$ points. This is so because if $\underline{x} \notin S$ and $A\underline{x} = \underline{0}$ a suitable change of the first component of $\underline{x}$ yields a point of S, if $A\underline{x} \neq 0$ both $A\underline{x}$ and $-A\underline{x}$ occur as columns of A and then at least one of the corresponding components can be changed to yield a point of S.

Remarks

(i) It is clear that for $\sigma(n,3)$ this idea does not work for $n < 9$.

(ii) If one could find a coloring of R_3^5, as was done above, with more than 81 blanks, the same subsequent construction would give a lower bound for $\sigma(9,3)$. Since trivially $\sigma(9,3) \geq 1036$ it is a priori clear that at most 127 blanks can occur (but this is a poor estimate).

(iii) From (2.3) for $n = p+1$, $k = p$ (p a prime) it follows that $\sigma(p+1,p) = p^{p-1}$ and therefore $\sigma(n,p) \leq p^{n-2}$ for $n > p$.

An immediate generalization of our method gives:

THEOREM 2: If p is a prime then

$$\sigma(2p+3,p) \leq (p-1)\,p^{2p}.$$

REFERENCES

[1] S.W. GOLOMB and E.C. POSNER, Rook domains, latin squares, affine planes and error-distributing codes, IEEE Trans. on Inf. Theory 10 (1964), 196-208.

[2] A.W. HALES, Cubes with zeros and ones, JPL Space Programs Summary 37-16 (1962), 35-36.

[3] J.G. KALBFLEISCH and R.G. STANTON, Itersection inequalities for the covering problem, to be published.

[4] J.G. KALBFLEISCH and R.G. STANTON, Covering problems for dichotomized matchings, Aequat. Math. 1 (1968), 103-112.

[5] H.J.L. KAMPS and J.H. VAN LINT, The football pool problem for 5 matches, J.C.T. 3 (1967), 315-325.

[6] S.K. ZAREMBA, A covering theorem for Abelian groups, J. London Math. Soc. 26 (1950), 70-71.

[7] S.K. ZAREMBA, Covering problems concerning Abelian groups, J. London Math. Soc. 27 (1952), 242-246.

How many sums of vectors can lie in a circle of radius $\sqrt{2}$

by

G. O. H. Katona
Budapest, Hungary

INTRODUCTION

Let $a_1, \dots, a_n$ be real numbers with the property $|a_i| \geq 1$ $(1 \leq i \leq n)$. Erdős [1] asked, what is the maximum number of sums $\sum_{i=1}^{n} \varepsilon_i a_i$ which can lie in an open interval of length h, where $\varepsilon_i = 0$ or 1. He proved, that this number is $\leq$ sum of the largest binomial coefficients of order n. The example $a_i = 1$ $(1 \leq i \leq n)$ shows that this estimation is the best possible. Kleitman [2] and Katona [3] independently proved the same for two dimensions and for $h = 1$:

If $a_1, \dots, a_n$ are two-dimensional vectors such that $|a_i| \geq 1$, then at most $\binom{n}{[n/2]}$ sums $\sum_{i=1}^{n} \varepsilon_i a_i$ can lie in an open circle with unit diameter. Now we consider the case of diameter $\sqrt{2}$.

THEOREM 1. If $a_1, \dots, a_n$ are two-dimensional vectors with the property $|a_i| \geq 1$ $(1 \leq i \leq n)$, then at most $\binom{n}{[n/2]} + \binom{n}{[n/2]+1}$ sums $\sum_{i=1}^{n} \varepsilon_i a_i$ can lie in a circle of diameter $\sqrt{2}$, where $\varepsilon_i = 0$ or 1.

In the proof we use a Sperner type theorem, which is formulated in a more general language. The method of the proof is what we used in [4] and [5].

DEFINITIONS AND THEOREM 2

Let G be a partially ordered set with rank function. The i-th level is the set of elements $g \in G$ with rank $r(g) = i$. A chain of length h is a sequence $g_1, \dots, g_h \in G$, where $r(g_{i+1}) = r(g_i) + 1 \quad (1 \le i \le h)$.

A chain is symmetrical if $r(g_1) + r(g_h) = n$, where

$$n = \max_{g \in G} r(g).$$

We say that a partially ordered set is a symmetrical chain set if we can split G into disjoint symmetrical chains. (It is defined in [4] under a different name.) It is easy to see, that the partially ordered set of the subsets of a finite set S is a partially ordered set with rank function $r(A) = |A|$ ($|A|$ is the number of elements of A.)

If G and H are partially ordered sets, then the direct sum G+H is the set of ordered pairs (g,h), $g \in G$, $h \in H$, with the ordering $(g_1,h_1) < (g_2,h_2)$ iff $g_1 \le g_2$ and $h_1 \le g_2$, but $(g_1,h_1) \neq (g_2,h_2)$. If G and H has rank function r and s, respectively, then we can define a rank function on G+H as follows:

$$t((g,h)) = r(g) + s(h).$$

It is easy to see if G is the partially ordered set of the subsets of a set S_1 and H is the same of a set S_2 $(S_1 \cap S_2 = \emptyset)$, then G+H is the partially ordered set of the subsets of $S_1 \cup S_2$.

Now we can formulate the following theorem:

THEOREM 2. Let G and H be symmetrical chain sets. If we have a set $(p_1,q_1), \dots, (p_m,q_m)$ of the elements of G+H, satisfying the following conditions:

no two different ones of them satisfy the conditions

$$p_i = p_j ,\quad q_i < q_j ,\quad s(q_i) < s(q_j) - 1$$

(C$_1$) or

$$p_i < p_j ,\quad q_i = q_j ,\quad r(p_i) < r(p_j) - 1 ,$$

no four different ones of them satisfy the conditions

(C$_2$)
$$p_i = p_j ,\quad q_\ell = q_i ,\quad p_k = p_\ell ,\quad q_k = q_j ,$$

$$q_i < q_j ,\quad p_\ell > p_i ,\quad q_k > q_\ell ,\quad p_k > p_j ,$$

then

$$m \le M_{[\frac{n}{2}]} + M_{[\frac{n}{2}]+1} ,$$

where M_i denotes the number of elements of the i-th level of $G+H$ and $n = \max_{(g,h)\in G+H} t((g,h))$. The estimation is the best possible.

PROOFS

The PROOF OF THEOREM 2 follows the ideas of the proof of the theorem in [4] and of Theorems 2, 3 in [5].

By the definition of the symmetrical chain sets, G and H are divisible into disjoint symmetrical chains. Denote by G' and H' the partially ordered sets which have ordering relations only along these chains, that is $g_1 < g_2$ can hold only if g_1 and g_2 lie on the same chain. Thus, the set of relations in $G'(H')$ is a part of that in $G(H)$. It follows that the set of relations in $G'+H'$ is a part of that in $G+H$. So, it is sufficient to prove the statement of the theorem for $G'+H'$ instead of $G+H$. However, the direct sum of two chains $g_0,\dots,g_a$ and $h_0,\dots,h_b$ is a rectangular lattice of pairs (g_i,h_j), where $(g_i,h_j) < (g_\ell,h_k)$ iff $i \le \ell, j \le k$ but $(i,j) \ne (\ell,k)$.

We will prove in the Lemma, that the maximum number of the elements of such a rectangular, under conditions (C$_1$) and (C$_2$) is the number of elements of the two maximal levels, that is the number of pairs (g_i, h_j) with

(1) $$i+j = \left[\frac{a+b}{2}\right]$$

and

(2) $$i+j = \left[\frac{a+b}{2}\right]+1.$$

However, by the symmetricity of the chains

(3) $$n_1 = r(g_0) + r(g_a) = 2r(g_0) + a$$
$$r(g_0) = \frac{n_1 - a}{2}$$

and

(4) $$s(h_0) = \frac{n_2 - b}{2}$$

follow, where $n_1 = \max_{g \in G} r(g)$ and $n_2 = \max_{h \in H} s(h)$.

Thus, in case (1), using (3) and (4) we get

$$t((g_i, h_j)) = r(g_0) + s(h_0) + i + j =$$
$$= \frac{n_1 - a}{2} + \frac{n_2 - b}{2} + \left[\frac{a+b}{2}\right] = \left[\frac{n_1 + n_2}{2}\right]$$

Similarly, in the case (2)

$$t((g_i, h_j)) = \left[\frac{n}{2}\right] + 1$$

holds. So, if we choose the elements in a given maximal way from every rectangle, then we obtain elements of the $\left[\frac{n}{2}\right]$-th and $\left[\frac{n}{2}\right]+1$-th levels. It is easy to see, that we obtain every element of $M_{\left[\frac{n}{2}\right]}$ and $M_{\left[\frac{n}{2}\right]+1}$ in this way. This completes the proof.

LEMMA. Let R be the set of pairs (i, j) $(1 \le i \le a,$ $1 \le j \le b,$ a, b, i, j integers), and $(p_1, q_1), \dots, (p_m, q_m)$ a subset of it, such that

no two different ones of them satisfy the conditions

$$p_i = p_j, \quad q_i < q_j - 1$$

(C_1') or

$$p_i < p_j - 1, \quad q_i = q_j ;$$

no four different ones of them satisfy the conditions

(C_2') $$p_i = p_j, \quad q_\ell = q_i, \quad p_k = p_\ell, \quad q_k = q_j$$

then the maximal m is given by the set of pairs satisfying

$$i+j = \left[\frac{a+b}{2}\right] \quad \text{or} \quad i+j = \left[\frac{a+b}{2}\right] + 1 .$$

PROOF. If $a \neq b$, assume $a > b$. By (C_1') at most two (p_i, q_i) can lie in every row (a row is the subset of elements in R with fixed second coordinate), thus the maximum is at most $2b$, however, it is easy to see that the maximal set given in the Lemma has exactly $2b$ element.

But if $a = b$, the given maximal set has $2a-1$ elements. We have to prove that we can not have $2a$ elements. We prove it in an indirect way. Let $(p_1, 0)$ and $(p_2, 0)$ be the elements chosen of the first row. We have two elements in the p_1-th column. By (C_1'), it must be $(p_1, 1)$. Similarly, we get $(p_2, 1)$, too. However, $(p_1, 0), (p_2, 0), (p_1, 1), (p_2, 1)$ form a configuration excluded in (C_2(in contradiction by our assumption. The proof is completed.

Let us return now to the PROOF OF THEOREM 1.

It is easy to see that we can reduce the problem to the case the first coordinates of the vectors are non-negative (transformating to $\varepsilon_i = \pm 1$, multiplying some a_i's by -1, and retransformating to $\varepsilon_i = 0,1$). Let S_1 and S_2 be the set of vectors a_i with nonnegative and negative second coordinates, respectively. We shall use Theorem 2 for G = partially ordered set of subsets of S_1 and H = partially ordered set of subsets of S_2. Let us fix an open circle of diameter $\sqrt{2}$ and consider the sums lying in the circle. We may correspond with each such sum a subset of $S_1 \cup S_2$, the subset of a_i's which have coefficient 1. We have only to verify, that the family of these subsets satisfies (C_1) and (C_2).

Indeed, if two different subsets $(p_i, q_i) = P_i \cup Q_i$, $(p_j, q_j) = P_j \cup Q_j$ of $S_1 \cup S_2$ satisfy e.g.

$$P_i = P_j, \quad Q_i \subset Q_j, \quad |Q_i| < |Q_j| - 1, \tag{5}$$

then for the corresponding sums

$$\begin{aligned} \left| \sum_{u=1}^{n} \varepsilon_u a_u - \sum_{u=1}^{n} \varepsilon'_u a_u \right| &= \left| \sum_{u \in P_j \cup Q_j} a_u - \sum_{u \in P_i \cup Q_i} a_u \right| = \\ &= \left| \sum_{u \in Q_j - Q_i} a_u \right| \end{aligned} \tag{6}$$

holds. The members of the sum are vectors with nonnegative first and negative second coordinates and with absolute value ≥ 1. The number of members is at least 2 by (5). It is easy to see, that the sum of such vectors has absolute value $\geq \sqrt{2}$. Thus the difference (6) is at least $\sqrt{2}$ which contradicts our assumption that both sums lie in the same open circle of diameter $\sqrt{2}$. The proof of holding of (C_1) is completed.

In order to prove the same for (C_2) we have to show that in the case

$$\begin{aligned} &P_i = P_j, \quad Q_\ell = Q_i \quad P_k = P_\ell \quad Q_k = Q_j \\ &Q_i \subset Q_j, \quad P_\ell \supset P_i, \quad Q_k \supset Q_\ell, \quad P_k \supset P_j \qquad \text{(proper subsets)} \end{aligned} \tag{7}$$

at least two of the sums

$$\sum_{u \in P_i \cup Q_i} a_u, \quad \sum_{u \in P_j \cup Q_j} a_u, \quad \sum_{u \in P_\ell \cup Q_\ell} a_u, \quad \sum_{u \in P_k \cup Q_k} a_u \tag{8}$$

differ with at least $\sqrt{2}$. The difference of the 4-th and 1-st sum is

$$\sum_{u \in (P_k - P_i) \cup (Q_k - Q_i)} a_u = \sum_{u \in P_\ell - P_i} a_u + \sum_{u \in Q_j - Q_i} a_u = v_2 + v_1 .$$

The difference of the 3-rd and 2-nd sum is

$$\sum_{u\in P_\ell - P_j} a_u - \sum_{u\in Q_j - Q_\ell} a_u = \sum_{u\in P_\ell - P_i} a_u - \sum_{u\in Q_j - Q_i} a_u = v_2 - v_1 .$$

Here v_1 (and v_2) is a (nonvoid) sum of vectors lying in the same quadrant, with absolute values ≥ 1. Thus $|v_1| \geq 1$, $|v_2| \geq 1$.

If the angle of v_1 and v_2 is $\leq \frac{\pi}{4}$, then $|v_1 + v_2| \geq \sqrt{2}$, conversely, if the angle $\geq \frac{\pi}{4}$ then $|v_2 - v_1| \geq \sqrt{2}$. In other words there are always two sums in (8) with absolute difference $\geq \sqrt{2}$. They cannot lie in the same circle what contradicts our supposition.

The conditions (C_1) and (C_2) are really satisfied in this case. We may apply Theorem 2, thus, the two middle levels of the partially ordered set of subsets of $S_1 \cup S_2$ give an optimal set. The two middle levels consists of the subsets with elements $\left[\frac{n}{2}\right]$ or $\left[\frac{n}{2}\right] + 1$. The number of such subsets is $\binom{n}{[n/2]} + \binom{n}{[n/2]+1}$.

The proof is completed.

CONCLUDING REMARKS

The estimation of Theorem 1 is the best possible in the sense that for $a_i = 1$ $(1 \leq i \leq n)$ the maximum is attained.

It is also true that the theorem does not hold for a larger number instead of $\sqrt{2}$, since the vectors $(0,1)$ and $(1,0)$ would provide a counterexample. However we have the following

CONJECTURE. Theorem 1 holds with 2 instead of $\sqrt{2}$ in the case $n > 2$.

REFERENCES

[1] ERDŐS, P.: On a Lemma of Littlewood and Offord, Bull. Amer. Math Soc. 51 (1945), 898-902.

[2] KLEITMAN, D.: On a Lemma of Littlewood and Offord on the distribution of certain sums, Math. Z. 90 (1965) 251-259.

[3] KATONA, G.: On a conjecture of Erdős and a stronger form of Sperner's theorem, Studia Sci. Math. Hungar. 1 (1966) 59-63.

[4] KATONA, G.O.H.: A generalization of some generalizations of Sperner' theorem, J. Combinatorial Theory (to appear).

[5] KATONA, G.O.H.: Families of subsets having no subset containing an other one with small difference, Nieuw Arch. Wiskunde (to appear

Matching problems

by

G. O. H. Katona and D. O. H. Szász

Budapest, Hungary

We have two types of lattice-figures: a "big" and a "small" one. We should like to cover the "big" figure with disjoint replicas of the "small" one. An old problem of this type is the well-known chessboard problem: is it possible to cover with 31 dominoes a chessboard deprived of two diagonal fields?

The following interesting matching problem of N.G. de Bruijn was published in 1962 in Matematikai Lapok [1]:

An n-dimensional rectangular parallelotop is to be decomposed into such congruent rectangular parallelotops, the edge-lengths of which are the given natural numbers $a_1, a_2, \dots, a_n$. Under which conditions can we say that such a decomposition exists if and only if there exists a decomposition with parallel parallelotops (i.e. the parallel edges of the parallelotops involved in the decomposition are equal).

(The solution of the problem was sent in by G. Hajós and the authors [2].)

In a not too general sense, matching problems deal with the

coverability and the number od different coverings of a lattice-figure - say B - with the replicas of another lattice-figure - say A. Rotation and symmetry are or are not allowed. Sometimes more types of A's are also allowed. The notion of coverability can be also generalized (see Definition B). The principal results of the paper give necessary and sufficient conditions for the coverabilit of a lattice parallelotop with

α) lattice-parallelotops of one or more given types;

β) lattice-figures consisting of two cubes (this is a generalization of the domino).

Here we summarize only the main definitions and results of our paper being in the press at the Journal of Combinatorial Theory.

The proofs can be found in that paper, and the numbering of definitions and theorems is also taken from there.

DEFINITIONS

Let us consider the set of n-dimensional lattice points (i.e. the points with integer coordinates).

An n-dimensional lattice-figure is an arbitrary subset of lattice-points.

There exists a natural correspondence between the lattice-points and lattice-fields (unit cubes). Thus, sometimes we shall use the more illustrative expression "lattice field" instead of "lattice-point".

We accept the usual concept of congruency, that is we allow of shift, rotation and symmetry.

For the sake of simplicity we suppose (unless we emphasize the contrary) that the parallelotop which we want to decompose will be situated in the non-negative octant, and that one of its vertices is the origin. (I.e. a parallelotop B with edge-lengths $b_1, b_2, \dots, b_n$ consists of the lattice-points $(x_1, x_2, \dots, x_n)$ satisfying the conditions $0 \le x_i < b_i$ $(1 \le i \le n)$.)

DEFINITION A.

We say that a parallelotop B can be filled up (covered) by the given lattice-figures $A_1, A_2, \dots, A_m$ if we can decompose B into disjoint subsets, each of which is congruent to one of A_i's; and in this case we write

$$(A_1, A_2, \dots, A_m) \mid B .$$

(If $m=1$, we write simply $A_1 \mid B$.)

If we use the above natural definition of coverability, then the necessary and sufficient conditions are valid only if all the edges of the parallelotop B are large enough. However, in the case of the next definition we can omit this.

DEFINITION B.

We say that a parallelotop B can be filled up (covered) in the weak sense by the given lattice-figures $A_1, A_2, \dots, A_m$ if there exist the parallelotops $A_1(1), \dots, A_1(r_1), A_2(1), \dots, A_2(r_2), \dots, A_m(1), \dots, A_m(r_m)$ and the integers $\nu_{11}, \dots, \nu_{1r_1}, \nu_{21}, \dots, \nu_{2r_2}, \dots, \nu_{m1}, \dots, \nu_{mr_m}$ such that

$$\sum_{\substack{1 \le i \le m \\ 1 \le j \le r_i \\ x \in A_i(j)}} \nu_{ij} = \begin{cases} 1 & \text{if} \quad x \in B \\ 0 & \text{if} \quad x \notin B , \end{cases}$$

where $A_i(j)$ $(1 \le j \le r_i)$ is congruent to A_i $(1 \le i \le m)$

In this case we write

$$(A_1, A_2, \dots, A_m) \mid^{*} B .$$

The number ν_{ij} is called the multiplicity of $A_i(j)$.

It is easy to see that $(A_1, A_2, \dots, A_m) | B$ implies $(A_1, A_2, \dots, A_m) |^* B$ and we can choose the integers ν_{ij} so that $\nu_{ij} = 1$ $(1 \le i \le m,\ 1 \le j \le r_i)$.

DEFINITION C.

We say that a parallelotop B can be filled up (covered) in a parallel manner by a given parallelotop A if we can decompose B into disjoint subsets, each of which is congruent to A and the parallel edges are equal. In this case we write $A |^P B$.

DEFINITION D.

If $a_1, a_2, \dots, a_n$ are given nonnegative integers $(\sum_{i=1}^{n} a_i > 0)$ then the lattice points $(x_1, x_2, \dots, x_n)$, $(y_1, y_2, \dots, y_n)$ form a knight-figure of type $a_1 \times a_2 \times \dots \times a_n$ if $|x_1 - y_1|, |x_2 - y_2|, \dots, |x_n - y_n|$ is a permutation of the integers $a_1, a_2, \dots, a_n$. The knight-figure of type $a_1 \times a_2 \times \dots \times a_n$ will be denoted by $K(a_1, \dots, a_n)$.

In Part 1 we give a necessary and sufficient condition for the validity of $(A_1, A_2, \dots, A_m) |^* B$ (Theorem 2), and for the validity of $(A_1, A_2, \dots, A_m) | B$ and $A | B$ if B is large enough (Theorem 3 and Theorem 4, resp.). Some special cases are also explained because of the simpler form of conditions (Theorem 1, 5, 6, 7). In the course of the proofs we need a generalization (Lemma 7) of the well-known marriage principle which may be interesting in itself.

In Part 2 we give a necessary and sufficient condition for the validity of $K(a,b) | B$ if B is large enough. (Theorem 12). For the case $K(a,1)$ a covering is constructed. Two simple n-dimensional generalizations (Theorem 13, 14) are also given.

1. COVERINGS WITH PARALLELOTOPS

The simplest but very interesting case is the case of parallelotop with edge-lengths $1,1,\dots,1,a$. The following theorem concerning this type is obviously a special case of general Theorem 2.

Theorem 1. [6], [7] Let A and B be n-dimensional parallelotops and the edge-lengths of A be $1,1,\dots,1,a$, then $A|B$ if and only if at least one edge of B is divisible by a.

DEFINITION 2.

If $e_1,\dots,e_n$ are natural numbers and B is an n-dimensional parallelotop, then $M(B,e_1,\dots,e_n)$ denotes the divisibility matrix: the j-th element of the i-th row is 1 if $e_i | b_j$ (where b_j is the j-th side of B) and 0 if $e_i \nmid b_j$.

DEFINITION 3.

We say that an $n\times n$ matrix M has not independent 0's (or 1's) if there are no n 0's (or 1's) in different rows and columns.

Theorem 2. $(A_1,A_2,\dots,A_m)|^*B$ holds if and only if, choosing in an arbitrary manner $k_i(\geq 1)$ edges of A_i, denoting by d_i their greatest common divisor and making n sets of the numbers d_i in an arbitrary manner, but using every d_i exactly in $n-k_i+1$ sets, finally, denoting by $e_1,\dots,e_n$ the greatest common divisors of the numbers in one set ($e_j=\infty$ if the j-th set is void), the matrix $M(B,e_1,\dots,e_n)$ has no n independent 0's.

Theorem 3. In the case of

$$b_i \geq 3^{nm\cdot 2^{nm}} \cdot a^{2^{nm}+2} \tag{5}$$

(where a is the maximum of edges of A_i's) $(A_1,\dots,A_m)|B$ holds if and only if (F_1) holds.

DEFINITION 4.

Let ε_1 and ε_2 be equal to 0 or 1. We call the logical sum of ε_1 and ε_2 the number

$$\varepsilon_1 \vee \varepsilon_2 = \begin{cases} 0 & \text{if} \quad \varepsilon_1 = \varepsilon_2 = 0 \\ 1 & \text{otherwise.} \end{cases}$$

Similarly, if t_1 and t_2 are row vectors with 0,1 coordinates then the coordinates of the logical sum of t_1 and t_2 are the logical sums of the corresponding coordinates.

Lemma 7. Let $M_1, M_2, \dots, M_m$ be $n \times n$ matrices with elements 0 and 1. If they have the property that choosing in an arbitrary manner $k_i \geq 1$ rows from M_i $(1 \leq i \leq m)$, denoting by w_i the logical sum of these rows and making n sets of the row vectors w_i in an arbitrary manner, but using every
(F_2) w_i exactly in $n-k_i+1$ sets, finally denoting by z_j $(1 \leq j \leq n)$ the logical sum of the w_i's lying in the j'th set (if the j'th set is void, then $w_i = (0,0,\dots,0)$) the matrix formed from $z_1, z_2, \dots, z_n$, as rows has not n independent 0's, then there is an index p $(1 \leq p \leq m)$ such that M_p has n independent 1's.

REMARK. This lemma is a generalization of the well-known marriage principle [3], which says:

MARRIAGE PRINCIPLE. Let M be an $n \times n$ matrix with elements 0 and 1. If choosing in an arbitrary manner k rows, the number of columns containing a 1 in of these rows (or the numbers of 1's of the logical sum of these rows) $\geq k$, then M has n independent 1's.

Now we consider some interesting special cases.

Theorem 4. In the case of

$$b_i \geq 3^{n \cdot 2^n} \cdot a^{2^n+1}$$

(where a is the maximum of edges of A)

$$A | B$$

holds if and only if choosing k $(1 \le k \le n)$ edges of A in an arbitrary manner
3) their greatest common divisor d is a divisor of at least k edges of B.

The problem of de Bruijn says that

Theorem 5. A has the property "$A|B$ if and only if $A|^{p}B$ " if and only if from any two edges of A one of them is the divisor of the other one.

In Theorem 5 we have shown that it is true only in a special case that we can fill up something only if we can fill it up in a "regular" wav. However, we may define the term "regular" in a wider sense.

DEFINITION 5.

A filling up $A|B$ is regular if we can reach to this filling up by cuts, where cut is the operation when we divide the whole parallelotop by an $n-1$-dimensional hyperplane.

Theorem 6. $(A_1, A_2, \dots, A_m) | B$ if and only if it is possible regularly, too.

Another interesting special case of Theorem 3 is the case when we have n-dimensional cubes with relative prime edges.

Theorem 7. Let $C_1, C_2, \dots, C_m$ be n-dimensional cubes with edges $c_1, c_2, \dots, c_m$ satisfying $(c_i, c_j) = 1$ $(i \ne j)$ and let

$$b_j > 3^{nm\, 2^{nm}} \cdot c^{2^{nm}+1} \qquad (1 \le j \le n)$$

where $c = \max(c_1, \dots, c_m)$. Then

$$(C_1, \dots, C_m) | B$$

holds if and only if the $m \times n$ matrix $M(B, c_1, \dots, c_m)$ has no m independent 0's.

We have obtained the following modified form of Theorem 7.

Theorem 7a. Under the condition of Theorem 7 $(C_1, \dots, C_m) | B$ if and only if

a) $m > n$

b) $m \leq n$ and there are $n-m+1$ edges of B divisible by all the numbers $c_1, \dots, c_m$, or we can fill up B by less than m of C_i's.

2. COVERINGS WITH KNIGHT FIGURES

Theorem 12. Let $(a,b) = d$.

1. If $\frac{a}{d} \equiv \frac{b}{d} \pmod 2$, then a rectangular with large enough sizes $(m \geq m_1(a,b), n \geq n_1(a,b))$ is coverable with knight-figures of type $a \times b$, if and only if m and n are divisible by $2d$.

2. If $\frac{a}{d} \not\equiv \frac{b}{d} \pmod 2$, then a rectangular with large enough sizes $(m \geq m_1(a,b), n \geq n_1(a,b))$ is coverable with knight-figures of type $a \times b$, if and only if either m or n is divisible by $2d$.

Theorem 13. An n-dimensional parallelotop R can be covered with knight-figures of type $a \times 0 \times \dots \times 0$ if and only if one of its sizes is divisible by $2a$.

Theorem 14. An n-dimensional parallelotop R can be covered with knight-figures of type $a \times \dots \times a$ if and only if its sizes are divisible by $2a$.

REFERENCES

[1] Problem 119., Mat. Lapok 12 (1961) 103.

[2] The solution of Problem 119., Mat. Lapok 13 (1962) 314-317.

[3] ORE, O.: Graphs and matching theorems, Duke Math. J. 22 (1955) 625-639.

[4] EGERVÁRY, J.: Mátrixok kombinatorikus tulajdonságairól. Mat. Fiz. Lapok 38 (1931) 16-28. Translation by H.W. KUHN: On combinatorial properties of matrices, George Washington Univ. Logistic Papers, 11 (1955).

[5] JULLIEN, P.: Essai sur la Théorie des Puzzles, Rev. Franc. Recherche Opérationnelle, 33 (1964) 375-384.

[6] HARPER, L.H. and ROTA, Gian-Carlo: Matching theory and introduction, preprint, The Rockefeller University.

[7] KLARNER, D.A.: Packing a Rectangle with Congruent N-ominoes, J. Combinatorial Theory 7 (1969). 107-115.

Methods for the general cell growth problem

by

D. A. Klarner

Eindhoven, The Netherlands

Let G be a graph with $\Gamma(G)$ the automorphism group of G, and let Σ be a subgroup of $\Gamma(G)$. Two subgraphs K and L of G are Σ-isomorphic if an automorphism of G in Σ maps K onto L; Σ-isomorphism is an equivalence relation on the subgraphs of G. The general cell growth problem asks for an algorithm for computing the number of equivalence classes containing connected, Σ-isomorphic subgraphs of G. In this paper we review the methods used by Eden, Read, and the author in the special case when G is the graph consisting of the lattice points in the plane with two points connected by an edge in G whenever the points are exactly at a unit from each other, and Σ is the set of automorphisms of G corresponding to translations of the plane or $\Sigma = \Gamma(G)$. In the third section we fill a gap in Read's work by describing the set of sections used to form sequences corresponding to animals with breadth r (Read only described the cases when $r \leq 5$). Also, we find the number of sections needed for a fixed r, so that we now have a measure of the amount of work required by Read's algorithm for computing the number of fixed animals with breadth r. In the last section, we show how Read's results can be reformulated in terms of the integral equation used by the author to prove a lower bound for the number of animals with n cells.

A PROBLEM CONCERNING ISOMORPHIC SUBGRAPHS:

Let G be a graph with $V(G)$ the vertices of G, and $E(G)$ (a subset of the 2-subsets of $V(G)$) the edges of G. A graph H is a subgraph of G if $V(H)$ is a subset of $V(G)$, and $E(H)$ is the set of all edges $\{x, y\}$ of G such that x, y are vertices of H. Let $\Gamma(G)$ be the group of automorphisms of G, and let Σ be a subgroup of $\Gamma(G)$. Two subgraphs H and K of G are Σ-isomorphic if there is an automorphism σ in Σ such that $V(K) = \{\sigma h : h \in H\}$; Σ-isomorphism is an equivalence relation on the subgraphs of G. An equivalence class containing Σ-isomorphic subgraphs of G, each subgraph having v vertices, is a (G, Σ, v)-graph. The number of (G, Σ, v)-graphs is the coefficient of x^v in the polynomial $Z(\Sigma : 1 + x, 1 + x^2, \ldots)$, where $Z(\Sigma : x_1, x_2, \ldots)$ is the cycle index for the permutation group Σ. (This is an immediate consequence of Pólya's enumeration theorem; for a clear statement of this theorem see N. G. de Bruijn [1].) An equivalence class containing connected, Σ-isomorphic subgraphs of G, each subgraph having v vertices, is a connected (G, Σ, v)-graph. The general problem of finding a useful algorithm for computing the number of connected (G, Σ, v)-graphs is unsolved, and seems to be very difficult. Our aim in this paper will be to present an exposition of the methods M. Eden [2], R. C. Read [12], and the author [7], [8], [9] have developed in attempts to solve the cell growth problem, a special case of the problem just formulated.

Let i, j be integers; a cell $[i, j]$ is a unit square in the plane having vertices (i,j), $(i+1,j)$, $(i,j+1)$, and $(i+1,j+1)$.

Let C be any set of cells; the components of C are the maximal subsets B of C such that $\cup B$ is a connected set with no finite set of cut points; thus, the set of components of C is a partition of C. An n-omino (an animal with n cells) is a set of n cells with exactly one component. Two point sets in the plane are congruent if an isometry of the plane maps one set onto the other, otherwise the sets are incongruent. Two n-ominoes X and Y are congruent or incongruent if the point sets $\bigcup X$ and $\bigcup Y$ are congruent or incongruent respectively. The cell growth problem asks for a useful algorithm for computing $t(n)$, the number of incongruent n-ominoes. The cell growth problem was first considered by Eden [3] in 1953, and Golomb [4] mentioned it later in his paper on polyominoes. Harary [5], [6] listed the cell growth problem as an unsolved problem in graphical enumeration without reference to Eden [2] who had already given upper and lower bounds for $t(n)$.

To see that the cell growth problem is a special case of the problem we posed at the start, we can give two formulations, the first in terms of an infinite graph, the second in terms of finite graphs. Let G be the graph whose vertices are the cells in the plane and whose edges are the pairs of cells having their centers at a unit distance from one another. The connected $(G, \Gamma(G), n)$-graphs are the equivalence classes containing congruent n-ominoes which Read called free animals with n cells. If Σ is the subgroup of G containing automorphisms of G that correspond to translations of the plane, the connected (G, Σ, n)-graphs are what Read called the fixed animals with n cells. Let n be a natural number and let G_n be the graph defined by $V(G) = \{(x,y): x,y = 1,\ldots,n\}$, and $\{(x,y),(u,v)\}$ is an edge in G_n whenever $|x-u| \equiv 1, |y-v| \equiv 0 \pmod{n}$ or $|x-u| \equiv 0, |y-v| \equiv 1 \pmod{n}$. It is convenient here to think of the vertices and edges of G_n arranged in a grid pattern on the surface of a torus. Let $\Sigma_n = \{\sigma_{a,b}: a,b = 0,\ldots,n-1\}$, where $\sigma_{a,b}(x,y) = (u,v), x \equiv u+a, y \equiv v+b \pmod{n}$, for $x, y, u, v = 1,\ldots,n$; Σ_n is a subgroup of $\Gamma(G_n)$. The connected $(G_n, \Gamma(G_n), n)$-graphs and the connected (G_n, Σ_n, n)-graphs correspond to the free and fixed animals with n cells respectively.

EDEN'S CONTRIBUTIONS:

An animal with n cells can be constructed by joining a cell to some cell in an animal with $n-1$ cells; thus, an animal can be "grown" in many ways by making a sequence of additions of cells to a single cell. The growth process indicated serves as a model for the two-dimensional growth of certain plants. Eden [2] proposed this model and asked two questions about it, one quantitative, the other qualitative. How many different animals with n cells are there, and if we grow an animal, what size and shape is an animal with cells most likely to have, for large n? These problems may be related, since to obtain an asymptotic estimate for the number of animals it may only be necessary to count animals which are typical in some sense. After devising a Monte Carlo procedure to be carried out on a computer, Eden obtained evidence to support the idea that growing an animal with n cells is most likely to result in a "solid" configuration "circular" in shape. He was able to prove that $t(n)$, the number of different animals with n cells, satisfies the inequalities

$$(1) \qquad (3.14)^n < t(n) < (6.75)^n,$$

for all sufficiently large n. It seems likely from an argument supplied by Eden [3] (but lacking a complete mathematical justification) that $(t(n))^{1/n}$ tends to 4 as n tends to infinity.

To prove $t(n) > (3.14)^n$, for all sufficiently large n. Eden proved a lower bound for the number of elements in a subset of the set of all fixed animals with n cells. Let $(n_1, \dots, n_k)$ be a composition of n with k positive parts, for $k = 1, \dots, n$; a fixed animal with n cells can be formed in $(n_1 + n_2 - 1)(n_2 + n_3 - 1) \dots (n_{k-1} + n_k - 1)$ ways by joining the bottom edge of a $1 \times n_{i+1}$ strip of cells to the top edge of a $1 \times n_i$ strip of cells, for $i = 1, \dots, k-1$. The total number of animals that can be formed in this way is

$$(2) \qquad b(n) = \sum (n_1 + n_2 - 1)(n_2 + n_3 - 1) \cdots (n_{k-1} + n_k - 1),$$

where the sum extends over all compositions $(n_1, n_2, \dots, n_k)$ of n into k positive parts, for $k = 1, \dots, n$. Of course, $8t(n) > b(n)$, and Eden was able to prove $b(n) > (3.14)^n$, for all sufficiently large n; however, it was shown in [7] and [8] that

$$b(n+3) = 5b(n+2) - 7b(n+1) + 4b(n), \tag{3}$$

for $n = 2,3,\dots,$ and it follows from this that $(b(n))^{1/n}$ tends to 3.20... as n tends to infinity. Thus, the best lower bound that can be obtained by Eden's method is $t(n) > (3.20)^n$, for all sufficiently large n.

Eden used an important idea in proving his upper bound $t(n) < (6.75)^n$, for all n. He showed that each animal with n cells, say X, corresponds to a unique binary sequence $W(X) = (x_1, \dots, x_{3n-1})$, with exactly $n-1$ of the x's equal to 1 and the rest equal to 0. From this it follows that

$$t(n) < \binom{3n-1}{n-1} < \left(\frac{27}{4}\right)^n. \tag{4}$$

Using the fact that $W(X)$ is a code word for a rooted plane tree, the estimate (4) can be improved slightly (see [10]).

READ'S CONTRIBUTIONS:

Now we will present the main ideas developed by Read [12] in his treatment of the cell growth problem. Read first focused his attention on animals $X = \{[x_1, y_1], \dots, [x_n, y_n]\}$ with <u>length</u> $\lambda(X) = 1 + \max_{i,j} |x_i - x_j|$ and <u>breadth</u> $\beta(X) = 1 + \max_{i,j} |y_i - y_j|$. An animal with n cells, length a, and breadth b is bounded with an $a \times b$ rectangle; the $1 \times b$ columns of the bounding rectangle are <u>sections</u> of the animal (we will give a precise definition presently), so an animal can be viewed as a sequence of sections. Read's idea was to begin with a set of sections and then describe how to construct all the fixed animals with breandth b as sequences of $1 \times b$ sections. Read's method is indicated by his treatment of the cases $b = 2, 3, 4,$ and 5, but it is not difficult to describe the general case, and we will do that here.

Let $(V, E : B, T)$ be a directed graph with V the set of vertices, $E \subseteq V \times V$ the set of edges, $B \subseteq V$ the set of initial vertices, and $T \subseteq V$ the set of terminal vertices. A j-tuple of vertices $(v_1, \ldots, v_j)$ is permissible if $v_1 \in B$, $v_j \in T$, and $(v_{i-1}, v_i) \in E$, for $i = 2, \ldots, j$. It is well known in the theory of graphs (see for example Ore [11]) that if $V = \{v_1, \ldots, v_n\}$, then the number of permissible j-tuples is the matrix product $[b_r][e_{rs}]^{j-1}[t_s]$, where $[b_r]$ is the row matrix defined by $b_r = 1$, or 0 if v_r is or is not an element of $B\,(r = 1, \ldots, n)$, $[e_{rs}]$ is the n by n matrix defined by $e_{rs} = 1$, or 0 if (v_r, v_s) is or is not an element of $E\,(r, s = 1, \ldots, n)$, and $[t_s]$ is the column matrix defined by $t_s = 1$, or 0 if v_s is or is not an element of $T\,(s = 1, \ldots, n)$. Let w be a mapping of V into a commutative ring; $w(v)$ is the weight of the vertex v, and $w(v) \ldots w(z)$ is the weight of the j-tuple of vertices $(v, \ldots, z)$. Again it is clear that the sum of the weights of all permissible j-tuples is the matrix product $[b_r w(v_r)][e_{rs} w(v_s)]^{j-1}[t_s]$. Using this observation, one obtains the theorem used by Read in his treatment of the cell growth problem.

THEOREM: Let I be the identity matrix having the same dimension as $[e_{rs}]$. The sum of the weights of all permissible j-tuples in $(V, E : B, T)$ is the coefficient of y^j in the matrix product

$$y[b_r w(v_r)]\{I - y[e_{rs} w(v_s)]\}^{-1}[t_s] = \tag{5}$$

$$= \sum_{j=1}^{\infty} [b_r w(v_r)][e_{rs} w(v_s)]^{j-1}[t_s]\, y^j .$$

The ingenious application of this Theorem to enumerating fixed animals is best appreciated by studying Read's paper. We will fill a gap in Read's work by discussing the enumeration of animals with arbitrary breadth k.

Let C be a column of cells in the plane (that is, for some integer i

$C = \{[i,j] : j = 0, \pm 1, \ldots\})$, let C^* be the half plane formed by the set of cells which includes the cells of C, but no cell to the right of C, and let X be an animal with n cells. If $X \cap C$ is non-empty, an equivalence relation π is defined on the cells of $X \cap C$ as follows: $x, y \in X \cap C$ belong to the same equivalence class in π if, and only if, x and y belong to the same component in $X \cap C^*$. The pair $(X \cap C, \pi)$ is a section; thus, associated with each animal X is a unique sequence of sections $S(X) = \{(X \cap C_1, \pi_1), \ldots, (X \cap C_j, \pi_j)\}$, where $C_1, \ldots, C_j$ are consecutive columns of the plane. A section (W, π), where W is a finite subset of some column C, is the initial section of some animal if, and only if, the components of W are the equivalence classes of π. Also, (W,π) is the terminal section of some animal if, and only if, $\pi = \{W\}$. Finally, if $(W_1 \pi_1)$ and (W_2, π_2) are sections of animals X_1 and X_2 with W_1 and W_2 subsets of consecutive columns, then (W_2, π_2) follows (W_1, π_1) in the sequence of sections of a third animal X if, and only if,every equivalence class in π_1 is followed by exactly one equivalence class in π_2. An equivalence class P_1 in π_1 is followed by an equivalence class P_2 in π_2 if every component of $P_1 \cup P_2$ contains an element belonging to P_1.

Let $S_k^j(n)$ be the set of fixed animals with length j and breadth k. Sections of animals in $S_k^j(n)$ split into equivalence classes in an obvious way, and the rules for determining the initial sections, terminal sections, and the following relation carry over to these classes. Let S_k be the set of equivalence classes of sections of animals in $S_k^j(n)$, $j, n = 1, 2, \ldots$. Thus, a directed graph $(V, E : B, T)$ can be defined with $V = S_k$, B the classes in S_k containing initial sections, T the classes in S_k containing terminal sections, and $(W,Z) \in E$ if the class W contains a section that can be followed by a section in the class Z. The permissible j-tuples of vertices in $(V, E : B, T)$ correspond one-one to the elements of $S_k^j(n)$, $n = 1, 2, \ldots$; in fact, if the weight of a class in S_k is defined to be $x^{|W|}$, where (W, π) is a representative section in this class, then the weight of a j-tuple corresponding to a class in $S_k^j(n)$

is x^n. Thus, for a given k, matrices $B_k(x)$, $E_k(x)$, and T_k can be defined such that

$$(6) \qquad y B_k(x)\{I - y E_k(x)\}^{-1} T_k = \sum_{n=1}^{\infty} \sum_{j} |S_k^j(n)| x^n y^j = S_k(x,y),$$

where I is the identity matrix with the same dimension as $E_k(x)$.

Furthermore, the number of fixed animals with n cells and breadth less than or equal to k is the coefficient of x^n in the power series $S_k(x,1) - S_{k-1}(x,1)$, for $k = 2,3,\ldots$.

We still have not described how to construct the elements of S_k, and this seems to be necessary to construct $E_k(x)$. It is not yet clear what the dimension of $E_k(x)$ is, and this is an interesting problem since it gives a measure of the amount of work required to compute $|S_k^j(n)|$.

To characterize the sections of animals with breadth k, consider a subset W of the cells in a $1 \times k$ column of cells. We want to determine which partitions π of W correspond to a section (W,π) of some animal. Let $(W_1,\ldots,W_i)$ be the components of W as they are encountered going from the bottom to the top of the column containing W. Clearly, a partition π of W corresponding to a section (W,π) of some animal must have W_r a subset of some equivalence class in π, for $r = 1,\ldots,i$. Now for each equivalence class $W_u \cup \ldots \cup W_v$ in π, with $u < \cdots < v$, we replace W_v and W_u in the sequence $(W_1,\ldots,W_i)$ with (W_v) and (W_u) respectively; also, if $u = v$, we replace W_u with (W_u).

If $W_{u_1} \cup \ldots \cup W_{v_1}$ and $W_{u_2} \cup \ldots \cup W_{v_2}$ are equivalence classes in π with $u_1 < \cdots < v_1$ and $u_2 < \cdots < v_2$, we cannot have $\ldots(W_{u_1} \cdots \ldots(W_{u_2} \cdots W_{v_1}) \ldots W_{v_2}) \ldots$ if (W,π) is the section of some animal unless $u_1 = u_2$, $v_1 = v_2$. The number of systems of parentheses on $(W_1,\ldots,W_i)$ satisfying this property, where $W_1,\ldots,W_i$ are now components of a subset of cells in some $1 \times k$ column of cells, is $\binom{2i}{i}/(i+1)$ and every system of parentheses of this kind gives rise to a partition π corresponding to the section of some animal.

Let R be a set of cells forming a $1 \times k$ rectangular column of cells, let W, W' be subsets of R with $W \cup W' = R$, and let $W_1, W_3, \ldots$ and $W_2, W_4, \ldots$ be the components of W and W' respectively as they are encountered going from the bottom to the top of R. Letting $|W_j| = k_j$, $(k_1, \ldots, k_i)$ is a composition of k, and the number of sections of animals having $W_1, W_3, \ldots$ or $W_2, W_4, \ldots$ as components is

(7) $$c(i) = \frac{1}{1+[i/2]}\binom{2[i/2]}{[i/2]} + \frac{1}{1+i-[i/2]}\binom{2i-2[i/2]}{i-[i/2]}.$$

for $i > 1$, and $c(1) = 1$. The number of compositions of k with exactly i parts is $\binom{k-1}{i-1}$ so the total number of sections of animals with breadth k is

(8) $$\sum_{i=1}^{k} \binom{k-1}{i-1} c(i) = b(k).$$

Letting $B(x) = b(1)x + b(2)x^2 + \ldots$, and $C(x) = c(1)x + c(2)x^2 + \ldots$, (8) implies $B(x) = C(x)/(1-x)$; also, since

(9) $$(1-2x-(1-4x)^{1/2})/2x = \sum_{i=1}^{\infty} \frac{1}{i+1}\binom{2i}{i} x^i .$$

we have

(10) $$C(x) = (1+x)^2(1-2x^2-(1-4x^2)^{1/2})/2x^3.$$

and hence

(11) $$B(x) = ((1-x) - 2x^2(1-x)^{-1} - (1+x)^{1/2}(1-3x)^{1/2})/2x^3.$$

from this it follows that $(b(n))^{1/n}$ tends to 3 as n tends to infinity. The first few values of $b(n)$ are 1, 3, 8, 20, 50, 126, 322, 834, 2187, and 5797, for $n = 1, 2, \ldots,$ and 10 respectively.

Let Λ be a permutation group defined on a finite set X, and for each λ in Λ, let $X_\lambda = \{x : \lambda(x) = x, x \in X\}$. Two elements x, y in X are Λ-equivalent if there is a permutation λ in Λ such that $\lambda(x) = y$; Λ-equivalence is an equivalence relation on X. The number of Λ-equivalence classes in X (the number of orbits of Λ in X) is $|\Lambda|^{-1} \sum X_\lambda$, $\lambda \in \Lambda$; this is Burnside's Theorem, and a proof can be found in de Bruijn [1]. Read used this formula to find the number of free animals with length a and breadth b, $a \neq b$, in terms of the number of fixed animals with these or smaller dimensions. This result has not been obtained in the case when $a = b$, and the problem seems to be difficult.

A CONNECTION WITH THE FREDHOLM EQUATIONS

Let $(a_1, a_2, \ldots)$ be a composition of n into positive parts $a_1, a_2, \ldots$ and let $F_{a_1 a_2 \cdots}$ be the set of all animals X such that the i-th row of the rectangle bounding X contains exactly a_i cells belonging to X, for $i = 1, 2, \ldots$. Let $f(a_1, a_2, \ldots)$ be the number of fixed animals in $F_{a_1 a_2 \cdots}$. We showed in [8] that

(12) $$f(a_1, a_2, \ldots, a_k) \geq f(a_1, a_2) f(a_2, a_3) \cdots f(a_{k-1}, a_k),$$

and

$$f(m+2, n+2) = f(m+1, n+2) + f(m+2, n+1) - f(m, n),$$

(13)

$$f(m, 0) = f(0, m) = 1, \quad f(m, 1) = f(1, m) = m,$$

for $m, n = 0, 1, \ldots$. Using (12) and (13) we obtained a lower bound for $s(n)$, the number of fixed animals with n cells; namely,

(14) $$s(n) \geq b(n) = \sum f(a_1,a_2) f(a_2,a_3) \dots f(a_{k-1},a_k),$$

where the sum extends over all compositions $(a_1, a_2, \dots, a_k)$ of n into k positive parts, $k = 1, \dots, n$. A theory for sums over compositions of n having the form of (14) was given in [9]; using this theory we were able to show that

(15) $$b(n) > (3.72\dots)^n,$$

for all sufficiently large n. We will give a synopsis of the theory which is developed in detail in [9].

Suppose $\{f(m,n): m,n = 1,2,\dots\}$ and $\{g(n): n = 1,2,\dots\}$ are given sets of numbers, and consider the set of numbers $b(n): n = 1,2,\dots$ defined by

(16) $$b(n) = \sum f(a_1,a_2) f(a_2,a_3) \cdots f(a_{i-1},a_i) g(a_i),$$

where the sum extends over all compositions $(a_1, a_2, \dots, a_i)$ of n into i positive parts, $i = 1, 2, \dots, n$, and $g(n)$ is the contribution to the sum when the number of parts of the composition is one.

The symbol $b_k^j(a,n)$ used with all or only some of the suffixes, is the partial sum obtained from (16) when the index of summation has been restricted to those compositions of n which have exactly k parts, no part greater than j, and the first part equal to a; if a suffix is dropped, the corresponding restriction on the index of summation is dropped as well. If

(17) $$F(x,y) = \sum_{m=1}^{\infty} \sum_{n=1}^{\infty} f(m,n) x^m y^n$$

is an analytic function of x for fixed y and of y for fixed x in neighborhoods of $x = 0$ and $y = 0$ respectively, and if

(18) $$G(x) = \sum_{n=1}^{\infty} y(n) x^n$$

is an analytic function in a neighborhood of $x = 0$, then each of the functions found by deleting any of the suffixes from

(19) $$B_k^j(x,y) = \sum_{n=1}^{\infty} \sum_{a=1}^{n} b_k^j(a,n)\, y^a x^n$$

is an analytic function of x for fixed y in a neighborhood of $x = 0$.

Furthermore,

(20) $$B_1(x,y) = G(x,y)$$

(21) $$B_{k+1}(x,y) = \frac{1}{2\pi i} \int_C s^{-1} F(x,y,s^{-1})\, B_k(x,s)\, ds, \quad k = 1,2,\ldots,$$

(22) $$B(x,y) = G(xy) + \frac{1}{2\pi i} \int_C s^{-1} F(xy,s^{-1})\, B(x,s)\, ds$$

where C is a contour in the s-plane which includes the singularities of $F(xy,s^{-1})s^{-1}$ but excludes those of $B_k(x,s)$ or $B(x,s)$ respectively. Since $B(x,y) = B_1(x,y) + B_2(x,y) + \ldots$, (22) follows from (20) and (21). The integral equation (22) can be solved by means of (20) and (21), but there is a second method which is also sometimes useful. We have

(23) $$B^j(x,y) = \sum_{n=1}^{j} g(n) x^n y^n + \frac{1}{2\pi i} \int_C \sum_{m=1}^{j} \sum_{n=1}^{j} f(m,n) x^m y^m B^j(x,s) s^{-n-1} ds,$$

where C is a small circle enclosing the origin in the s-plane; this follows from (22) and the definition of $B^j(x,y)$. Now (23) implies

(24) $$B^j(x,y) = \sum_{n=1}^{j} g(n)x^n y^n + \sum_{n=1}^{j}\left\{\sum_{m=1}^{j} f(m,n)x^m y^m\right\} \partial^n B^j(x,0)/n!\ ,$$

where $\partial^n B^j(x,0)$ is the n-th partial derivative with respect to s of $B^j(x,s)$ at $s = 0$. Since

(25) $$B^j(x,y) = \sum_{m=1}^{j} \partial^m B^j(x,0)\, y^m/m!\ ,$$

we can equate coefficients of y^m in (24) to obtain the following system of equations linear in the quantities $\partial B^j(x,0)$, $\partial^2 B^j(x,0)/2!$, ... :

(26) $$\partial^m B^j(x,0)/m! = g(m)x^m + \sum_{n=1}^{j} f(m,n)x^m\, \partial^n B^j(x,0)/n!\ ,$$

for $m = 1, 2, \ldots, j$. Now we can solve (26) using Cramer's rule to find $\partial^m B^j(x,0)/m!$ in terms of x and the numbers $\{f(m,n): m,n = 1,\ldots,j\}$ and $\{g(m): m = 1,\ldots,j\}$; these functions and (24) provide a means of calculating $B^j(x,y)$, which is evidently a rational function.

Returning to the situation described in the second paragraph of the third section of this paper, let $f(m,n) = w(v_m)$ if $(v_m, v_n) \in E$, and $f(m,n) = 0$ otherwise; also $g(m) = w(v_m)$ if $v_m \in T$, and $g(m) = 0$ otherwise. Now the sum of the weights of all permissible k-tuples $(v_a, v_{a_2}, \ldots, v_{a_k})$ is

(27) $$\sum_{n=k}^{\infty} b_k(a,n) = \sum f(a,a_2)f(a_2,a_3)\cdots f(a_{k-1},a_k)g(a_k),$$

where the second sum extends over all compositions $(a, a_2, \ldots, a_k)$ of n into exactly k parts, for $n = k, k+1, \ldots$. In particular, if $(V, E : B, T)$ is the directed graph used to find sequences of sections corresponding to animals with breadth r, the weight of a section $(W,\pi) \in V$ is $z^{|W|}$. Thus, $f(m,n)$, $g(n)$, and $b_k(a,n)$ are defined so that $F(x,y) = F(x,y:z)$, $G(x) = G(x:z)$, and $B_k(x,y) = B_k(x,y:z)$. If we set

(28) $$B(x,y) = B(x,y:z,t) = \sum_{k=1}^{\infty} t^k B_k(x,y:z),$$

then it follows from

(29) $$\begin{aligned} tB_1(x,y:z) &= tG(xy:z) \\ t^{k+1}B_{k+1}(x,y:z) &= \frac{1}{2\pi i}\int_C s^{-1}F(xy,s^{-1}:z)t^k B_k(x,s:z)\,ds, \end{aligned}$$

that

(30) $$B(x,y:z,t) = tG(xy:z) + \frac{1}{2\pi i}\int_C s^{-1}F(xy,s^{-1}:z)B(x,s:z,t)\,ds,$$

where C is a small circle enclosing the origin in the s-plane. Now since $F(x,y:z)$ is a polynomial in x, y, z, the kernel of the integral in (30) has finite rank and so the integral equation can be solved by means of the method summarized in equations (23), (24), (25) and (26). We are only interested in finding the functions $\partial^a B(x,0:z,t)/a!$ where $v_a \in B$ because

(31) $$\sum_{v_a \in B} \partial^a B(1,0:z,t)/a! = S_r(z,t),$$

where $S_r(z,t)$ is the function defined by (6). This shows that Read's result can be formulated in terms of a linear integral equation instead of using matrices.

REFERENCES

[1] N. G. de BRUIJN, Polya's Theory of counting, Chapter 5 of Applied Combinatorial Mathematics, E. Beckenbach, Editor.

[2] M. EDEN, A two-dimensional growth process, Proceedings of the Fourth Berkeley Symposium on Mathematical Statistics and Probability, Vol. IV (Berkeley, California, 1961), pp. 223, 239.

[3] M. EDEN, private correspondence: Professor Eden's letter reads as follows: "The cell growth problem was original with me and I began to work on it in 1953 at Princeton. It is related to the Polyomino problem that Golomb has worked on. I cannot say whether he derived the problem from my work or initiated it independently.

Regarding my estimate of 4^n as an upper bound I have worried about making it more precise for quite a long time. The mathematical missing step involves the expected value for the number of constraints induced as the code words are generated. The expected values given in my figure are based on the assumption of equal likelihood of outcome. I feel it should be possible to show that this assumption is justified, but I certainly did not do so in the paper.

In order to get some feel for the problem I ran a number of computer simulations and if I recall correctly I obtained an estimate of $(4.02)^k$ from this simulation.

I am afraid that I have not done anything serious on the problem since the Berkeley Symposium and I am not aware of any further references since your own."

[4] S. W. GOLOMB, checkerboards and polyominoes, Amer. Math. Monthly, 67 (1954) pp. 275-282.

[5] F. HARARY, Unsolved problems in the enumeration of graphs, (Magyar Tud. Akad. Mat. Kutató Int. Közl.) Publ. Math. Inst. Hungar. Acad. Sci., 5 (1960), pp. 63-95.

[6] F. HARARY, "Combinatorial problems in graphical enumeration", Chap. 6, Applied Combinatorial analysis, edited by F. Beckenbach (New York, 1964), pp. 185-217.

[7] D. A. KLARNER, Some results concerning polyominoes, Fibonacci Quarterly 3 (1965), pp. 9-20.

[8] D. A. KLARNER, Cell growth problems, Can. J. Math., Vol. 19 (1967), pp. 851-863.

[9] D. A. KLARNER, a combinatorial formula involving the Fredholm integral equation, Journal of Combinatorial Theory, 5 (1968), pp. 1-16.

[10] D. A. KLARNER, Correspondences between plane trees and binary sequences, to appear in Journal of Combinatorial Theory.

[11] O. ORE, The theory of graphs, Amer. Math. Society Colloquium Pub., Vol. XXXVIII, 1962.

[12] R. C. READ, Contributions to the cell growth problem, Can. J. Math., Vol. 14 (1962), pp. 1-20.

On the sum of elements of ± 1 matrices

by

J. Komlós and M. Sulyok

Budapest, Hungary

Let us consider a quadratic $n \times n$ matrix with elements ± 1. We can multiply arbitrary rows and columns by -1 and our aim is to minimize the absolute value of the sum of elements of the matrix.

In other words we want to draw the number of $+1$ to that of -1 as near as possible.

Put

$$S_n = \max_{(a_{ij})} \min_{x_i, y_j} \left| \sum_{i,j=1}^{n} a_{ij} x_i y_j \right| ,$$

where x_i, y_j take the values ± 1 in all possible ways and the maximum is extended for all $n \times n$ matrices (a_{ij}) with elements ± 1.

ERDŐS has remarked that by applying the central limit theorem it is easy to show that

$$S_n \leqq \sqrt{c n \log n} .$$

Moser and Moon guessed that for all n

$$S_n \leqq 2 .$$

In this paper we prove this conjecture for $n > n_0$. More precisely (for $n > n_0$)

$$S_n = \begin{cases} 1 & \text{if } n \text{ is odd} \\ 2 & \text{if } n \text{ is even.} \end{cases} \tag{1}$$

(Actually we shall prove that $S_n \leqq 2$, but it is equivalent to (1) as for odd n $S_n \geqq 1$, and the example $\begin{pmatrix} a_{11} = -1 & \\ a_{ij} = 1 & \text{for } i+j > 2 \end{pmatrix}$ shows that $S_n \geqq 2$ for even n.)

In paragraph 1 we shall prove relation (1), and in paragraph 2 we consider similar questions.

§ 1.

Throughout this paper n_0 means an integer great enough, and when a non-integer number stands e.g. for index, one has to understand its integer part.

As the sum of a row $\underline{r}$ we shall mean the sum of the elements of this row and denote it by $\| \underline{r} \|$.

a) First we give several lemmas.

LEMMA 1. There are at least $\frac{n-1}{n} \cdot 2^n$ changes of signs of the columns for which the absolute value of the sum of any row is less than $\sqrt{5n \log n}$.

LEMMA 2. For $n > n_0$ there are more than $\frac{2^n}{n}$ changes of signs of the columns, for which there exist $\sqrt{7n \log n}$ rows whose sums are between 1 and $\sqrt{100 \log n}$.

LEMMA 3. We are given an $m \times n$ matrix with elements ± 1, $(m, n > n_0)$, where in every row there are at least $\frac{n}{3}$ 1's. Then we can find a $q \times q$ minor which has only 1 as element, where

$$q = \left[\frac{\log m}{2} \right] \quad \text{if} \quad \log m < \frac{n}{4} .$$

We shall often use the following simple

REMARK. If there are p rows the absolute values of whose sums are between 0 and t, then by changes of signs of some of these rows we can reach that the sum of all elements in these p rows be between 0 and t.

The remark can be reformulated in the following form. If we have p numbers $a_1, a_2, \ldots, a_p$ with $0 \leqq a_i \leqq t$, then we can find p numbers $y_1, \ldots, y_p$ with values ± 1, for which

$$0 \leqq \sum_{i=1}^{p} a_i y_i \leqq t .$$

The proof of this statement is so simple that we can omit it.

b) PROOF OF LEMMA 1.

Let us consider only one row and denote its elements by $b_1, b_2, \ldots, b_n$. We want to count at how many changes of signs of the columns of the matrix will the absolute value of the sum of this row be greater than $\sqrt{5n \log n}$; I.e. between the 2^n numbers

$$\sum_{k=1}^{n} b_k y_k$$

(the y's are $\pm$ 1's) how many numbers have absolute value greater than $\sqrt{5n \log n}$

It is obvious that we can assume that $b_1 = b_2 = \cdots = b_n = 1$. So this number is equal to

$$\sum_{|k-\frac{n}{2}| > \frac{\sqrt{5n \log n}}{2}} \binom{n}{k}$$

We apply the inequality

$$\binom{n}{k} < \frac{2^n}{\sqrt{n}} e^{-\frac{x^2}{2}}$$

where

$$x = \frac{k - \frac{n}{2}}{\sqrt{\frac{n}{4}}} ,$$

which is true for $n \geqq 1$ and $|x| < \frac{\sqrt{n}}{2}$.

The monotonity of the binomial coefficients implies

$$\sum_{\left|k-\frac{n}{2}\right| > x\sqrt{\frac{n}{4}}} \binom{n}{k} < n \cdot \frac{2^n}{\sqrt{n}} e^{-\frac{x^2}{2}} = 2^n \cdot \sqrt{n}\, e^{-\frac{x^2}{2}} ,$$

if $n \geqq 1$, and $|x| < \frac{\sqrt{n}}{2}$.

Thus, (as $\sqrt{5 \log n} < \frac{\sqrt{n}}{2}$ for large n), the number considered above is less than

$$2^n \cdot \sqrt{n}\, e^{-\frac{(\sqrt{5 \log n})^2}{2}} = \frac{2^n}{n^2} .$$

We have n rows, so the number of changes of signs of columns at which the absolute value of at least one row is greater than $\sqrt{5 n \log n}$, is less than $n \cdot \frac{2^n}{n^2} = \frac{2^n}{n}$, which proves lemma 1.

c) PROOF OF LEMMA 2. We use the so-called first-moment method.

Let us take the possible 2^n changes of signs of the columns in some in some order and define ε_{ik} for $i = 1, 2, \dots, n$, $k = 1, 2, \dots, 2^n$ as follows:

$\varepsilon_{ik} = 1$ if at the k-th change of signs the sum of the i-th row is between 1 and $\sqrt{100 \log n}$ and $\varepsilon_{ik} = 0$ otherwise.

Clearly

$$\sum_{k=1}^{2^n} \varepsilon_{ik} = \sum_{\frac{1}{2} \leq s - \frac{n}{2} \leq \frac{\sqrt{100 \log n}}{2}} \binom{n}{s} > \frac{2^n}{\sqrt{\frac{n}{28 \log n}}}$$

for $i = 1, 2, \ldots, n$. Therefore,

$$\sum_{k=1}^{2^n} \sum_{i=1}^{n} \varepsilon_{ik} > \frac{n \cdot 2^n}{\sqrt{\frac{n}{28 \log n}}} .$$

Thus for at least $\dfrac{2^n}{2\sqrt{\frac{n}{28 \log n}}}$ incides k

$$\sum_{i=1}^{n} \varepsilon_{ik} \geqq \frac{n}{2\sqrt{\frac{n}{28 \log n}}} = \sqrt{7n \log n} .$$

But $\dfrac{2^n}{2\sqrt{\frac{n}{28 \log n}}} > \dfrac{2^n}{n}$, which proves our lemma.

d) THE PROOF OF LEMMA 3. t 1-s in a row will be called a t-line. By our assumption the number of t-lines in the matrix is at least

$$m \binom{\left[\frac{n}{3}\right]}{t} .$$

If there were no $q \times q$ minors of 1-s, then in every q columns there would be at most $q-1$ q-lines.

Then we should have the inequality

$$m \binom{\left[\frac{n}{3}\right]}{q} \leqq (q-1) \binom{n}{q} .$$

An easy calculation shows that this inequality contradicts our assumption on q.

e) THE PROOF OF RELATION (1).

In view of lemmas 1 and 2, there is a change of signs at which the absolute value of any row is less than $\sqrt{5n \log n}$, further there eixst $\sqrt{7n \log n}$ rows whose sums are between 1 and $\sqrt{100 \log n}$. We start from this change of signs (after this change, the matrix will be denoted by M), and modify it as follows. As we have $\sqrt{7n \log n}$ rows whose sums assume only $\sqrt{100 \log n}$ values, there is at least one value between 1 and $\sqrt{100 \log n}$, which is assumed at least $\sqrt{\frac{7n}{100}}$ times.

Let us consider the smallest integer k between 1 and $\sqrt{100 \log n}$ which is taken (as sum of rows) at least $a = \sqrt{\frac{n}{10000}}$ times. Let these rows be denoted by $\underline{r}_1, \underline{r}_2, \ldots, \underline{r}_a$. Clearly among our $\sqrt{7n \log n}$ sums of rows at least $b = \sqrt{6n \log n}$ are greater than, or equal to k. (as $\sqrt{7n \log n} - a \cdot k > \sqrt{6n \log n}$).

f) Let us study the matrix formed by the rows $\underline{r}_1, \underline{r}_2, \ldots, \underline{r}_a$. Lemma 3 implies the existence of a $q \times q$ "block" of 1-s, where

$$q = \left[\frac{\log \frac{n}{10000}}{2} \right] > \frac{\log n}{3} .$$

(The condition $\log a < \frac{n}{4}$ is valid for $n > n_0$.

Let us consider the q columns of matrix M including this block. Multiplying $\left[\frac{k-1}{2}\right]$ of them by -1, the sums of the rows including our block will be 1 or 2 (n is odd or even, respectively); these rows will be denoted by $\underline{s}_1, \underline{s}_2, \ldots, \underline{s}_q$.

g) Now we have a matrix which has q rows (say $\underline{s}_1, \underline{s}_2, \ldots, \underline{s}_q$) with sum 1 (or 2), other $b-q$ rows (say $\underline{t}_1, \underline{t}_2, \ldots, \underline{t}_{b-q}$) with sum between 1 and $\sqrt{100 \log n} + k = K_1$, and other $n-b$ rows (say $\underline{v}_1, \underline{v}_2, \ldots, \underline{v}_{n-b}$), the absolute values of whose sums are at most $\sqrt{5n \log n} + k = K_2$.

Taking into account the relations (for $n > n_0$):

1^0 $$-K_2 \leqq \| u_i \| \leqq K_2 \qquad (i = 1, 2, \ldots, n-b)$$

2° $\qquad \| t_i \| \geqq 1 \qquad (i = 1, 2, \dots, b-q)$

and $\qquad 1 \cdot (b-q) > K_2$

3° $\qquad \| s_i \| = 1 \quad \text{(or 2)} \qquad (i = 1, 2, \dots, q)$

and $\qquad 1 \cdot q > K_1 \qquad (2 \cdot q > K_1),$

and applying the Remark three times, we get the desired result.

§ 2.

a) First we remark that it is easy to modify the proof to the case of non-quadratic matrix; it is sufficient to request for an $m \times n$ matrix the condition

$$\max(m, n) > n_0$$

(probably this condition can be omitted).

b) As we have seen in the proof, in an ± 1 matrix one can change the signs of the columns in such a way that the absolute value of the sum of any row be less than $\sqrt{cn \log n}$. Erdős and ourselves have shown (with enumeration of matrices) that almost all $n \times n$ matrices have the property that changing the signs of the columns in any way, the absolute value of sum of at least one row will be grater than $\frac{\sqrt{n}}{2}$. (Davidson remarked that for Hadamard-matrices this minimum of row-sums is $\sqrt{n}$).

The exact value of this extremum

$$\left(A_n = \max_{(a_{ik})_{n \times n}} \ \min_{y_k = \pm 1} \ \max_{i=1}^{n} \left| \sum_{k=1}^{n} a_{ik} y_k \right| ; \right.$$

$$\left. c_1 \sqrt{n} < A_n < c_2 \sqrt{n \log n} \right)$$

is unknown.

c) Tusnády and van Lint asked what is the maximum of the sum of the elements of the matrix, more exaclty, the number

$$B_n = \min_{(a_{ik})} \max_{x_i, y_k = \pm 1} \left| \sum_{i,k=1}^{n} a_{ik} x_i y_k \right| .$$

One can easily show, using the first-moment method, that

$$c_1 n^{3/2} < B_n < c_2 n^{3/2} .$$

Enumeration problems and context-free languages

by

W. Kuich

Vienna, Austria

ABSTRACT:

The enumeration of the number of distinct words of length n that are contained in a context-free language provides a useful tool in solving enumeration problems in combinatorics.

1. PRELIMINARIES

We first provide the reader with basic facts concerning context-free languages.

An alphabet is a finite nonempty set. The set of all words, including the empty word ε, over an alphabet Σ is denoted by Σ^*.

A context-free grammar is a 4-tuple $G = (V, \Sigma, P, \sigma)$, where (i) V is an alphabet (the vocabulary); (ii) $\Sigma \subseteq V$ is an alphabet (the terminals); (iii) P is a finite set of productions of the form $\xi \longrightarrow v$, where ξ is in $V - \Sigma$ and v is in V^*; (iv) σ is in $V - \Sigma$ (the start symbol). Elements of $V - \Sigma$ are called nonterminals.

The relation $\underset{l}{\Longrightarrow}$ over words of V^* is defined as follows. For x, y in V^*, $x \underset{l}{\Longrightarrow} y$, if there exist ξ in $V - \Sigma$, u in Σ, v, w in V^* such that $x = u\xi v$, $y = uwv$ and $\xi \longrightarrow w$ is in P. The relation $\overset{*}{\underset{l}{\Longrightarrow}}$ is the transitive closure of $\underset{l}{\Longrightarrow}$, i.e. $x \overset{*}{\underset{l}{\Longrightarrow}} y$, if there exist $z_0, \dots, z_r$ such that $z_0 = x$, $z_r = y$, and $z_i \underset{l}{\Longrightarrow} z_{i+1}$ for $i = 0, \dots, r-1$. Such a sequence $z_0, \dots, z_r$ of words is called a leftmost derivation or a leftmost generation (from z_0 to z_r) of length r and is written $z_0 \underset{l}{\Longrightarrow} \cdots \underset{l}{\Longrightarrow} z_r$.

A subset L of Σ^* is called a context-free language if $L = L(G)$ for some context-free grammar $G = (V, \Sigma, P, \sigma)$ where $L(G) = \{w \text{ in } \Sigma^* \mid \sigma \overset{*}{\underset{l}{\Longrightarrow}} w\}$. $L(G)$ is said to be the language generated by G.

A context-free grammar $G = (V, \Sigma, P, \sigma)$ is ambiguous if there is some word in $L(G)$ generated by at least two different leftmost derivations from σ. A grammar which is not ambiguous is unambiguous.

Let L be a context-free language and let $u(n)$ be the structure function of L, i.e. $u(n)$ is the number of distinct words of length n contained in L. Then the function $f(z)$

(1) $$f(z) = \sum_{n=1}^{\infty} u(n) z^n$$

is called the structure generating function of L.

Assuming that the unambiguous context-free grammar

(2) $G = (V, \Sigma, P, \gamma_1)$ with nonterminals $V - \Sigma = \{\gamma_1, \dots, \gamma_k\}$

and terminals $\Sigma = \{x_1, \dots, x_\ell\}$

contains no productions of the form $\gamma_i \longrightarrow \gamma_j$ or $\gamma_i \longrightarrow \varepsilon$, it is possible to construct a system of equations whose unique solution is the structure generating function of $L(G)$.

Let $\phi_{i,1}, \ldots, \phi_{i,m_i}$ be all the elements of V^* such that $\gamma_i \longrightarrow \phi_{i,j}$, $(1 \le j \le m_i)$, is an element of P and write

$$\gamma_i = \phi_{i,1} + \cdots + \phi_{i,m_i}$$

for all γ_i in $V - \Sigma$. Replacement of each occurrence of γ_j and x_m by the power series y_j and the variable z, respectively, yields a system of equations

(3) $$y_i = H_i(y_1, \ldots, y_k; z), \quad (1 \le i \le k)$$

which has a unique solution

(4) $$y_i = f_i(z), \qquad f_i(0) = 0 \qquad (1 \le i \le k).$$

Besides $f_1(z)$ is the structure generating function of $L(G)$ (compare Theorem 2 of Kuich [2]).

2. ENUMERATION OF CODED OBJECTS

Consider a denumerable set S of element s called objects. Assume that each object s is mapped by a rank function r on a positive integer $r(s)$, called the rank of s, in such a manner that at most finitely many objects have rank n for a given positive integer n.

The generating function of S according to the rank function r is defined as

$$g(z) = \sum_{n=1}^{\infty} c(n) z^n,$$

where $c(n)$ is the number of distinct objects of rank n, i.e. $c(n)$ is the cardinality of the set $r^{-1}(n)$.

A coding of the set S compatible with the rank function r is the following:

Let Σ be an alphabet and let L be a subset of Σ^*. If it is possible to find a one-to-one onto mapping

$$h : S \longrightarrow L$$

such that for each s of S $h(s) = w$ implies $r(s) = \lambda(w)$ ($\lambda(w)$ denotes the length of the word w), then the coding of the set S by L is said to be compatible with the rank function r. Such a coding demands that objects of rank n are mapped on words of length n.

If L is a context-free language, then $c(n) = u(n)$, where $u(n)$ is the structure function of L. Hence the following theorem is obvious:

Theorem of Enumeration by Codes:

Let $L(G)$ be a context-free language generated by the context-free grammar $G = (V, \Sigma, P, \gamma_1)$ of (2). Let S be a denumerable set of objects s with associated rank function r. Assume that it is possible to code the objects of S by $L(G)$ compatibly with the rank function r.

Then the generating function $g(z)$ of S according to the rank function r is equal to the structure generating function $f(z)$ of $L(G)$.

3. THE ENUMERATION OF (k+1)-VALENT PLANTED PLANE TREES

This section brings an application of the Theorem of Enumeration by Codes.

Plane trees are trees which are embedded in the plane. A planted plane tree is a plane tree in which one point with valency one is distinguished and is called root. The line incident with the root is called stem.

We may suppose a positive sense of rotation in the plane to be fixed. Then the embedding of a tree in the plane induces a cyclic order on the lines incident with a certain point. Two planted plane trees are map-isomorphic if there exists a one-to-one mapping of the points of one tree onto those of the other which preserves (i) adjacency; (II) the root; and (III) the cyclic order.

A $(k+1)$-valent (planted plane) tree is a (planted plane) tree whose points assume only valencies one or $k+1$. We want to count the number of non-map-isomorphic $(k+1)$-valent planted plane trees with n points.

The set S of objects is the set of non-map-isomorphic $(k+1)$-valent planted plane trees. The rank $r(s)$ of a $(k+1)$-valent planted plane tree is the number of its points without the root.

The context-free language $L = L(G_k)$ is generated by the context-free grammar $G_k = (V_k, \Sigma_k, P_k, \sigma)$, where

(i) $\quad V_k = \{\sigma\} \cup \Sigma_k$

(ii) $\quad \Sigma_k = \{a_0, a_k\}$

(iii) $\quad P_k = \{\sigma \longrightarrow a_0, \; \sigma \longrightarrow a_k \underbrace{\sigma\sigma\cdots\sigma}_{k\text{-times}}\}$

The coding of S by $L(G_k)$ compatible with the rank function r is as follows (proofs are in Kuich [3]):

The word corresponding to a $(k+1)$-valent planted plane tree is obtained by walking in the plane around the tree in the positive sense of rotation, starting at the root. Whenever passing for the first time a point (except the root) of valency one or $k+1$, we put down the symbol a_0 or a_k respectively. Concatenation of all the symbols we put down yields the word corresponding to the given $(k+1)$-valent planted plane tree.

The system (3) associated with the context-free grammar G_k, has the form

$$y = z + zy^k .$$

The transformation

$$\begin{cases} w(x) = y/z \\ x = z^k \end{cases}$$

yields the equation

$$1 - w(x) + xw(x)^k = 0$$

which is solved by Pólya, Szegő [4], p. 125, No. 211.

Retransformation yields the structure generating function

$$f(z) = \sum_{n=0}^{\infty} \frac{(kn)!}{n![(k-1)n+1]!} z^{kn+1}$$

which is equal to the generating function of the set of non-map-isomorphic $(k+1)$-valent planted plane trees according to the number of points less one.

Hence there exist

$$\frac{(kn)!}{n![(k-1)n+1]!}$$

non-map-isomorphic $(k+1)$-valent planted plane trees with $kn+2$ points, $n = 0,1,2,\ldots$. This result generalizes a result achieved by Harary, Prins, Tutte [1] for trivalent planted plane trees, i.e. in the case $k = 2$.

REFERENCES

[1] HARARY, F., PRINS, G., TUTTE, W. T.: The Number of Plane Trees. - Indigationes Mathematicae 26 (1964), pp. 319-329.

[2] KUICH, W.: On the Entropy of Context-free Languages I. - Structure Generating Functions. - IBM Lab Vienna, Techn. Rep. TR 25.092, April 1969, to be published.

[3] KUICH, W.: Languages and the Enumeration of Planted Plane Trees. - IBM Lab Vienna, Lab. Rep. LR 25.6.010, July 1969, to be published.

[4] PÓLYA, G., SZEGŐ, G.: Aufgaben und Lehrsätze aus der Analysis. - Springer (1925), Berlin; reprinted, Dover (1945), New York.

Ein Kantennumerierungsproblem bei kubischen Graphen

von

R. Lang
Ilmenau, DDR

Im folgenden sei $G=(X,U)$ ein endlicher, ungerichteter Graph mit n Knotenpunkten und m Kanten. Eine Kantennumerierung von G ist eine eineindeutige Abbildung φ der Kantenmenge U auf die Menge $\mathcal{A}_m = \{1,2,3,\ldots,m\}$. Von den zahlreichen auf G möglichen Kantennumerierungen interessieren naturgemäß nur solche, die bestimmten Bedingungen genügen.

Eine Rechtfertigung dafür, Kantennumerierungsprobleme zu untersuchen, folgt aus der Tatsache, daß sich u.a. das Vierfarbenproblem als ein Kantennumerierungsproblem formulieren läßt. Der Vierfarbenvermutung ist bekanntlich folgende Vermutung äquivalent: Jeder planare, kubische, zweifachkantenzusammenhängende Graph zerfällt in drei Linearfaktoren.

Es sei nun G ein solcher Graph und φ eine Kantennumerierung von G. Jedem Knotenpunkt P von G ist durch φ eindeutig ein Zahlentripel (a,b,c) mit $a<b<c$ zugeordnet, wobei a,b,c die Nummern der drei mit P inzidierenden Kanten bezeichnen. Als Eckenschema E_φ von G bzgl. φ bezeichnen wir die folgende Zusammenstellung aller Zahlentripel von G bei vorgegebener Kantennumerierung φ.

Knotenpunkt	I	II	III
P_1	.	.	.
⋮			
P_l	a	b	c
⋮			
P_n	.	.	.

Da G ohne Schlingen ist, kommt in E_φ jede Zahl i, $1 \leq i \leq m$, genau zweimal, und zwar in verschiedenen Zeilen des Eckenschemas vor. Befindet sich überdies jede Zahl i, $1 \leq i \leq m$, beide Male in der gleichen Spalte, dann besitzt G drei paarweise fremde Linearfaktoren: Jede Spalte von E_φ repräsentiert einen Linearfaktor. Besitzt G umgekehrt eine Zerlegung in drei Linearfaktoren L_1, L_2 und L_3, dann kann man eine Numerierung φ von G angeben derart, daß in E_φ jede Zahl i, $1 \leq i \leq m$, zweimal in der gleichen Spalte vorkommt: Man braucht dazu nur die Kanten von L_1 mit den Nummern $1, 2, \ldots, \frac{n}{2}$, die Kanten von L_2 mit den Nummern $\frac{n}{2} + 1, \ldots, n$ und die Kanten von L_3 mit den Nummern $n+1, \ldots, \frac{3n}{2}$ zu versehen.

Daraus folgt unmittelbar folgender Satz, der die Äquivalenz des vierfarbenproblems mit einem Kantennumerierungsproblem aussagt.

SATZ 1: Die Flächen einer normalen Landkarte G lassen sich genau dann mit vier Farben zulässig färben, wenn es eine Kantennumerierung φ von G gibt derart, daß im Eckenschema E_φ keine Zahl in verschiedenen Spalten vorkommt.

Anstatt nun - wie das bei der Vierfarben-Vermutung der Fall ist - nach einer Kantennumerierung von G zu fragen, bei der im zugehörigen Eckenschema "größtmögliche Ordnung" herrscht, fragen wir im folgenden nach einer Kantennumerierung von G bei der im zugeordneten Eckenschema "größtmögliche Unordnung" herrscht, und zwar in dem Sinne, daß im Eckenschema keine Zahl - mit Ausnahme der Zahlen 1 und m - zweimal in der gleichen Spalte vorkommt.

Wir nennen eine Kantennumerierung φ eines kubischen Graphen zulässig, wenn keine Nummer i, $1 < i < m$, zweimal in der gleichen Spalte des Eckenschemas E_φ vorkommt. Wir werden auch sagen, eine einzelne Kante sei (bei gegebener Numerierung φ) zulässig numeriert, wenn sie entweder die Nummer 1 oder die Nummer m hat oder ihre Nummer nicht beide Male in der gleichen Spalte von E_φ auftritt. Eine Kantennumerierung φ von G ist genau dann zulässig, wenn jede Kante von G zulässig numeriert ist.

Aus dieser Definition ergeben sich unmittelbar folgende Eigenschaften zulässiger Kantennumerierungen:

1. Ist c die Nummer einer beliebigen Kante von G mit den Endpunkten P und Q und sind a, b, d, e die Nummern der übrigen mit P und Q inzidenten Kanten (Abb. 1), dann gilt

$$\mathrm{Min}(a,b,c,d,e) < c < \mathrm{Max}(a,b,c,d,e) \quad \text{für } c \neq 1, m .$$

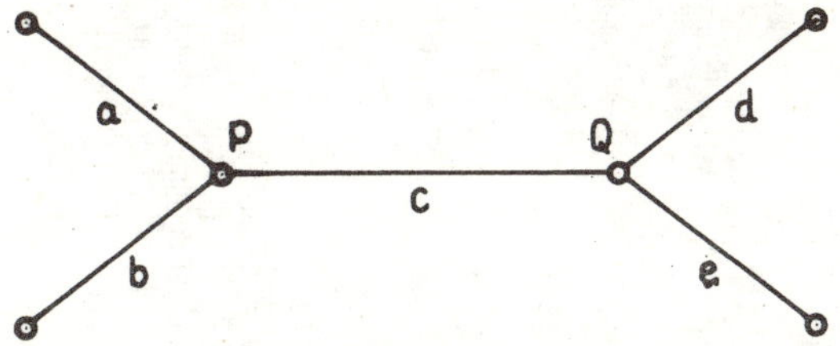

Abb. 1.

2. c liegt niemals sowohl zwischen den Nummern der mit P inzidenten Kanten als auch zwischen den Nummern der mit Q inzidenten Kanten; d.h. die Kante (P,Q) hat bezüglich der von P ausgehenden Kanten oder bezüglich der von Q ausgehenden Kanten eine extremale Nummer.

Das Ziel dieser Arbeit ist, diejenigen kubischen Graphen zu charakterisieren, die eine zulässige Kantennumerierung gestatten.

Wir beweisen den

SATZ 2: Jeder kubische, zweifach-kantenzusammenhängende Graph G mit mehr als zwei Knotenpunkten besitzt eine zulässige Kantennumerierung.

Beweis: Nach dem Satz von PETERSEN (Siehe [2], S. 190) kann ein Graph G, der den Voraussetzungen des Satzes genügt, in einen linearen und einen quadratischen Faktor zerlegt werden. Wir denken uns G in der Ebene so gezeichnet, daß bei einer gegebenen Zerlegung von G den Kreisen des quadratischen Faktors Kreise in der Ebene entsprechen (Abb. 2). Ferner denken wir uns die Kanten des Linearfaktors rot und die des quadratischen Faktors blau gefärbt. (In den Abbildungen werden die Kanten des Linearfaktors gestrichelt gezeichnet.) Eine rote Kante von G, die zwei Knotenpunkte des gleichen blauen

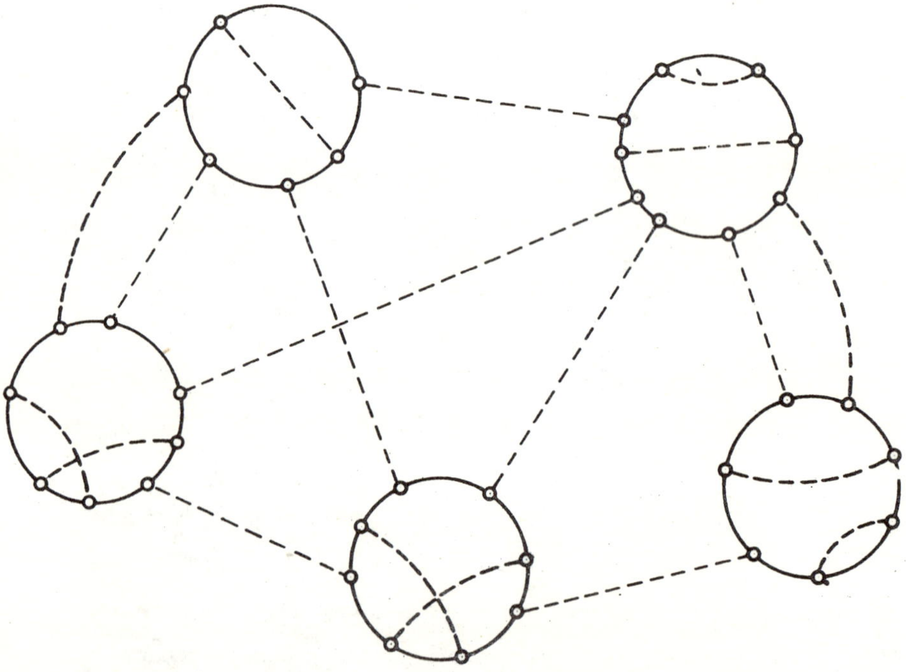

Abb. 2.

Kreises verbindet, nennen wir eine innere rote Kante; eine rote Kante, die zwei Knotenpunkte von verschiedenen blauen Kreisen verbindet, heiße äußere rote Kante.

Wir bezeichnen den quadratischen Faktor von G mit G'. Wir unterscheiden nun zwei Fälle, je nachdem ob G' zusammenhängend ist oder nicht.

I. G' ist zusammenhängend.

Der Graph G hat somit einen Hamiltonkreis, den wir dazu benutzen werden, um eine Kantennumerierung von G herzustellen. Es sei A ein beliebiger Knotenpunkt von G und a eine mit A inzidierende blaue Kante, die nicht in einem Zweieck liegt. Mit A als Anfangspunkt und a als erster Kante ist eine Durchlaufsrichtung des Hamiltonkreises festgelegt, und mit Hilfe dieser Durchlaufsrichtung des Hamiltonkreises kann man jeder blauen Kante eine Richtung zuordnen, die mit der Durchlaufsrichtung des Hamiltonkreises übereinstimmt. Dadurch ist es möglich, Anfangs- und Endpunkte bei den blauen Kanten zu unterscheiden. Den roten Kanten werden während der Numerierung Richtungen zugewiesen. Wir nehmen zunachst eine Numerierung der Kanten von G nach folgender Vorschrift vor:

NUMERIERUNGSVORSCHRIFT 1:

1. Die Kante a erhält die Nummer 1.

2. Es sei $v=(P,Q)$ die zuletzt numerierte blaue Kante und $N(v)$ ihre Nummer.

a) Sind zwei mit dem Endpunkt Q von v inzidierende Kanten noch unnumeriert, dann erhält die mit Q inzidierende rote Kante die Nummer $N(v)+1$ und diejenige Richtung, daß Q Anfangspunkt der roten Kante wird, und die mit Q inzidierende noch unnumerierte blaue Kante erhält die Nummer $N(v)+2$.

b) Inzidiert mit Q nur eine noch nicht numerierte Kante (diese ist notwendig eine blaue Kante), so erhält diese die Nummer $N(v)+1$.

c) Sind alle mit Q inzidierenden Kanten numeriert, so sind alle Kanten von G numeriert.

Wir zeigen als erstes, daß alle Kanten von G, die nicht mit A inzidieren, zulässig numeriert sind.

Sei v eine beliebige blaue Kante, die nicht mit A inzidiert. $N(v)$ steht im Eckenschema E_φ der durch Numerierungsvorschrift 1 erhaltenen Numerierung φ von G in der Zeile des Anfangspunktes von v in der dritten Spalte, denn die beiden anderen mit dem Anfangspunkt von v inzidierenden Kanten haben kleinere Nummern erhalten. In der Zeile des Endpunktes von v steht $N(v)$ sicherlich nicht in der dritten Spalte, denn die im Hamiltonkreis auf v folgende Kante hat eine Nummer, die größer als $N(v)$ ist. Daraus folgt, daß v zulässig numeriert ist.

Sei v eine beliebige rote Kante, die nicht mit A inzidiert. Ihre Nummer $N(v)$ steht in der Zeile des Anfangspunktes von v in der zweiten Spalte, in der Zeile des Endpunktes von v in der ersten Spalte; v ist also zulässig numeriert.

Von den mit A inzidierenden Kanten sind die beiden blauen Kanten per definitionem zulässig numeriert, denn sie haben die Nummern 1 und m erhalten. Die Nummer der mit A inzidierenden roten Kante steht jedoch im Eckenschema zweimal in der zweiten Spalte; diese Kante ist also nicht zulässig numeriert. Es macht sich eine Umnumerierung erforderlich. (Siehe Abb. 3).

UMNUMERIERUNGSVORSCHRIFT 2:

Es sei v die mit A inzidierende rote Kante und $N(v)$ ihre Nummer, B sei der Anfangspunkt von v. Ferner sei u diejenige blaue Kante,

deren Endpunkt B ist. Ihre Nummer ist $N(u) = N(v) - 1$.
Wir vertauschen die Nummern der Kanten u und v.

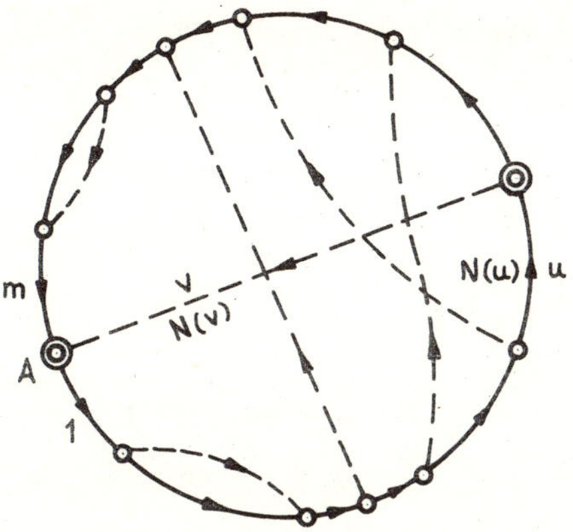

Abb. 3.

Man sieht unmittelbar, daß nun alle Kanten von G zulässig numeriert sind, denn die Nummer der Kante v steht nun in der Zeile von A in der zweiten Spalte, in der Zeile von B in der ersten Spalte. Die Zeile des Knotenpunktes B ist unverändert. Die neue Nummer der Kante u steht nun in der Zeile des Anfangspunktes von u in der dritten Spalte, wo auch die alte Nummer der Kante u stand. Alle anderen Zeilen des Eckenschemas blieben unverändert.

Wir haben somit eine zulässige Numerierung φ von G erhalten.

II. G' ist nicht zusammenhängend.

$G_1, G_2, \ldots, G_s$ seien die Komponenten von G'. Mit X_i bezeichnen wir die Knotenpunktmenge der Komponente G_i.

Wegen des zweifachen Kantenzusammenhanges von G gibt es in jeder Knotenpunktmenge X_i wenigstens zwei Knotenpunkte, die in G mit außeren roten Kanten inzidieren.

Mit $\overline{G}_i$ $(1 \leq i \leq s)$ bezeichnen wir denjenigen Untergraphen von G, der aus G_i entsteht, indem zu G_i alle inneren roten Kanten von G, die zwei Knotenpunkte von X_i verbinden, und alle äußeren roten Kanten von G, die mit einem Knotenpunkt aus X_i inzidieren, enschließlich der nicht zu X_i gehörenden Endpunkte der äußeren roten Kanten, hinzugefügt werden.

Um zu zeigen, daß alle Kanten von G zulässig numeriert werden können, führen wir eine Induktion durch.

a) Wir betrachten $\bar{G}_1$ und zeichnen zwei äußere roten Kanten a und b von $\bar{G}_1$ aus. Ihre zu X_1 gehörenden Endpunkte seien A und B (siehe Abb. 4). Wir werden auf $\bar{G}_1$ eine Kantennumerierung durchführen, so daß alle blauen und alle inneren roten Kanten von $\bar{G}_1$ zulässig numeriert sind, wobei die Kante a die kleinste und die Kante b die größte Nummer erhalten. Die Knotenpunkte A und B zerlegen den zu $\bar{G}_1$ gehörenden blauen Kreis in zwei Teile, und zwar in einen Weg K_1^1 von A nach B und einen Weg K_2^1 von A nach B. Die zu diesen Wegen gehörenden blauen Kanten werden so gerichtet, daß beim Durchlaufen der Wege von A nach B die Kanten in positiver Richtung durchlaufen werden

NUMERIERUNGSVORSCHRIFT 3:

1. Die Kante a bekommt die Nummer 1 und diejenige Richtung, daß A Anfangspunkt von a wird.

2. Von A aus werden die blauen und die mit den Knotenpunkten von X_1 inzidierenden roten Kanten von K_1^1 nach Numerierungsvorschrift 1, beginnend mit der Nummer 2, numeriert, bis die in B einlaufende blaue Kante von K_1^1 eine Nummer erhaltan hat; dann brechen wir die Numerierung ab und setzen die Numerierung mit der mit A inzidierenden blauen Kante von K_2^1 fort, wieder nach Numerierungsvorschrift 1, bis alle Kanten von $\bar{G}_1$ numeriert sind. Die Kante b erhält die größte Nummer.

In Abbildung 4 wird die Numerierungsvorschrift 3 demonstriert.

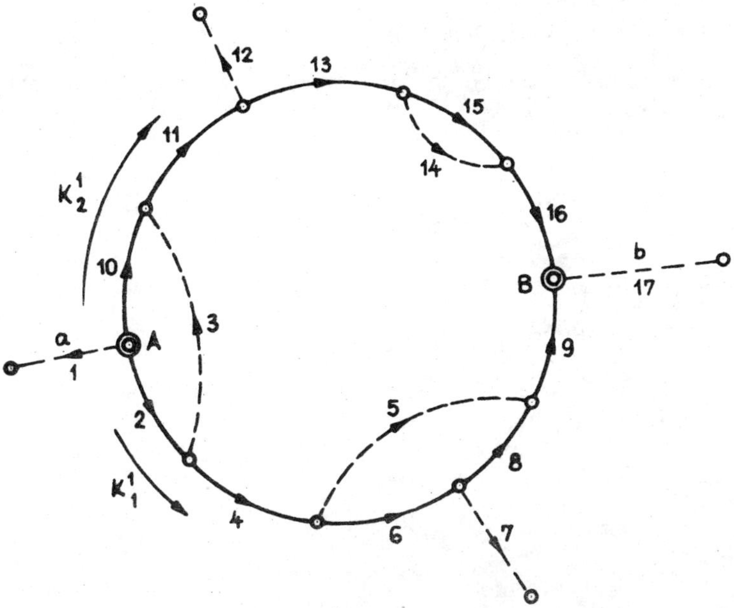

Abb. 4.

Zu Abbildung 4 gehört das folgende Eckenschema:

A	1	2	10
P_1	2	3	4
P_2	4	5	6
P_3	6	7	8
P_4	5	8	9
P_5	3	10	11
P_6	11	12	13
P_7	13	14	15
P_8	14	15	16
B	9	16	17

Man überzeugt sich leicht davon, daβ alle blauen und alle inneren roten Kanten von $\bar{G}_1$, auch die mit A and B inzidierenden, zulässig numeriert sind. Die Zulässigkeit der Numerierung der äuβeren roten Kanten wird später überprüft.

b) Angenommen, von den s Untergraphen $\bar{G}_1, \bar{G}_2, \dots, \bar{G}_s$ seien r bereits zulässig numeriert, und zwar $\bar{G}_1, \bar{G}_2, \dots, \bar{G}_r$ mit $1 \leq r \leq s$, d.h. alle blauen Kanten, alle inneren roten Kanten und alle äußeren roten Kanten, die zwischen Knotenpunkten von bereits numerierten Untegraphen verlaufen, seien zulässig numeriert. Dabei seien genau die Nummern $1, 2, 3, \dots, L$ verbraucht worden.

Wir zeigen: Dann können wir von den Untegraphen $\bar{G}_{r+1}$, $\bar{G}_{r+2}, \dots, \bar{G}_s$ einen auswählen - wir wählen die Numerierung der Untergraphen so, daß dieser ausgewählte der Untergraph $\bar{G}_{r+1}$ ist - und eine zulässige Kantennumerierung mit aufeinanderfolgenden Nummern für die Untegraphen $\bar{G}_1, \bar{G}_2, \dots, \bar{G}_r, \bar{G}_{r+1}$ angeben.

Wir unterscheiden zwei Fälle:

α) Unter den Untegraphen $\bar{G}_{r+1}, \bar{G}_{r+2}, \dots, \bar{G}_s$ gibt es einen, bei dem genau eine äußere rote Kante bereits numeriert ist.

Ohne Beschränkung der Allgemeinheit nehmen wir an, dieser sei $\bar{G}_{r+1}$. Die bereits numerierte äußere rote Kante sei die Kante a mit der Nummer $N(a)$. Von den blauen Kanten von $\bar{G}_{r+1}$ ist noch keine numeriert (denn unser Numerierungsverfahren ist so eingerichtet, daß mit einer blauen Kante alle blauen Kanten derselben Komponente von G' numeriert werden). Der Untegraph $\bar{G}_{r+1}$ möge k unnumerierte Kanten besitzen. Diesen Kanten werden wir die Nummern $N(a)+1, N(a)+2, \dots, N(a)+k$ zuordnen. Möglicherweise steht aber ein Teil dieser Nummern nicht mehr zur Verfügung, da er bereits für die Numerierung anderer Kanten benutzt wurde. Wir können aber die benötigten Nummern freisetzen, indem wir auf den bereits numerierten Kanten die folgende Umnumerierung durchführen:

NUMMERNTRANSFORMATION $T[N(a); k]$;

Hat die bereits numerierte Kante u die Nummer $N(u)$, dann wird $N(u)$ ersetzt durch die neue Nummer $N^*(u)$ mit

$$N^*(u) = \begin{cases} N(u), & \text{wenn } 1 \leq N(u) \leq N(a) \\ N(u)+k, & \text{wenn } N(a)+1 \leq N(u) \leq L. \end{cases}$$

Man sieht unmittelbar, daß eine Kante, die vor der Nummerntransformation T zulässig numeriert war, auch nach der Nummerntransformation zulässig numeriert ist. Denn sind etwa f, g, h drei beliebige bereits numerierte Kanten und $N(f)$, $N(g)$, $N(h)$ ihre Nummern, dann gilt: aus $N(f) < N(g) < N(h)$ folgt stehts $N^*(f) < N^*(g) < N^*(h)$. Daraus folgt, daß sich bei Anwendung der Transformation T die Reihenfolge der Kantennummern in den Zeilen des Eckenschemas nicht ändert.

Nach Ausführung der Nummerntransformation T (die Sterne lassen wir anschließend der Einfachheit halber wieder weg), wählen wir eine äußere rote Kante von $\bar{G}_{r+1}$, die noch ohne Nummer ist (wegen des zweifachen Kantenzusammenhanges gibt es eine solche), nennen diese Kante b und numerieren

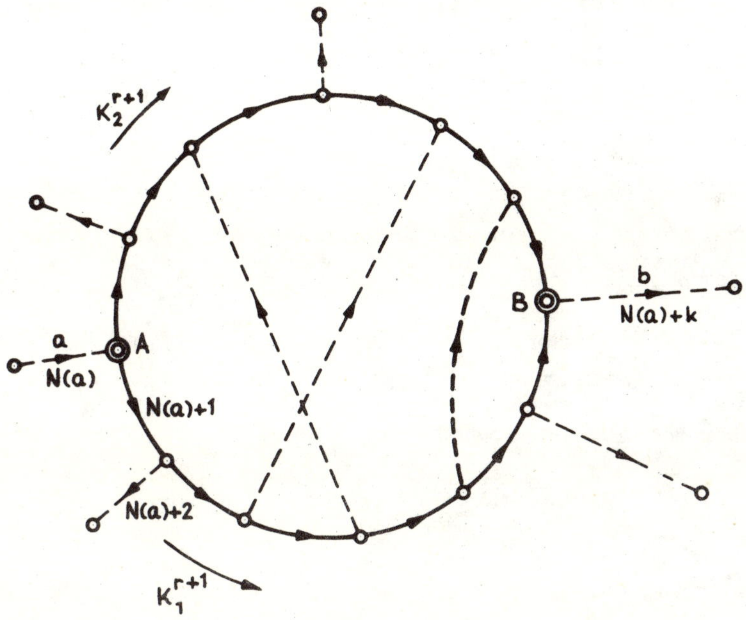

Abb. 5.

$\bar{G}_{r+1}$ nach der Numerierungsvorschrift 3 mit a und b als den ausgezeichneten äußeren roten Kanten, und zwar nun mit den Nummern $N(a)+1, N(a)+2, \ldots$ $\ldots, N(a)+k$. Die Kante b bekommt die Nummer $N(a)+k$ (Abb. 5).

Wie bei der Numerierungsvorschrift 3 sind alle blauen und alle inneren roten Kanten von $\bar{G}_{r+1}$ zulässig numeriert. Überdies ist nun die Kante a zulässig numeriert, denn wenn $N(a) = 1$, dann ist a per definitionem zulässig numeriert, wenn $N(a) \neq 1$, dann steht $N(a)$ im Eckenschema in der Zeile des Endpunktes von a in der ersten Spalte, in der Zeile des Anfangspunktes aber in der zweiten Spalte.

Damit haben wir eine zulässige Numerierung der Kanten der Untegraphen $\bar{G}_1, \bar{G}_2, \bar{G}_3, \ldots, \bar{G}_r, \bar{G}_{r+1}$ mit den fortlaufenden Nummern $1, 2, \ldots, L+k$ erhalten.

β) Unter den Untergraphen $\bar{G}_{r+1}, \bar{G}_{r+2}, \ldots, \bar{G}_s$ gibt es keinen, bei dem genau eine äußere rote Kante bereits numeriert ist.

Dann gibt es unter ihnen aber einen Untergraphen - ohne Beschränkung der Allgemeinheit können wir annehmen, daß dieser $\bar{G}_{r+1}$ ist - mit mindestens zwei bereits numerierten äußeren roten Kanten. Wäre das nämlich nicht der Fall, dann würde es in G keine äußere rote Kante (P, Q) geben mit $P \in X_i$ $(1 \leq i \leq r)$ und $Q \in X_j$ $(r+1 \leq j \leq s)$, denn eine solche Kante hätte bei der Numerierung von G_i eine Nummer erhalten. Das bedeutet aber: G wäre nicht zusammenhängend, im Widerspruch zur Voraussetzung.

b und a seien diejenigen bereits numerierten äußeren roten Kanten von $\bar{G}_{r+1}$ mit größter bzw. zweitgrößter Nummer $N(b)$ und $N(a)$. Ihre zu X_{r+1} gehörenden Endpunkte seien B und A (Abb. 6).

Zur Numerierung der Kanten von $\bar{G}_{r+1}$ werden wir die Nummern $N(a)+1, N(a)+2, \ldots, N(a)+1$ verwenden, wobei wir annehmen, daß $\bar{G}_{r+1}$ noch 1 unnumerierte Kanten besitzt. Zunächst wenden wir auf alle Nummern der bereits numerierten Kanten die Nummerntransformation $T[N(a); 1]$ an.

Dadurch werden die Nummern $N(a)+1, \ldots, N(a)+1$ freigesetzt. K_1^{r+1} und K_2^{r+1} seien die beiden von A nach B führenden Wege mit blauen Kanten.

Wir unterscheiden zwei Fälle:

FALL 1: In X_{r+1} gibt es einen Knotenpunkt, der mit einer noch unnumerierten roten Kante inzidiert.

Dann gibt es auf K_1^{r+1} oder auf K_2^{r+1} wenigstens einen solchen Knotenpunkt D mit der Eigenschaft: Beim Durchlaufen von K_1^{r+1} oder K_2^{r+1} von A nach B ist D der erste Knotenpunkt, der angetroffen wird, der mit einer noch unnumerierten roten Kante inzidiert. C sei der Anfangspunkt der in D einmündenden blauen Kante. (Eventuell ist $C = A$).

Von C nach B führen nun zwei blaue Wege: W_1^{r+1}, der die Kante (C,D) enthält, und W_2^{r+1}. Wir legen die Orientierung der blauen Kanten von $\bar{G}_{r+1}$ neu fest: Die Richtung jeder blauen Kante von $\bar{G}_{r+1}$ stimme mit der Durchlaufsrichtung desjenigen der beiden Wege W_1^{r+1} und W_2^{r+1}, dem sie angehört, überein.

Nun numerieren wir nach der

NUMERIERUNGSVORSCHRIFT 4:

Wir numerieren nach Numerierungsvorschrift 3 zunächst W_1^{r+1}, beginnend mit der Kante (C,D), welche die Nummer $N(a)+1$ erhält, dann W_2^{r+1}, bis alle Kanten von $\bar{G}_{r+1}$ numeriert sind.

Man kann leicht überprüfen, daß alle Kanten von $\bar{G}_{r+1}$ zulässig numeriert sind. (Abb. 6)

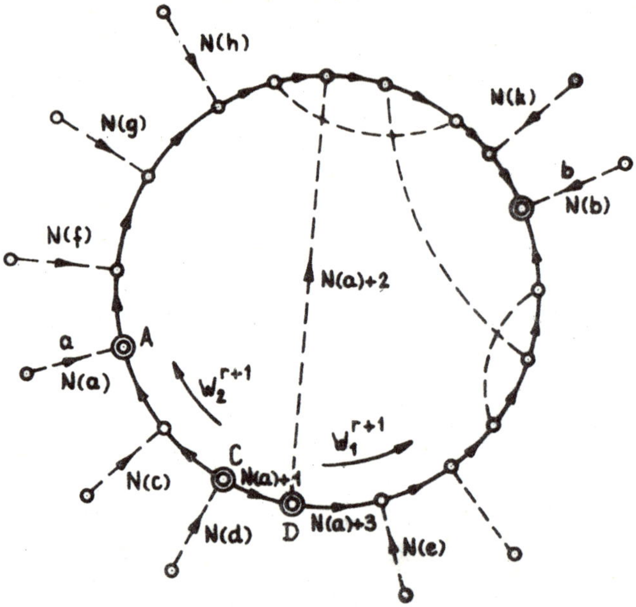

Abb. 6.

FALL 2: Alle roten Kanten von $\bar{G}_{r+1}$ sind bereits numerierte äußere rote Kanten (Abb. 7).

In diesem Falle numerieren wir die Kanten von $\bar{G}_{r+1}$ nach der folgenden

NUMERIERUNGSVORSCHRIFT 5:

Wir durchlaufen von B ausgehend den blauen Kreis von $\bar{G}_{r+1}$, geben der Kanten die Orientierung entsprechend der Durchlaufsrichtung und numerieren die blauen Kanten der Reihe nach mit den Nummern $N(a)+1, N(a)+2, \ldots, N(a)+1$.

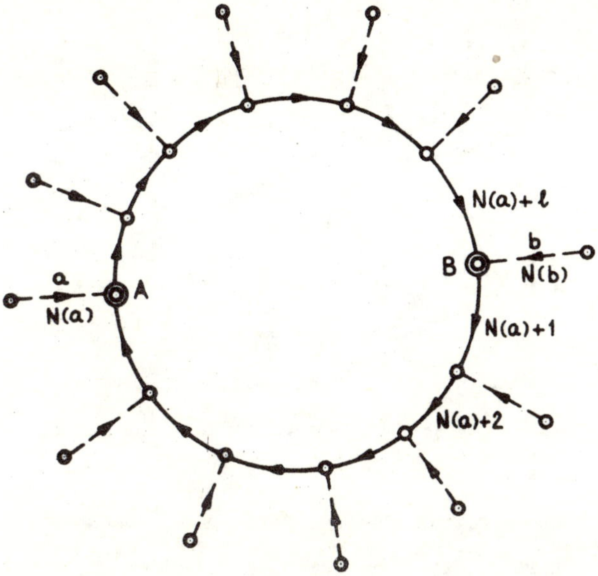

Abb. 7.

Auch in diesem Fall ist leicht einzusehen, daß alle Kanten von $\bar{G}_{r+1}$ zulässig numeriert sind.

Da G' nur endlich viele Komponenten enthält, bricht das Numerierungsverfahren nach endlich vielen Schritten ab.

Nach Beendigung des Numerierungsverfahrens sind alle Kanten von G zulässig numeriert, d.h. es liegt eine zulässige Kantennumerierung φ von G vor. Damit ist der Beweis des Satzes 2 beendet.

BEMERKUNGEN:

1. Die Voraussetzung des Satzes 2, daß G mehr als zwei Knotenpunkte haben muß, ist notwendig; denn der Graph von Abb. 8 besitzt keine zulässige Kantennumerierung.

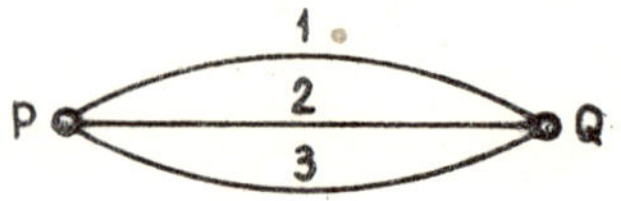

P	1	(2)	3
Q	1	(2)	3

Abb. 8.

2. Die Voraussetzung des Satzes, daß G zweifach-kantenzusammenhängend ist, kann noch etwas abgeschwächt werden. Es gnügt zu fordern, daß es in G einen Weg gibt, der alle Brücken von G enthält. Nach dem Satz von PETERSEN-ERRERA ([1], S. 209) besitzt dann G einen Linearfaktor, der notwendig alle Brücken von G enthält. Die Numerierung der Kanten von G kann mittels des oben angegebenen Algorithmus in der durch Abb. 9 angegebenen Reihenfolge geschehen.

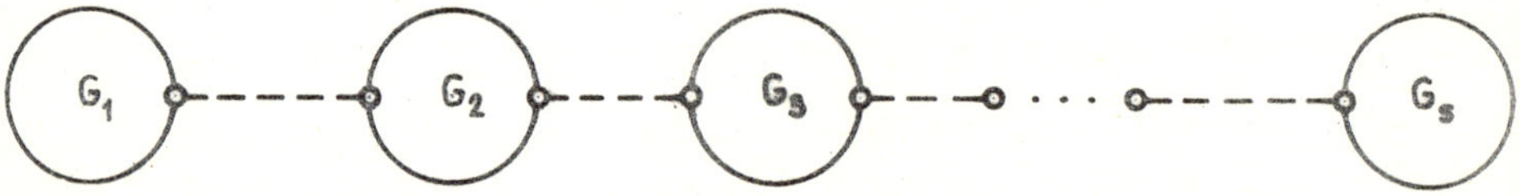

Abb. 9.

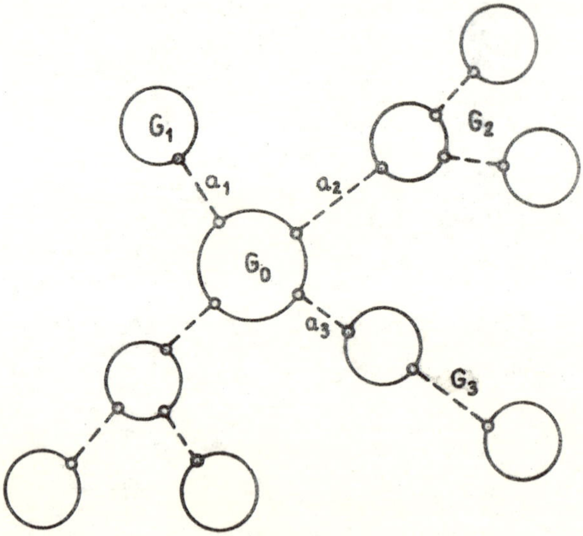

Abb. 10.

Gibt es jedoch in G keinen Weg, der alle Brücken von G enthält, dann läßt sich zeigen, daß G keine zulässige Kantennumerierung besitzt. Denn dann gibt es einen Untergraphen G_0, von dem aus drei Brücken a_1, a_2, a_3 zu den zugehörigen Ufern G_1, G_2, G_3 führen (Abb. 10).

Es genügt offenbar zu zeigen, daß derjenige Untergraph G^* von G, der aus $G_0, G_1, G_2, G_3, a_1, a_2, a_3$ besteht, keine zulässige Kanten-numerierung gestattet.

Angenommen, φ sei eine zulässige Kantennumerierung des Graphen G^*, r_i sei die kleinste und R_i die größte Nummer der Kanten in G_i $(1 \leq i \leq 3)$. Wenigstens eine Brücke, sagen wir a_1, hat eine von 1 und m verschiedene Nummer. Wir betrachten r_1 und R_1. Ist $r_1 \neq 1$ oder $R_1 \neq m$, dann ergibt sich ein Widerspruch, denn dann steht r_1 oder R_1 im Ecken-schema E_φ zweimal in der gleichen Spalte. Ist $r_1 = 1$ und $R_1 = m$, dann steht aber entweder r_2 oder R_2 zweimal in derselben Spalte von E_φ, im Widerspruch zur Voraussetzung.

Damit kann die Klasse der kubischen Graphen, die eine zulässige Kantennumerierung gestatten, vollständige charakterisiert werden.

SATZ 3: Ein zusammenhängender kubischer Graph G mit mehr als zwei Knotenpunkten besitzt genau dann eine zulässige Kantennumerierung, wenn es in G einen Weg gibt, der alle Brücken von G enthält.

3. Setzt man bei G zusätzlich Planarität und die Existenz einer Whitneyschen Transversalen (das ist eine einfach-geschlossene Linie, die durch jede Fläche von G genau einmal hindurch geht und dabei keinen Knoten-punkt enthält) voraus, dann läßt sich die Existenz einer zulässigen Kanten-numerierung von G sehr leicht nachweisen [3].

4. Betrachtet man allgemeiner reguläre Graphen r-ten Grades ohne Schlingen, so entspricht jeder Kantennumerierung φ ein Eckenschema

mit r Spalten. φ heißt zulässig, wenn außer den Zahlen 1 und m keine Zahl zweimal in derselben Spalte vorkommt. Es ist möglich, die Richtigkeit der folgenden beiden Sätze nachzuweisen:

SATZ 4: Jeder zusammenhängende, schlichte reguläre Graph von geradem Grad ist zulässig numerierbar.

SATZ 5: Jeder zweifach-kantenzusammenhängende, schlichte, reguläre Graph von ungeradem Grad, der einen Linearfaktor hat, ist zulässig numerierbar.

Beim Beweis des Satzes 4 wird wesentlich die Existenz einer Eulerlinie benutzt, beim Beweis von Satz 5 wird die Tatsache benutzt, daß jede Komponente von G' (G' entsteht aus G, indem alle Kanten des Linearfaktors aus G entfernt werden) eine Eulersche Linie besitzt.

LITERATUR:

[1] BERGE, C., "Théorie des graphes et ses applications", Paris 1958.

[2] KÖNIG, D., "Theorie der endlichen und unendlichen Graphen", Leipzig 1936

[3] LANG, R., "Über Kantennumerierungen regulärer Graphen" Wiss. Zeitschrift der TH Ilmenau (im Druck)

Permutations with strongly restricted displacements

by

D. H. Lehmer

Berkeley, USA

The study of classes of permutations of the n marks

$$1, 2, 3, \ldots, n$$

subject to various restraints is a time honored problem in combinatorics. When the restraints refer to iterated permutations we have the study of finite groups. There is a lot of information available in the cases when the restraints refer to the cycle structure of the permutations [1], the differences between successive marks [2], runs up and down [1] etc. The present paper is a modest contribution to the study of classes of permutations in which the positions of the marks after permutation are restricted, especially those cases in which the restrictions are natural and strong.

A class of permutations of this latter type can be specified by an $n \times n$ matrix $A = A_n = \{a_{ij}\}$, of zeros and ones, in which

$$a_{ij} = \begin{cases} 1 \text{ if the mark } i \text{ is permitted to occupy} \\ \text{the } j^{\text{th}} \text{ place} \\ 0 \text{ otherwise} \end{cases}$$

Thus

$$A = \begin{bmatrix} 0 & 1 & 0 & 0 \\ 1 & 0 & 1 & 0 \\ 1 & 1 & 0 & 1 \\ 0 & 0 & 1 & 1 \end{bmatrix}$$

specifies the restriction satisfied by the two permutations

$$\begin{pmatrix} 1 & 2 & 3 & 4 \\ 2 & 1 & 4 & 3 \end{pmatrix} \quad \text{and} \quad \begin{pmatrix} 1 & 2 & 3 & 4 \\ 3 & 1 & 2 & 4 \end{pmatrix}$$

The number of permutations specified by any $(0,1)$ matrix A is given by the permanent function

(1) $$\operatorname{perm}(A) = \sum_{(\pi)} a_{1,\pi(1)} a_{2,\pi(2)} \cdots a_{n,\pi(n)}$$

This includes even those matrices A which, strictly speaking, have no actual permutation associated with them, namely those with a row of zeros so that some mark is not allowed to participate in the new assignment, and those with a column of zeros in which some position in the permutation goes unfilled.

From one viewpoint the permanent function solves our problem of enumerating position-wise restricted permutations. On the other hand, the calculation of permanents is not nearly as easy as that of determinants and so we can regard (1) as merely a restatement of the enumeration problem. We can even evaluate permanents of $(0,1)$ matrices by counting the number of restricted permutations in some independent way.

Let the row sums of A be denoted by

$$r_i = \sum_{j=1}^{n} a_{ij}$$

There are two conspicuous cases of interest: those in which $n-r_i$ is uniformly small in i, and those in which r_i is uniformly small. These cases refer to weakly and strongly restricted permutations respectively. The trivial cases are the matrix J consisting of all ones and the unit matrix I (or any other permutation matrix) giving $n!$ and 1 permutations respectively. If $A = J - I$ we have the famous Problème des Rencontres in which

$$\mathrm{perm}(A) \sim n!/e.$$

Other weakly restricted permutations have been studied by Mendelsohn [3], [4], who obtained similar asymptotic expressions in these cases. He also investigated two or three strongly restricted cases. These are more difficult and his results are slightly in error. Two examples occur in Hall [5].

The restrictions imposed on the class of permutations by the matrix A are twofold: (a) each mark is confined to a set of positions and (b) each mark is responsible for not interfering with his fellow marks. If we overlook condition (b) we obtain the trivial inequality

$$\mathrm{perm}(A) \leq \prod_{i=1} r_i$$

In the case of strongly restricted permutations where

$$r_i \leq K$$

for some integer K not depending on n we have

$$[\mathrm{perm}(A_n)]^{1/n} \leq K.$$

Let R be a strong restriction, that is a general restrictive statement confining the movement of marks (such as, "no mark may move left or right more than two places"). We denote by $N(R,n)$ the total number of permutations resulting from the imposition of R on n marks. If the limit

$$\lim_{n\to\infty} \{N(R,n)\}^{1/n} = \mu(R)$$

exists we call $\mu = \mu(R)$ the "index of mobility of R".

This is appropriate since for large n the effect of R is asymptotically the same as though each mark could move independently to any one of μ places (including staying put).

In the case of strongly restricted permutations the geometric mean of the row sums of A_n will tend to a limit $P = P(R)$ as $n \to \infty$. The "index of interference" of R is defined by

$$\Theta(R) = \mu(R)/P(R)$$

and it is a measure of the effect of conditions (b). Of course, the smaller Θ the more interference is encountered.

We intend to introduce natural restrictions R, to discuss briefly the methods available to us, and to give more or less explicit expressions for $N(R,n)$, $\mu(R)$ and $\Theta(R)$.

For future reference it is convenient to list the restrictions to be considered.

List of Strong Restrictions R

Restriction	Description
$R_1^{(k)}$	No mark shall move more than k places left or right.
$R_2^{(k)}$	When the marks are deployed in a circle no mark shall move more than k positions clockwise or counterclockwise.
$R_3^{(k)}$	When in a circle each mark shall move clockwise only, but not more than k places.
$R_4^{(k)}$	When deployed on a line, the mark n goes to the first place. All other marks move right not more than k places.
$R_5^{(k)}$	No mark shall move more than k places left or right but each mark must move.

It should be noted that $N(R_2^{(k)}, n) = N(R_3^{(2k}, n)$ since in each case the rows of the specification matrix A consist in all n cyclic permutations of

$$1 \;\; 1 \;\; 1 \;\; 1 \;\; . \;\; . \;\; . \;\; 1 \;\; 0 \;\; 0 \;\; 0 \;\; . \;\; . \;\; . \;\; 0$$

where the number of ones is $2k+1$.

We list here six methods of approach to the problems corresponding to the above restrictions.

1. Cross classification (Inclusion and Exclusion, Eratosthenes sieve
2. Back tracking
3. Separation of many cases
4. Inductive expansion of the permanent
5. Diagraph method
6. Inequalities for permanents.

Method 1, which is so powerful for the weakly restricted case, turns out to be rather inefficient in our case. To illustrate its use, consider the problem of evaluating $N(R_1^{(2)}, 5)$. The total population consists in the permutations of 1,2,3,4,5 . There are six illegal properties of these that are of interest:

1. 1 is in place 5	4. 4 is in place 1
2. 1 is in place 4	5. 5 is in place 2
3. 2 is in place 5	6. 5 is in place 1

We ask for the number of permutations having no property.

The number of permutations having 1,2,3 and 4 properties is 4!, 3!, 2! and 1! respectively. The numbers of simultaneous consistent properties are

1 at a time - 6

2 at a time - 11

3 at a time - 6

4 at a time - 1

Hence

$$N(R_1^{(2)}, 5) = 5! - 6.4! + 11.3! - 6.2! + 1.1! = 31 .$$

In contrast, method 2, back tracking, is quite efficient for strongly restricted permutations. Using a computer, especially a captive one, the initial values of $N(R,n)$ can be computed for the first 10 or 20 values of n without much trouble. The method consists in building up restricted permutations, as vectors, one component at a time abandoning in a single step the millions of permutations that cannot be completed. For details see [6].

Both methods 1 and 2 are useful in obtaining the initial values of what will turn out to be a recurring series for $N(R,n)$. The other methods except 6 have to do with the determination of the recurrence relation for $N(R,n)$.

The simplest example of method 3 is that of showing that

(2) $$N(R_1^{(1)}, n) = N(R_1^{(1)}, n-1) + N(R_1^{(1)}, n-2) .$$

There are two cases of permutations restricted by $R_1^{(1)}$.

Case I. The mark n remains fixed

Case II. The mark n goes to place $n-1$

In Case I the number of permutations is $N(R_1^{(1)}, n-1)$.

In Case II the mark $n-1$ is forced to the n place.

Cancelling the last two positions, the number of Case II permutations is $N(R_1^{(1)}, n-2)$. Thus (2) is established. From the initial conditions

$$N(R_1^{(1)}, 1) = 1 \qquad N(R_1^{(1)}, 2) = 2 \qquad \text{we have}$$

$$N(R_1^{(1)}, n) = F_{n+1}$$

the Fibonacci number of index $n+1$. It follows that

$$\mu(R_1^{(1)}) = \frac{1+\sqrt{5}}{2} = 1.6180, \quad \Theta(R_1^{(1)}) = \frac{\mu(R_1^{(1)})}{3} = 0.5393$$

Although each mark, except the first and last, is given permission to make one of three moves, only about 50 percent of the time is it successful in its choice because of strong neighborly interference. In the case of $R_2^{(1)}$ the argument is more complicated since the image of n can be $n, n-1$ or 1. There are 3 main cases, two with two subcases each. The answer is

$$N(R_2^{(1)}, n) = 2 + L_n$$

where L_n is the nth term of the associated Fibonacci series

$$1, 3, 4, 7, 11, 18, \ldots$$

As might be expected, the two indices μ and Θ are the same as for $R_1^{(1)}$. $N(R_2^{(1)} n)$ is really a third order recurring series

$$N(R_2^{(1)}, n) = 2N(R_2^{(1)}, n-1) - N(R_2^{(1)} n-3) \quad (n>5)$$

It is generated by

$$\sum_{n=1}^{\infty} N(R_2^{(1)}, n)x^n = x + 2x^2 + x^3 \frac{6-3x-5x^2}{1-2x+x^3}$$

For $k>1$ the separation of cases method becomes much more elaborate as the results for $k=2,3$ given below indicate. There is the possibility here of carrying out the case splitting automatically. For $R_4^{(k)}$ the restriction is sufficiently strong to enable a complete solution of the problem by separation of cases and induction. The result is the recurrence

$$N(R_4^{(k)}, n) = 2N(R_4^{(k)}, n-1) - N(R_4^{(k)}, n-1-k).$$

Method 4, inductive expansion of $\text{perm}(A_n)$, in somewhat akin to method 3 and perhaps is more effective. It can be used to obtain the indices μ and Θ without giving $N(R,n)$ explicitly.

The method is as follows: If we expand $\text{perm}(A_n)$ by the first row (or column) we obtain, for a general n,

$$N(R,n) = \text{perm}(A_n) = \text{perm}(A^{(1)}_{n-1}) + \text{perm}(A^{(2)}_{n-1}) + \dots + \text{perm}(A^{(s)}_{n-1}) \tag{3}$$

Each of these permanents in turn can be so expanded and the process continued until no "new" matrices occur. For each occurring matrix we define a formal power series

$$f_t(x) = \sum_{n=1}^{\infty} \text{perm}(A^{(t)}_n)x^n \qquad (t = 0,1,\dots)$$

where $A^{(0)}_n = A_n$. The relations of the type (3) now become linear equations in the f_t with coefficients that are simple polynomials in x . These equations may now be solved for $f_0(x)$ as a rational function of x . By a theorem of Lagrange, $\text{perm}\ A_n$ is then a combination of the nth powers of the distinct roots of the denominator of $f_0(x)$. The largest root then becomes $\mu(R)$ which by its very nature is positive.

To illustrate the procedure let us consider $N(R^{(2)}_5, n)$ in which each mark must move but not more than 2 places left or right. Here we have

$$A_n = \begin{bmatrix} 0 & 1 & 1 & 0 & 0 & . & . & . & 0 & 0 & 0 \\ 1 & 0 & 1 & 1 & 0 & . & . & . & 0 & 0 & 0 \\ 1 & 1 & 0 & 1 & 1 & . & . & . & 0 & 0 & 0 \\ . & . & . & . & . & . & . & . & . \\ 0 & 0 & 0 & 0 & 0 & . & . & . & 1 & 0 & 1 \\ 0 & 0 & 0 & 0 & 0 & . & . & . & 1 & 1 & 0 \end{bmatrix}$$

If we set

$$B_n = \begin{bmatrix} 1 & 1 & 1 & 0 & 0 & . & . & . & 0 & 0 & 0 \\ 1 & 0 & 1 & 1 & 0 & . & . & . & 0 & 0 & 0 \\ 0 & 1 & 0 & 1 & 1 & . & . & . & 0 & 0 & 0 \\ 0 & 1 & 1 & 0 & 1 & . & . & . & 0 & 0 & 0 \\ . & . & . & . & . \\ 0 & 0 & 0 & 0 & 0 & . & . & . & 1 & 0 & 1 \\ 0 & 0 & 0 & 0 & 0 & . & . & . & 1 & 1 & 0 \end{bmatrix} \qquad C_n = \begin{bmatrix} 1 & 0 & 1 & 0 & 0 & . & . & . & 0 & 0 & 0 \\ 1 & 1 & 1 & 1 & 0 & . & . & . & 0 & 0 & 0 \\ 0 & 1 & 0 & 1 & 1 & . & . & . & 0 & 0 & 0 \\ 0 & 0 & 1 & 0 & 1 & . & . & . & 0 & 0 & 0 \\ 0 & 0 & 1 & 1 & 0 & . & . & . & 0 & 0 & 0 \\ . & . & . & . & . & . & . & . & . & . & . \\ 0 & 0 & 0 & 0 & 0 & . & . & . & 1 & 1 & 0 \end{bmatrix}$$

Then we find for $n \geq 4$

$$\begin{aligned} \operatorname{perm}(A_n) &= \operatorname{perm}(B_{n-1}) + \operatorname{perm}(C_{n-1}) \\ \operatorname{perm}(B_n) &= \operatorname{perm}(A_{n-1}) + \operatorname{perm}(B_{n-1}) \\ \operatorname{perm}(C_n) &= \operatorname{perm}(A_{n-3}) + \operatorname{perm}(B_{n-1}) \end{aligned} \tag{3'}$$

Setting

$$f_0(x) = \sum_{n=1}^{\infty} \operatorname{perm}(A_n)x^n = x^2 + 2x^3 + 4x^4 + \dots$$

$$f_1(x) = \sum_{n=1}^{\infty} \operatorname{perm}(B_n)x^n = x + x^2 + 2x^3 + 4x^4 + \dots$$

$$f_2(x) = \sum_{n=1}^{\infty} \operatorname{perm}(C_n)x^n = x + x^2 + 2x^3 + 2x^4 + \dots$$

we see that (3') implies

$$\begin{aligned} f_0 - xf_1 - xf_2 &= -x^2 \\ -xf_0 + (1-x)f_1 &= x \\ -x^3f_0 - xf_1 + f_2 &= x + x^3 \end{aligned}$$

Solving for f_0 we find

$$f_0(x) = \sum N(R_5^{(2)}, n)x^n = \frac{x^2(1+x+x^2-x^3)}{1-x-x^2-x^3-x^4+x^5}$$

Here the denominator factors into

$$(x+1)(x^2-(1+\sqrt{2})x+1)(x^2-(1-\sqrt{2})x+1)$$

whose largest root

$$\frac{1}{2}\left\{1+\sqrt{2}+\sqrt{2\sqrt{2}-1}\right\} = 1.88320$$

is the index of mobility for $R_5^{(2)}$.

This method can be made the basis of a proof of the fact that $\mu(R)$, the index of mobility, is an algebraic number.

Method 5 which was suggested to me by Professor DAVID

BLACKWELL [7] is aimed at finding the index of mobility $\mu(R)$ rather than a precise formula for $N(R,n)$.

Suppose R is such that no mark moves more than k places left or right. Since n will tend to infinity, we may suppose that n is much larger than k. Suppose further that we are in the process of constructing a permutation subject to R and that the partially built permutation is

$$(4) \qquad \begin{pmatrix} 1 & 2 & \dots & h-1 \\ \pi(1) & \pi(2) & \dots & \pi(h-1) \end{pmatrix}$$

where h is also much larger than k. Our immediate problem is to select a value for $\pi(h)$. Our choice of $\pi(h)$ will depend not only on R but also on the actual values already chosen for $\pi(h-1), \pi(h-2),\dots,\pi(h-k)$, since we must not select a value of $\pi(h)$ that has already been used. In some cases our choice of $\pi(h)$ may be forced by the fact that certain values have not yet been chosen. Hence there are a number of states of incompletion of the permutation (4). We may number these states, in any order, $1,2,3,\dots,s$. We now build the matrix

$$M = \{m_{ij}\}$$

where

$$m_{ij} = \begin{cases} 1 & \text{if, subject to } R\text{, there is a value of } \pi(h) \text{ which, beginning with an incomplete permutation in state } i\text{, will produce one state } j. \\ 0 & \text{otherwise} \end{cases}$$

The matrix M which depends only on R is the incidence matrix of a directed graph, with loops allowed in which the states of incompletion are the vertices and two vertices P_i and P_j are adjacent if there is a choice of $\pi(h)$ which converts state i into state j.

The general element $m_{ij}^{(n)}$ of M^n is the number of walks of length n from an incomplete permutation of length h and state i to one of length $h+n$ and state j. The growth of $m_{ij}^{(n)}$ is as the nth power of the largest eigenvalue

λ_0 of M and so $\mu(R) = \lambda_0$. The characteristic polynomial of M gives a recurrence relation for $N(R,n)$.

To illustrate let us return to $R_1^{(1)}$. For the incompleted permutation (4) there are 4 states of incompletion

1. neither $h-1$ nor h has been used in the second row

2. $h-1$ but not h has been used

3. h but not $h-1$ has been used

4. both have been used.

If we are in state 1 our choice must be $\pi(h) = h-1$ and this produces a new incompleted permutation of state 1. Thus $m_{11} = 1$, $m_{12} = m_{13} = m_{14} = 0$. Filling in the other entries of M in like manner we find

$$M = \begin{bmatrix} 1 & 0 & 0 & 0 \\ 0 & 1 & 1 & 0 \\ 0 & 1 & 0 & 0 \\ 0 & 0 & 0 & 1 \end{bmatrix}$$

The characteristic polynomial of M is

$$(\lambda-1)^2(\lambda^2-\lambda-1)$$

whose largest root is $\lambda_0 = (1+\sqrt{5})/2 = \mu(R_1^{(1)})$ as before. The factor $(\lambda+1)^2$ is extraneous. The difficulty with this method is the large number of possible states and the consequent size of the matrix M. For example, in the case $R_1^{(3)}$ there are 20 states and the characteristic polynomial of M factors into

$$(\lambda-1)(\lambda^6+\lambda^5+\lambda^4-\lambda^3+\lambda^2+\lambda+1)(\lambda^{13}-\lambda^{12}-3\lambda^{11}-3\lambda^{10}-13\lambda^9-21\lambda^8-19\lambda^7-$$
$$-3\lambda^6+7\lambda^5+9\lambda^4+5\lambda^3+3\lambda^2+3\lambda+1)$$

The sextic factor is extraneous. The largest root of the irreducible 13-ic is

$$\lambda_0 = 3.061773$$

The product of the first and last factors gives us the 14th degree denominator of the generating function for $N(R_1^{(3)}, n)$ listed below.

Method 6 merely reasserts the proposition that

$$N(R, n) = \operatorname{perm}(A_n)$$

and so any inequality available for permanents may be of use. One of these uses was called to my attention by Dr. J. HAMMERSLEY [7]. Consider $R_2^{(k)}$. Each row and each column of the corresponding matrix A_n has $2k+1$ units the other elements being zero. If we replace each unit by $1/(2k+1)$ we obtain a so-called doubly stochastic matrix whose permanent according to the van der Waerden conjecture is at least $n!/n^n$. From this we have

$$\operatorname{perm}(A_n) \geq (2k+1)^n n!/n^n > [(2k+1)/e]^n .$$

Hence

$$\mu(R_2^{(k)}) > (2k+1)/e$$

and

$$\Theta(R_2^{(k)}) > e^{-1} = 0.36788$$

follow from the conjecture. The same inequalities hold for $R_1^{(k)}$.

We conclude with some explicit results about strongly restricted permutations. Instead of tabulating $N(R,n)$ it suffices to list for each R the rational generating function

$$G(R,x) = 1 + \sum_{n=1}^{\infty} N(R,n)\, x^n .$$

These are as follows

Restriction R	Generator G(R,x)
$R_1^{(1)}$	$1/(1-x-x^2)$
$R_1^{(2)}$	$(1-x)/(1-2x-2x^3+x^5)$
$R_1^{(3)}$	$\dfrac{1-x-2x^2-2x^4+x^7+x^8}{1-2x-2x^2-10x^4-8x^5+2x^6+16x^7+10x^8+2x^9-4x^{10}-2x^{11}-2x^{13}-x^{14}}$
$R_2^{(2)}$	$\dfrac{1-2x-x^2+2x^3+10x^4+60x^5-75x^6-150x^7-158x^8-122x^9-340x^{10}+213x^{11}+113x^{12}+48x^{13}-64x^{14}-40x^{15}-3x^{16}}{1-3x+2x^3+2x^4+6x^5-2x^6-4x^7-2x^8-2x^9+x^{10}+x^{11}}$
$R_3^{(1)}$	$(1+x)/(1-x)$
$R_3^{(2)}$	$(1-x+3x^3-2x^4-3x^5)/(1-2x+x^3)$
$R_3^{(3)}$	$(1-x+2x^3+13x^4-3x^5-6x^6-10x^7)/(1-2x+x^4)$
$R_4^{(k)}$	$(1-x-x^2+x^{k+1})/(1-2x+x^k)$
$R_5^{(1)}$	$1/(1-x^2)$
$R_5^{(2)}$	$(1-x)/(1-x-x^2-x^3-x^4+x^5)$

To obtain the recurring series for any $N(R,n)$ one can obtain the initial values by division and then use the coefficients in the denominator for the recurrence.

Thus for $N(R_1^{(1)}, n)$ we have the initial values

$$1, 1, 2, 6, 9, 13, 20, \ldots$$

and the recurrence

$$N(R_2^{(1)}, n) = 2N(R_2^{(1)}, n-1) - N(R_2^{(1)}, n-3) \quad (n \geq 6).$$

The corresponding indices of mobility μ and interference Θ are tabulated below

Restriction R	$\mu(R)$	$\Theta(R)$
$R_1^{(1)}, R_2^{(1)}, R_3^{(2)}$	1.61803	0.53934
$R_1^{(2)}, R_2^{(2)}, R_3^{(4)}$	2.33355	0.46671
$R_1^{(3)}, R_2^{(3)}, R_3^{(6)}$	3.06177	0.43740
$R_3^{(1)}$	1.00000	1.00000
$R_3^{(3)}$	1.83929	0.45982
$R_4^{(k)}$	$2 - 2^{-k}$	$(2-2^{-k})/(k+1)$
$R_5^{(2)}$	1.88320	0.47080

ERRATA in [3]

p. 31. formula (3) should read $\phi(n,3) = f^{(2)}_{n+1} + f^{(2)}_{n-1} + 2$

formula (4) should read $\phi(n,4) = 2(3f^{(3)}_{n} - 2f^{(3)}_{n-1} - f^{(3)}_{n-2} + 1)$

p. 32. Corrected tabular values of $\phi(n,3)$ and $\phi(n,4)$ are

$\phi(5,3) = 13$

$\phi(6,3) = 20$

$\phi(7,3) = 31$

$\phi(8,3) = 49$

$\phi(9,3) = 78$

$\phi(10,3) = 125$

$\phi(8,4) = 264$

$\phi(9,4) = 484$

$\phi(10,4) = 888$

REFERENCES

[1] JOHN RIORDAN, Introduction to Combinatorial Analysis, New York, 1958 (Wiley).

[2] D. H. LEHMER, "Combinatorial problems with digital computers," Canadian Math. Congress Proc. p. 160-173, 1959.

[3] N. S. MENDELSOHN, "Permutations with confined displacements," Canadian Math. Bulletin, V.4, p. 29-38, 1961.

[4] N. S. MENDELSOHN, "The asymptotic series for a certain class of permutation problems," Canadian Jour. Math. B. 8, p. 234-244, 1956.

[5] MARSHALL HALL, Combinatorial Theory, Waltham, 1967 (Blaisdell), p. 22-24.

[6] Applied Combinatorial Mathematics, New York, 1964 (Wiley), p. 26, 27.

[7] Private conversation.

Some combinatorial problems of multinomial sequential estimation

(abstract)

by

Yu. V. Linnik

Leningrad, USSR

ABSTRACT

Consider a "first hit" plan of statistical estimation for a multinomial walk on the lattice points of the domain: $x_1 \geqq 0, x_2 \geqq 0, \ldots, x_k \geqq 0$. This plan S is determined by its boundary ∂S (consisting of lattice points). It is supposed to be closed, complete (in the sense of sufficient statistics) and minimal. In the binomial case $(k = 2)$ and given plan size: $sz(S) = r$ the quantity of such plans is enumerated by means of a recurrence formula; it is proved that such plans are completely determined by the average value $E_p \tau$ of the stopping time τ known for all values of the propability $\underline{p}$ of the binomial walk. The corresponding problems for multinomial walk $(k \geqq 2)$ are not solved.

Generalized factors of graphs

by

L. Lovász

Budapest, Hungary

§ 1. A generalized factor (GF) of a graph $\underline{G}$ means a system of edge-disjoint paths of $\underline{G}$. Here path means a more general configuration than usually: it is either an open path (these are the usual simple paths) or a closed path which is a circuit on which a beginning- and endpoint is selected. We shall deal with the question of the existence of certain GFs. (Naturally, the same problems arise if we allow only open paths. In one of our detailed problems, no results are known for open paths, in the other case the two types of problems are equivalent. This will be pointed out in the same place.)

Let now $\mathcal{R}$ be a GF. We can define two values in every point $\underline{x}$ which are "valencies" of $\mathcal{R}$:

(a) the indegree $\underline{i}_{\mathcal{R}}(\underline{x})$, the number of paths of $\mathcal{R}$ ending in $\underline{x}$;

b) the throughdegree $\underline{t}_{\mathcal{R}}(\underline{x})$, the number of paths of $\mathcal{R}$ going through $\underline{x}$.

Here we have to mention that a closed path with endpoint $\underline{x}$ increases the value of $\underline{i}_{\mathcal{R}}(\underline{x})$ by 2 (it ends in $\underline{x}$ twice) but it is not counted in $t_{\mathcal{R}}(\underline{x})$ (it does not go through $\underline{x}$).

Two trivial remarks can be made here:

1) A subgraph $\underline{F}$ of $\underline{G}$ is a generalized factor consisting of edges, i.e. of paths without inner points. Thus $\underline{t}_{\underline{F}}(\underline{x})=0$, $\underline{i}_{\underline{F}}(\underline{x})$ is the valency of the vertex $\underline{x}$ in the subgraph $\underline{F}$.

2) On the other hand, every GF $\mathcal{R}$ determines a subgraph $\bar{\mathcal{R}}$, the union of its paths. Obviously we have

$$\underline{i}_{\bar{\mathcal{R}}}(\underline{x}) = \underline{i}_{\mathcal{R}}(\underline{x}) + 2\cdot\underline{t}_{\mathcal{R}}(\underline{x}).$$

Two further notations: $\underline{S}_{\underline{G}}$ denotes the set of points of the graph $\underline{G}$. If $\underline{A},\underline{B}\subseteq\underline{S}_{\underline{G}}$ then $\underline{G}(\underline{A},\underline{B})$ denotes the set of those edges of $\underline{G}$ which join a vertex of $\underline{A}$ to a vertex of $\underline{B}$.

GALLAI [1] answered the following question: Given a pair $\underline{f}(\underline{x})$, $\underline{g}(\underline{x})$ of integers in every vertex $\underline{x}$, which is the maximum cardinality of a GF satisfying $\underline{i}_{\mathcal{R}}(\underline{x})\leqq\underline{f}(\underline{x})$, $\underline{t}_{\mathcal{R}}(\underline{x})\leqq\underline{g}(\underline{x})$? The answer was:

$$\text{(1)}\qquad \max|\mathcal{R}| = \min \sum_{\underline{x}\in\underline{B}}\{\underline{f}(\underline{x})+\underline{g}(\underline{x})\} + \underline{G}(\underline{A},\underline{A}) + \sum_{\substack{\underline{D}\text{ is a}\\ \text{component of}\\ \underline{G}(\underline{C},\underline{C})}} \left[\tfrac{1}{2}\cdot\{\underline{G}(\underline{A},\underline{D})+\sum_{\underline{x}\in\underline{D}}\underline{f}(\underline{x})\}\right]$$

where the minimum is taken over all partitions $\underline{A}\cup\underline{B}\cup\underline{C}$ of $\underline{S}_{\underline{G}}$.

GALLAI mentioned in his paper that his formula, being a generalization of Tutte's factorization theorem [3], probably can be reduced to it. In the present paper, we are first going to carry out this reduction. Then by similar, but more complicated, ideas we answer the following question: When does a GF exist if an interval is prescribed for its indegree and another for its throughdegree? In particular: When does a GF exist with prescribed indegree and throughdegree? Unfortunately, at this latter problem I have to assume that the prescribed indegree is positive and the graph has no loops or multiple edges (see § 3.).

§ 2. The formula (1) is rather complicated. Our reduction shall show that this is only because Tutte's condition for the existence of a factor is complicated; in other words, the reduction will consist of natural transformations of the problem. We shall not use the knowledge of formula (1), but we ask:

(A) What is the maximum cardinality of $\mathcal{R}$?

...and try to find the answer.

There is a trivial estimation for $|\mathcal{R}|$:

$$|\mathcal{R}| = \frac{1}{2}\sum_{\underline{x}} \underline{i}_{\mathcal{R}}(\underline{x}) \leq \frac{1}{2}\sum_{\underline{x}} \underline{f}(\underline{x}).$$

Hence

$$\max|\mathcal{R}| \leq \frac{1}{2}\sum_{\underline{x}} \underline{f}(\underline{x}).$$

FIRST TRANSFORMATION:

(B) When does the equality hold here?

In other words:

(B') When does a GF $\mathcal{R}$ exist for which $\underline{i}_{\mathcal{R}}(\underline{x}) = \underline{f}(\underline{x})$, $\underline{t}_{\mathcal{R}}(\underline{x}) = \underline{g}(\underline{x})$?

Suppose we know the answer to question (B). Then the answer to question (A) follows like this: We add a new vertex $\underline{y}$ to the graph $\underline{G}$ and join it to every vertex of $\underline{G}$ by a large number of edges. We obtain a new graph $\underline{G}_0$. The functions $\underline{f}(\underline{x})$, $\underline{g}(\underline{x})$ are extended to this new vertex $\underline{y}$, $\underline{f}(\underline{y})$ is arbitrary for this time, $\underline{g}(\underline{y}) = 0$. We determine the least value $\underline{f}_0$ of $\underline{f}(\underline{y})$ for which the answer to question (B) (concerning $\underline{G}_0$) is positive. Then

(2) $$\max |\mathcal{R}| = \frac{1}{2}\{\sum_{\underline{x}} \underline{f}(\underline{x}) - \underline{f}_0\}$$

In order to prove formula (2), let us consider a GF $\mathcal{R}$ of $\underline{G}$. We add to $\mathcal{R}$ $\underline{f}(\underline{x}) - \underline{i}_{\mathcal{R}}(\underline{x})$ new paths for every $\underline{x} \in \underline{S}_G$, each of them being a single $\underline{xy}$-edge. Let $\mathcal{R}'$ denote the obtained GF; then $\underline{i}_{\mathcal{R}'}(\underline{x}) = \underline{f}(\underline{x})$, $\underline{t}_{\mathcal{R}'}(\underline{x}) = \underline{t}_{\mathcal{R}}(\underline{x}) \leq \underline{g}(\underline{x})$ if $\underline{x} \in \underline{S}_G$ and $\underline{i}_{\mathcal{R}'}(\underline{y}) = \sum_{\underline{x}} \underline{f}(\underline{x}) - 2 \cdot |\mathcal{R}|$. Since putting $\underline{f}(\underline{y}) = \underline{i}_{\mathcal{R}'}(\underline{y})$ the answer to question (B)(concerning G_0) is positive, we have

$$\underline{f}_0 \leq \underline{i}_{\mathcal{R}}(\underline{x}) = \sum_{\underline{x}} \underline{f}(\underline{x}) - 2 \cdot |\mathcal{R}|,$$

$$|\mathcal{R}| \leq \frac{1}{2}\{\sum_{\underline{x}} \underline{f}(\underline{x}) - \underline{f}_0\}.$$

Conversely, if $\mathcal{R}'$ is a GF of $\underline{G}_0$ and $\underline{i}_{\mathcal{R}'}(\underline{x}) = \underline{f}(\underline{x})$, $\underline{t}_{\mathcal{R}'}(\underline{x}) = \underline{g}(\underline{x})$ for $\underline{x} \in \underline{S}_G$, $\underline{i}_{\mathcal{R}'}(\underline{y}) = \underline{f}_0$, $\underline{t}_{\mathcal{R}'}(\underline{y}) = 0$ (by the definition of $\underline{f}_0$ such a GF exists), then we omit the paths which have an endpoint in $\underline{y}$. The remaining paths form a GF $\mathcal{R}$ of the original graph $\underline{G}$, and for this GF we have $\underline{i}_{\mathcal{R}}(\underline{x}) \leq \underline{i}_{\mathcal{R}'}(\underline{x}) = \underline{f}(\underline{x})$, $\underline{t}_{\mathcal{R}}(\underline{x}) \leq \underline{t}_{\mathcal{R}'}(\underline{x}) \leq \underline{g}(\underline{x})$, furthermore

$$|\mathcal{R}| = |\mathcal{R}'| - \underline{f}_0 = \frac{1}{2}\{\sum_{\underline{x}} \underline{f}(\underline{x}) + \underline{f}_0\} - \underline{f}_0 = \frac{1}{2}\{\sum_{\underline{x}} \underline{f}(\underline{x}) - \underline{f}_0\}.$$

This proves formula (2).

SECOND TRANSFORMATION:

(C) When has a graph $\underline{G}_0$ a subgraph $\underline{F}$ with the property that $\underline{i}_F(\underline{x}) \in \{\underline{f}(\underline{x}), \underline{f}(\underline{x})+2, \ldots, \underline{f}(\underline{x}) + 2 \cdot \underline{g}(\underline{x})\}$ in every vertex?

It is easy to see that the positive answer to question (B) implies the positive answer to question (C): If $\underline{R}$ is a GF for which $\underline{i}_{\underline{R}}(\underline{x}) = \underline{f}(\underline{x})$, $\underline{t}_{\underline{R}}(\underline{x}) \leqq \underline{g}(\underline{x})$, then $\underline{F} = \overline{\underline{R}}$ is a subgraph of the required property. Conversely, suppose $\underline{F}$ is a subgraph of $\underline{G}_0$ showing that the answer to question (C) is positive. We have to construct a GF $\underline{R}$ for which $\underline{i}_{\underline{R}}(\underline{x}) = \underline{f}(\underline{x})$, $\underline{t}_{\underline{R}}(\underline{x}) \leqq \underline{g}(\underline{x})$. Choose in $\underline{F}$ a GF $\underline{R}$ in such a way that $\underline{i}_{\underline{R}}(\underline{x}) \leqq \underline{f}(\underline{x})$, $\underline{t}_{\underline{R}}(\underline{x}) \leqq \underline{g}(\underline{x})$ and $|\underline{R}|$ is maximal. We have to show that then $\underline{i}_{\underline{R}}(\underline{x}) = \underline{f}(\underline{x})$ in every vertex.

Let $\underline{A}$ be the set of those vertices for which $\underline{i}_{\underline{R}}(\underline{x}) = \underline{f}(\underline{x})$. We have to show that every vertex belongs to $\underline{A}$.

First we consider the non-isolated vertices of the subgraph $\underline{F} - \overline{\underline{R}}$. If such a vertex $\underline{x}$ belongs to $\underline{A}$, then $\underline{t}_{\underline{R}}(\underline{x}) < \underline{g}(\underline{x})$, and the valency of $\underline{x}$ in $\underline{F} - \overline{\underline{R}}$ is even.

We examine what the maximality of $|\underline{R}|$ means.

(i) No open path of $\underline{F} - \overline{\underline{R}}$ can be added to $\underline{R}$ without breaking the conditions $\underline{i}_{\underline{R}}(\underline{x}) \leqq \underline{f}(\underline{x})$, $\underline{t}_{\underline{R}}(\underline{x}) \leqq \underline{g}(\underline{x})$. Hence $\underline{F} - \overline{\underline{R}}$ cannot contain a path the inner points of which belong to $\underline{A}$ but the endpoints do not. Consequently, $\underline{F} - \overline{\underline{R}}$ does not contain any path connecting two vertices of $\underline{S}_{\underline{G}_0} - \underline{A}$.

(ii) No closed path can be added to $\underline{R}$. Hence, $\underline{F} - \overline{\underline{R}}$ cannot contain a circuit which contains just one vertex of $\underline{S}_{\underline{G}_0} - \underline{A}$. By combining this and (i) we obtain that the vertices of any circuit of $\underline{F} - \overline{\underline{R}}$ belong to $\underline{A}$.

From (i) we deduce now, that a connected component of $\underline{F} - \overline{\underline{R}}$ contains at most one point of $\underline{S}_{\underline{G}_0} - \underline{A}$. As we have seen, the other vertices of the component are of even valency. Hence every vertex of it is of even valency. This implies that every vertex, which is non-isolated in $\underline{A}$ is contained in a circuit of $\underline{F} - \overline{\underline{R}}$, and thus by (ii) it belongs to $\underline{A}$.

The isolated vertices of $\underline{F} - \overline{\underline{R}}$ belong obviously to $\underline{A}$, thus $\underline{i}_{\underline{R}}(\underline{x}) = \underline{f}(\underline{x})$ is proved for any $\underline{x} \in \underline{S}_{\underline{G}}$.

THIRD TRANSFORMATION:

(D) When does a subgraph $\underline{F}_0$ exist for which $\underline{i}_{\underline{F}_0}(\underline{x}) = \underline{f}(\underline{x}) + 2 \cdot \underline{g}(\underline{x})$?

Suppose we can answer question (D) concerning any graph. Then we ask this about the graph obtained from the graph of question (C) by adding $\underline{g}(\underline{x})$ loops to every vertex $\underline{x}$. Obviously, the answer is positive just then if the answer to question (C) is so.

The question (D) was answered by Tutte [3], who proved that a graph $\underline{G}$ has an $\underline{h}$-factor if and only if for any partition $\underline{A} \cup \underline{B} \cup \underline{C}$ of we have

$$\frac{1}{2}\sum_{\underline{x} \in \underline{A} \cup \underline{C}} \underline{h}(\underline{x}) \leq \frac{1}{2}\sum_{\underline{x} \in \underline{B}} \underline{h}(\underline{x}) + |\underline{G}(\underline{A},\underline{A})| + \sum_{\substack{\underline{D} \text{ is a} \\ \text{component of} \\ \underline{G}(\underline{C},\underline{C})}} [\frac{1}{2}\{|\underline{G}(\underline{A},\underline{D})| + \sum_{\underline{x} \in \underline{D}} \underline{h}(\underline{x})\}]$$

If we apply the "inverse transformations" on this condition, we obtain GALLAI's formula (1).

§ 3. Let us repeat the problem of § 2. after the first transformation: When does a GF $\underline{R}$ exist for which $\underline{i}_{\underline{R}}(\underline{x}) = \underline{f}(\underline{x})$, $\underline{t}_{\underline{R}}(\underline{x}) \leq \underline{g}(\underline{x})$? It is a natural requirement to avoid this asymmetry in the indegree and throughdegree; this leads to the following problem:

When does a GF $\underline{R}$ exist for which $\underline{f}_1(\underline{x}) \leq \underline{i}_{\underline{R}}(\underline{x}) \leq \underline{f}_2(\underline{x})$, $\underline{g}_1(\underline{x}) \leq \underline{t}_{\underline{R}}(\underline{x}) \leq \underline{g}_2(\underline{x})$ ($\underline{f}_1, \underline{f}_2, \underline{g}_1, \underline{g}_2$ are given as integer-valued functions on the set $\underline{S}_{\underline{G}}$ of vertices of a given graph $\underline{G}$)?

If we want to follow the solution of the problem discussed in § 2, the proof of the "second transformation" (which is the first step here) is much more difficult. In fact it can be carried out only if we suppose

(i) the graph $\underline{G}$ does not contain loops or multiple edges;

(ii) $0 < \underline{f}_1(\underline{x})$ in every vertex $\underline{x}$.

On the other hand, we obtain more than in § 2.: the necessary condition for the existence of a GF will be sufficiant for the existence of a GF consisting of open paths only.

Put $\underline{H}(\underline{x}) = \{\underline{u}+2\underline{v} : \underline{f}_1(\underline{x}) \leqq \underline{u} \leqq \underline{f}_2(\underline{x}),\ \underline{g}_1(\underline{x}) \leqq \underline{v} \leqq \underline{g}_2(\underline{x})\}$. In case $\underline{f}_1(\underline{x}) < \underline{f}_2(\underline{x})$, $\underline{H}(\underline{x})$ is an interval, in case $\underline{f}_1(\underline{x}) = \underline{f}_2(\underline{x})$ it is an arithmetical sequence of difference 2. Now we prove:

LEMMA: The necessary and sufficient condition for the existence of a GF $\mathcal{R}$ for which $f_1(x) \leqq i_{\mathcal{R}}(x) \leqq f_2(x)$, $g_1(x) \leqq t_{\mathcal{R}}(x) \leqq g_2(x)$ is the existence of a subgraph $\underline{F}$ for which $\underline{i}_{\underline{F}}(\underline{x}) \in \underline{H}(\underline{x})$ in every vertex. If the condition is satisfied then the GF can be constructed only from simple paths. ((i) and (ii) are assumed.)

P r o o f: The necessity of the condition is obvious. To prove the sufficiency let $\underline{F}$ be a subgraph for which $\underline{i}_{\underline{F}}(\underline{x}) \in \underline{H}(\underline{x})$ in every vertex. That means that $\underline{i}_{\underline{F}}(\underline{x}) = \underline{u}(\underline{x}) + 2\cdot\underline{v}(\underline{x})$, where $\underline{f}_1(\underline{x}) \leqq \underline{u}(\underline{x}) \leqq \underline{f}_2(\underline{x})$, $\underline{g}_1(\underline{x}) \leqq \underline{v}(\underline{x}) \leqq \underline{g}_2(\underline{x})$.

Let us now add $\underline{u}(\underline{x})-1$ new vertices to every point $\underline{x}$ of $\underline{F}$ and join them to $\underline{x}$. These new points and edges together with $\underline{F}$ form a new graph $\underline{F}'$. Apply now the following theorem to $\underline{F}'$:

THEOREM ([4]): For any graph of n vertices there exists a GF $\mathcal{R}$ of the graph for which $|\mathcal{R}| \leqq \frac{n}{2}$ and $\overline{\mathcal{R}}$ is the whole graph. In other words: A graph of n vertices can be covered by $\left[\frac{n}{2}\right]$ edge-disjoint paths.

If we consider the GF $\mathcal{R}'$ of $\underline{F}'$, the existence of which is guaranteed by this theorem, we can observe several interesting properties of it. First of all, an "old" vertex $\underline{x}$ of $\underline{F}'$ has valency $\underline{u}(\underline{x})+2\cdot\underline{v}(\underline{x})+\{\underline{u}(\underline{x})-1\} = 2(\underline{u}(\underline{x})+\underline{v}(\underline{x}))-1$, the "new" vertices have valency 1, consequently every vertex of $\underline{F}'$ has odd valency. Hence, every vertex of $\underline{F}'$ must be the endpoint of at least one path of $\mathcal{R}'$. Since $|\mathcal{R}'| \leqq \frac{1}{2}\cdot\{$the number of vertices of $\underline{F}'\}$, no vertex can be the endpoint of more than one path of $\mathcal{R}$. That means $i_{\mathcal{R}}(\underline{x}) = 1$ for every vertex $\underline{x}$ of $\underline{F}'$; consequently, $\mathcal{R}'$ consists of open paths.

Omit now the "new" vertices and edges. The remaining arcs of the paths of $\mathcal{R}'$ form a GF $\mathcal{R}''$ of $\underline{F}$, for which $\overline{\mathcal{R}}'' = \underline{F}$, $\underline{i}_{\mathcal{R}''}(\underline{x}) \leqq \underline{u}(\underline{x})$, $\underline{i}_{\mathcal{R}''}(\underline{x}) \equiv \underline{u}(\underline{x}) \pmod 2$. If we have the inequality here, then we can split up a path of $\mathcal{R}''$ going through $\underline{x}$ into two parts and in this way we can construct a GF $\mathcal{R}$ for which $\underline{i}_{\mathcal{R}}(\underline{x}) = \underline{u}(\underline{x})$ in every point. Since then $\underline{t}_{\mathcal{R}}(\underline{x}) = = \frac{1}{2}\{\underline{i}_F(\underline{x}) - \underline{i}_{\mathcal{R}}(\underline{x})\} = \underline{v}(\underline{x})$ we are ready with the proof of the lemma.

We have still to answer the following question: When does a subgraph $\underline{F}$ of $\underline{G}$ eixst for which $\underline{i}_F(\underline{x}) \in \underline{H}(\underline{x})$ in every vertex $\underline{x}$? Applying results of [2], we obtain

THEOREM: Let a graph $\underline{G}$ be given without loops and multiple edges. Let $\underline{f}_1(\underline{x})$, $\underline{f}_2(\underline{x})$, $\underline{g}_1(\underline{x})$, $\underline{g}_2(\underline{x})$ be four functions on the set $\underline{S}_G$ such that $0 < \underline{f}_1(\underline{x}) \leqq \underline{f}_2(\underline{x})$, $0 \leqq \underline{g}_1(\underline{x}) \leqq \underline{g}_2(\underline{x})$. Then $\underline{G}$ has a GF $\mathcal{R}$ with the properties $\underline{f}_1(\underline{x}) \leqq \underline{i}_{\mathcal{R}}(\underline{x}) \leqq \underline{f}_2(\underline{x})$, $\underline{g}_1(\underline{x}) \leqq \underline{t}_{\mathcal{R}}(\underline{x}) \leqq \underline{g}_2(\underline{x})$ if and only if the following condition is fulfilled: Considering an arbitrary partition $\underline{A} \cup \underline{B} \cup \underline{C}$ of $\underline{S}_G$ and counting those components $\underline{D}$ of the graph spanned by $\underline{C}$ for which $\underline{f}_1(\underline{x}) = \underline{f}_2(\underline{x})$ if $\underline{x} \in \underline{D}$ and $\sum_{\underline{x} \in \underline{D}} \underline{f}_1(\underline{x}) \not\equiv |\underline{G}(\underline{D}, \underline{B})| \pmod 2$, we obtain at most

$$\sum_{\underline{x} \in \underline{A}} \{\underline{f}_2(\underline{x}) + 2 \cdot \underline{g}_2(\underline{x})\} - \sum_{\underline{x} \in \underline{B}} \{\underline{f}_1(\underline{x}) + 2 \cdot \underline{g}_1(\underline{x})\} + |\underline{G}(\underline{B}, \underline{C})| + 2 \cdot |\underline{G}(\underline{B}, \underline{B})| .$$

REFERENCES

[1] GALLAI, T.: "Maximum-minimum Sätze und verallgemeinerte Faktoren von Graphen" Acta Math. Acad. Sci. Hung. 12 (1961) pp. 131-173.

[2] LOVÁSZ, L.: "The factorization of graphs" to appear in the Proceedings of the Calgary International Conference on Combinatorial Structues, Calgary, 1969.

[3] TUTTE, W. T.: "The factors of graphs" Canad. Journal Math. 4 (1952) pp. 314-328.

[4] LOVÁSZ, L.: "On covering of graphs" Theory of graphs, Proc. of the Coll. held in Tihany, Akad. Kiadó, Budapest 1968. pp. 231-236.

Directed graphs with the unique path property

by

N. S. Mendelsohn

Winnipeg, Canada

ABSTRACT

A class of directed graphs in which between any two of the vertices of such a graph there is a unique path of length t are studied. Such graphs have an adjacency matrix A which is a $0,1$ matrix satisfying the matrix equation $A^t = J$, where, as usual, J is a matrix each of whose entries is 1. For each integer r there are graphs with r^t vertices and r^{t+1} edges satisfying this unique path property (UPP) of order t. In general there is a large class of such graphs which are non-isomorphic.

I. J. Good [3] and N. G. de Bruijn [1] discovered a specific subclass of graphs with the UPP of order t. We denote a graph of this class by $G_t^{(r)}$ which can be described as follows. The vertices of $G_t^{(r)}$ are all the t-tuples $(\alpha_1, \alpha_2, \dots, \alpha_t)$ where the α_i range over the elements of the r-set $\{0, 1, 2, 3, \dots, r-1\}$. A directed edge joins $(\alpha_1, \alpha_2, \dots, \alpha_t)$ to $(\alpha_2, \alpha_3, \dots, \alpha_t, \beta)$. Hence each vertex of $G_t^{(r)}$ has both its in-degree and out-degree equal to r.

In this paper some properties for $G_r^{(r)}$ are developed. Furthermore, many of these hold for the more general graphs satisfying UPP of order t.

For any $n \leq t$, each vertex of $G_t^{(r)}$ is on at most one elementary circuit (directed polygon) of exact length n and the number of such circuits is independent of t. In fact, the exact number $f(n)$ of such circuits is given by

$$f(n) = \frac{1}{n} \sum_{d|n} \mu\left(\frac{n}{d}\right) r^d .$$

This property transfers to the more general graphs. Similarly, properties of Eulerian and Hamiltonian directed circuits transfer. It is established that $G_t^{(1)}$, $G_1^{(2)}$, $G_1^{(3)}$, $G_1^{(4)}$, $G_2^{(2)}$, $G_2^{(3)}$, $G_3^{(2)}$ are the only planar Good - de Bruijn graphs and it is conjectured that this property transfers to the general UPP graphs. Finally, an important connection between UPP graphs and a class of universal algebras is established.

GOOD'S MOTIVATION FOR $G_t^{(r)}$.

Let S be the set of r symbols $\{0, 1, 2, \dots, r-1\}$. Can a cyclic sequence of r^{t+1} of these symbols $\lambda_1, \lambda_2, \lambda_3, \dots, \lambda_{r^{t+1}}$ be constructed such that the r^{t+1} blocks of $t+1$ consecutive elements in this cyclic sequence (note λ_1 comes after $\lambda_{r^{t+1}}$) be all possible sequences of length $t+1$? I. J. Good [3], solved this problem as follows. Construct a graph with r^t vertices. Each vertex is a t-ple $(\alpha_1, \alpha_2, \dots, \alpha_t)$ of elements where each α_i belongs to the set $\{0, 1, 2, \dots, r-1\}$. A directed edge is drawn from $(\alpha_1, \alpha_2, \dots, \alpha_t)$ to $(\alpha_2, \alpha_3, \dots, \alpha_t, \beta)$ and the edge may be labelled $(\alpha_1, \alpha_2, \alpha_3, \dots, \alpha_t, \beta)$. Each vertex has out-degree r and in-degree r, and the graph is connected (since as will be shown below each vertex is joined to any other vertex by exactly one path of length t). Hence an Eulerian path can be drawn and the labelled edges which appear consecutively in the path determine the required sequence.

PROPERTIES OF $G_t^{(r)}$.

THEOREM 1. Between any two vertices of $G_t^{(r)}$ there is exactly one path of length t.

Proof. Let $(\alpha_1, \alpha_2, \dots, \alpha_t)$ and $(\beta_1, \beta_2, \dots, \beta_t)$ be the two vertices. The unique path of length t is $(\alpha_1, \alpha_2, \dots, \alpha_t) \to$
$\to (\alpha_2, \alpha_3, \dots, \beta_1) \to \dots \to (\alpha_t, \beta_1, \beta_2, \dots, \beta_{t-1}) \to (\beta_1, \beta_2, \dots, \beta_t)$.

COROLLARY 1. If A is the adjacency matrix of $G_t^{(r)}$ and J is a matrix with entries exclusively 1, then $A^t = J$.

COROLLARY 2. Between any two vertices there is at most one path of length u if $u < t$.

THEOREM 2. Let $n \leq t$. In $G_t^{(r)}$ there is at most one cycle of length n through any given vertex. If such a vertex is on a cycle of length d where $d|n$ then the cycle of length n through the given vertex is the cycle of length d repeat $\frac{n}{d}$ times. The number $f(n)$ of elementary cycles of length is independent of t and is given by the formula $f(n) = \frac{1}{n} \sum_{d|n} \mu(\frac{n}{d}) r^d$, where μ is the usual Möbius function.

Proof. From the vertex $(\alpha_1, \alpha_2, \alpha_3, \dots, \alpha_t)$ we can get in n steps to the vertex $(\alpha_{n+1}, \alpha_{n+2}, \dots, \alpha_t, \beta_1, \beta_2, \dots, \beta_n)$ where $\beta_1, \beta_2, \dots$ $\dots, \beta_n$ are arbitrary. If the path is a cycle we have the equations:

$$\begin{array}{lll} \alpha_1 = \alpha_{n+1} & \qquad & \alpha_{t-n+1} = \beta_1 \\ \alpha_2 = \alpha_{n+2} & & \alpha_{t-n+2} = \beta_2 \\ \vdots & & \vdots \\ \alpha_{t-n} = \alpha_t & & \alpha_t = \beta_n \end{array}$$

Note that the β's uniquely determine all the α's Pictorially this can be seen by looking at the sequence $\alpha_1, \alpha_2, \dots, \alpha_t, \beta_1, \beta_2, \dots, \beta_n$.

Each of the α's is equal to the element n steps to its right and hence is equal to a unique β. Furthermore, no two distinct set of β's determine the same set of α's. Since there are r^n choices for the sequence $\beta_1, \beta_2, \ldots, \beta_n$ it follows that there are r^n vertices on cycles of length n, and each of these vertices is on exaclty one such cycle. These include vertices on cycles of length d where $d|n$. We note further that if a point $(\alpha_1, \alpha_2, \ldots, \alpha_t)$ is on a cycle of length r and on a cycle of length s where $r+s = n$ then
$\alpha_i = \alpha_{i+r}$ $(i = 1,2,3,\ldots)$, $i+r \leq t$ and $\alpha_i = \alpha_{i+s}$ $(i=1,2,3,\ldots)$ $i+s \leq t$.
But, if $s>r$, and $i+s \leq t$, then $\alpha_{i+s-r} = \alpha_{i+s-r+r} = \alpha_{i+s} = \alpha_i$.
If $i+s>t$, and $r+s \leq t$, then $i-r>0$ so that $\alpha_{i-r} = \alpha_{i-r+r} = \alpha_i$
and $\alpha_{i-r} = \alpha_{i-r+s}$. Hence, again $\alpha_i = \alpha_{i+s-r}$. Hence, continuing we obtain that if $i+q \leq t$ $\alpha_i = \alpha_{i+q}$ where q is the g.c.d. of r and s. Hence, the vertex $(\alpha_1, \alpha_2, \ldots, \alpha_t)$ is on a d-cycle. But the d-cycle repeated r/d times is an r-cycle and as a point is on at most one r-cycle it follows that the r-cycle and s-cycle through $(\alpha_1, \alpha_2, \ldots, \alpha_t)$ are multiples of the d-cycle. This implies that if a point is on an n-cycle it is either on an elementary n-circuit or on an elementary d-circuit when $d|n$. Let $f(n)$ be the number of vertices on elementary circuits of length exaclty n. Then $\sum_{d|n} df(d) = r^n$. By Möbius inversion, we obtain $f(n) = \frac{1}{n} \sum_{d|n} \mu\left(\frac{n}{d}\right) r^d$.

LONG PATHS IN $G_t^{(r)}$.

A path of length greater than t in $G_t^{(r)}$ is called a long path. If the path is a cycle it is called a long cycle. In comparison with short cycles the number of long cycles does depend on t. However, information on long elementary circuits is scant. There are three known cases of the exact number of long elementary circuits. For $n = t+1$ the formula used for $n \leq t$ is still valid but the proof is somewhat more complicated. In a private communication P. Bryant has informed me that he has worked out the case for $n = t+2$ and $r = 2$. In this case the formula turns out to be

$f(t+2) = \frac{1}{t+2} \sum_{d|t+2} \mu\left(\frac{t+2}{d}\right) 2^d - \varphi(t+2)$ where φ is the Euler totient function. The longest elementary circuit is the Hamiltonian circuit of length r^t. This is treated in the next section.

Since the adjacency matrix A of $G_t^{(r)}$ has r 1's in each row and $A^t = J$, it follows that $A^{t+k} = r^k J$. Hence, there are exactly r^k paths of lenght $t+k$ from any point to any other point. In particular, there are exactly r^k cycles of length $t+k$ through any point in $G_t^{(r)}$. A non-matrix proof of this property goes as follows. First note that each vertex is of out-degree r. Hence one can get from a vertex A to a vertex B (which could be A) as follows. Proceed from A along any one of its out edges to a vertex A_1 adjacent from A. Proceed from A_1 along any one of its out edges to a vertex A_2 adjacent from A_1. After k steps one reaches a vertex A_k. Then proceed from A_k to B along the unique path of length t. Obviously this leads to r^k distinct paths.

Although there are r^k cycles of length $t+k$ through any vertex, it does not imply the existence of an elementary circuit of length $t+k$. A. Lempel [9] proved that elementary circuits exist for all $t+k \leq r^n$. However, even when such elementary circuits do exist, it does not follow that there is such a circuit through any given point. For instance, in $G_3^{(2)}$ every cycle of length 5 through $(0,1,0)$ is the union of a 3-cycle and a 2-cycle.

HAMILTONIAN AND EULER CIRCUITS

In this section, as throughout the paper, a circuit is taken to be a directed circuit. Since $G_t^{(r)}$ contains r^t vertices the number of cycles of length r^t through any given vertex is r^{r^t-t}. This large number is enough to prove the existence of a Hamiltonian circuit since it is possible by an extremely arduous combinatorial argument to establish the existence of fewer than r^{r^t-t} cycles of length r^t which are not elementary. However, we proceed differently, since the alternate approach actually gives an efficient algorithm for constructing the Hamiltonian circuit as well as establishing an interesting connection between Euler circuits and Hamiltonian circuits.

THEOREM 3. Consider the graphs $G_t^{(r)}$ and $G_{t+1}^{(r)}$. To each Euler circuit in $G_t^{(r)}$ there is a unique Hamiltonian circuit in $G_{t+1}^{(r)}$, and conversely.

Proof. The edge of $G_t^{(r)}$ joining $(\alpha_1, \alpha_2, \dots, \alpha_t)$ to $(\alpha_2, \alpha_3, \dots, \alpha_t, \alpha_{t+1})$ is labelled $(\alpha_1, \alpha_2, \dots, \alpha_t, \alpha_{t+1})$. In a directed path two edges $(\alpha_1, \alpha_2, \dots, \alpha_t, \alpha_{t+1})$ and $(\beta_1, \beta_2, \dots, \beta_t, \beta_{t+1})$ are consecutive, if and only if, $\beta_1 = \alpha_2$, $\beta_2 = \alpha_3, \dots, \beta_t = \alpha_{t+1}$. Hence, an Euler circuit leads to a cyclic sequence $\alpha_1, \alpha_2, \dots, \alpha_{r^{t+1}}$ such that all blocks of length $t+1$ are distinct. Hence, the consecutive blocks of length $t+1$ may be taken as the vertices of $G_{t+1}^{(r)}$ and since $(\alpha_1, \alpha_2, \dots, \alpha_{t+1})$ and $(\alpha_2, \alpha_3, \dots, \alpha_{t+2})$ etc. are joined in $G_{t+1}^{(r)}$ this leads to a Hamiltonian circuit. The argument may be reversed and from a Hamiltonian circuit in $G_{t+1}^{(r)}$ we may obtain an Euler circuit in $G_t^{(r)}$.

The algorithm for constructing an Euler circuit is quite trivial and extremely efficient, and this transfers to the Hamiltonian circuits. In [1] N.G. de Bruijn computed the number of Hamiltonian circuits for $r = 2$. The number for general r was obtained by de Bruijn and van Aardenne-Ehrenfest in [2]. This number is $\frac{(r!)^{r^{t-1}}}{r^t}$.

PLANARITY OF GOOD - DE BRUIJN GRAPHS

Figures 1, 2, 3 illustrate that $G_2^{(2)}$, $G_3^{(2)}$ and $G_2^{(3)}$ are planar. The graphs $G_1^{(r)}$ are complete directed graphs K_r (with loops at every vertex. Hence $G_1^{(1)}$, $G_1^{(2)}$, $G_1^{(3)}$, $G_1^{(4)}$ are planar but $G_1^{(r)}$ is non-planar for $r \geq 5$. The graph $G_t^{(1)}$ consists of a single vertex and is also planar.

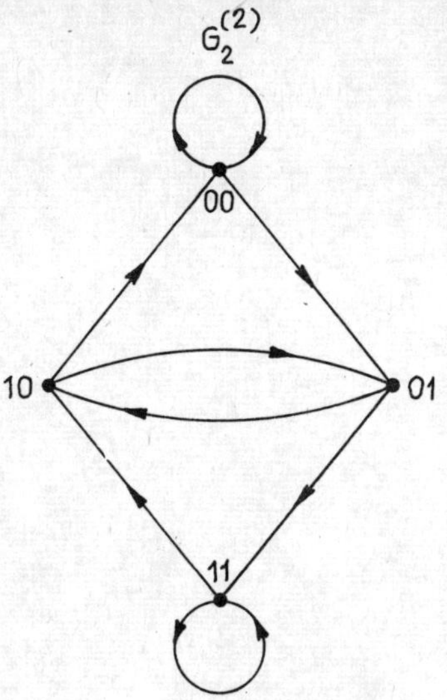

Figure 1.

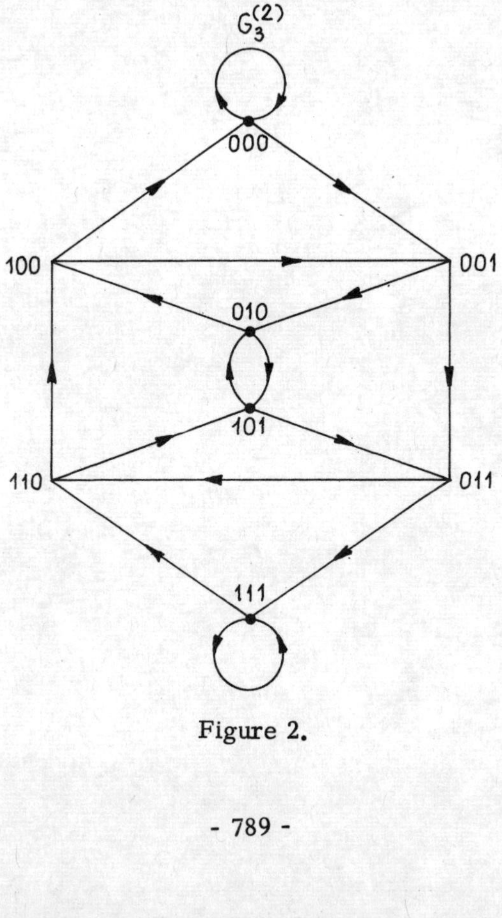

Figure 2.

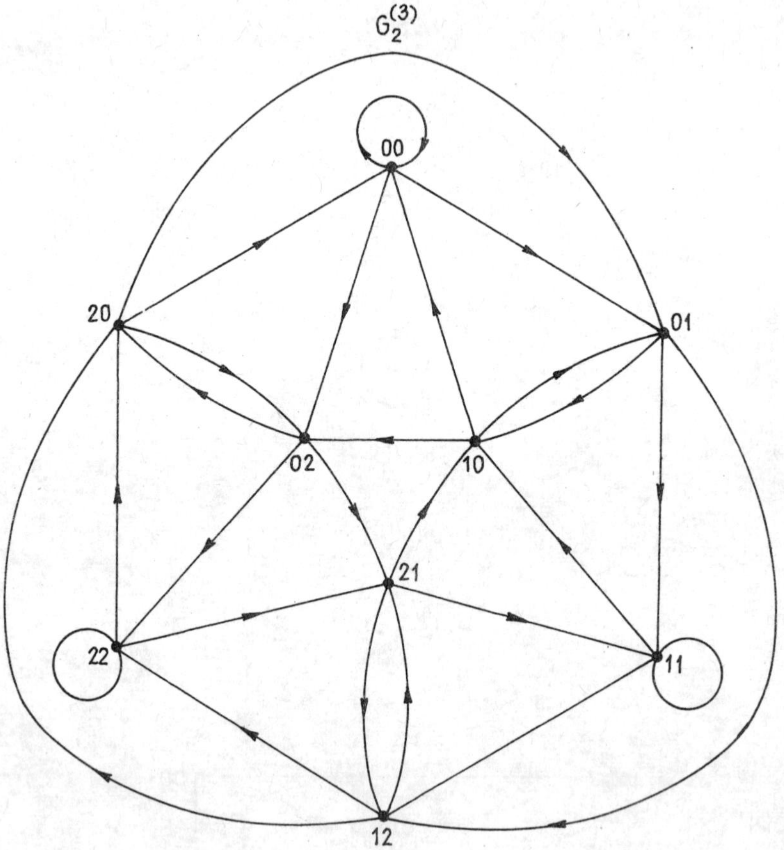

Figure 3.

In all other cases we can show that the graph is non-planar. It may be noted that we need only prove that $G_t^{(r)}$ is non-planar for the cases $r = 4,\ t \geq 2,\ r = 3,\ t = 3$ and for $r = 2,\ t \geq 4$. This follows from the fact that $G_t^{(s)}$ is a subgraph of $G_t^{(r)}$ if $s < r$.

Let $\bar{G}_t^{(r)}$ be the graph obtained from $G_t^{(r)}$ by ignoring the orientation of the edges and then removing the loops and double edges from $G_t^{(r)}$. The graph $\bar{G}_t^{(r)}$ is planar, if and only if, $G_t^{(r)}$ is planar. Now if E is the number of edges and V the number of vertices of $\bar{G}_t^{r}$ we have $V = r^t$, $E = r^{t+1} - \frac{1}{2} r^2 - \frac{1}{2} r$. This follows from the fact that $G_t^{(r)}$ has r^t vertices, r^{t+1} edges, r loop edges and $\frac{1}{2} r(r-1)$ 2-cycles. Now, a necessary condition for a connected graph without loop vertices and without 2-cycles to be planar is $E \leq 3V - 6$. For the graph $\bar{G}_t^{(r)}$ this condition becomes $r^{t+1} - 3r^t - \frac{1}{2} r^2 - \frac{1}{2} r + 6 \leq 0$. This condition is not satisfied if $r \geq 4$, $t \geq 2$. The condition is satisfied for $r = 3,\ t = 3$ and for $r = 2, t$ arbitrary. Hence, we require a different proof for the non-planarity of $G_3^{(3)}$ and $G_t^{(2)}, t \geq 4$.

In $G_3^{(3)}$ the following topological Kuratowski $K_{3,3}$ exists. Take $A = 012,\ B = 221,\ C = 110,\ A' = 122,\ B' = 211,\ C' = 101$. We have AA', $A'B$, BB', $B'C$, CC' and $C'A$ are edges of $G_3^{(3)}$. Also AB', CA' and $C'B$ are joined by the following paths

$$012 \rightarrow 120 \rightarrow 202 \rightarrow 021 \rightarrow 211$$

$$110 \leftarrow 111 \rightarrow 112 \rightarrow 122$$

$$101 \leftrightarrow 210 \rightarrow 102 \rightarrow 022 \rightarrow 221$$

In $\bar{G}_4^{(2)}$ we have the following topological $K_{3,3}$

$A = 0010 \qquad B = 1011 \qquad C = 1010; \qquad A^* = 0101, \qquad B^* = 1101, \qquad C^* = 0100$

In $\bar{G}_4^{(2)}$, AA^*, AC^*, BA^*, BB^*, CA^*, CB^*, CC^* are edges while AB^* and BC^* are joined by the paths

$$0010 \leftarrow 0001 \rightarrow 0011 \rightarrow 0111 \rightarrow 1110 \rightarrow 1101;$$

$$1011 \rightarrow 0110 \rightarrow 1100 \rightarrow 1000 \leftarrow 0100.$$

The graphs $\bar{G}_t^{(2)}$, $t>4$ all have $K_{3,3}$ as a topological subgraph. The details are left out here, but will appear in a paper by D. M. Johnson and the author in the Proceedings of the Calgary International Conference on Combinatorial Structures and their applications.

GRAPHS WITH UNIQUE PATH PROPERTY OF ORDER t.

The Good-de Bruijn graphs $G_t^{(r)}$ are directed graphs whose incidence matrices satisfy the equation $A^t = J$. From elementary matrix theory the following is easily obtained. Any $0,1$ matrix of order m satisfying $A^t = J$ has these properties:

(1) $m = r^t$ for some integer r.

(2) Each row and each column of A has r 1's

(3) The main diagonal of A has exactly r 1's

The directed graph corresponding to such a matrix shares the following with $G_t^{(r)}$:

(1) It contains r^t vertices.

(2) Each vertex has in-degree and out-degree equal r.

(3) There are exactly r loop vertices.

The number of solutions of $A^t = J$ with r fixed is in general very large. For $t = 2$, $r = 3$ there are 6 non-isomorphic solutions given by the following matrices.

$$A_1 = \begin{vmatrix} 111 & 000 & 000 \\ 000 & 111 & 000 \\ 000 & 000 & 111 \\ 111 & 000 & 000 \\ 000 & 111 & 000 \\ 000 & 000 & 111 \\ 111 & 000 & 000 \\ 000 & 111 & 000 \\ 000 & 000 & 111 \end{vmatrix} \qquad A_2 = \begin{vmatrix} 111 & 000 & 000 \\ 000 & 111 & 000 \\ 000 & 000 & 111 \\ 111 & 000 & 000 \\ 000 & 011 & 100 \\ 000 & 100 & 011 \\ 111 & 000 & 000 \\ 000 & 011 & 100 \\ 000 & 100 & 011 \end{vmatrix}$$

$$A_3 = \begin{vmatrix} 111 & 000 & 000 \\ 000 & 111 & 000 \\ 000 & 000 & 111 \\ 111 & 000 & 000 \\ 000 & 111 & 000 \\ 000 & 000 & 111 \\ 111 & 000 & 000 \\ 000 & 011 & 100 \\ 000 & 100 & 011 \end{vmatrix} \qquad A_4 = \begin{vmatrix} 111 & 000 & 000 \\ 000 & 111 & 000 \\ 000 & 000 & 111 \\ 111 & 000 & 000 \\ 000 & 111 & 000 \\ 000 & 000 & 111 \\ 101 & 010 & 000 \\ 010 & 101 & 000 \\ 000 & 000 & 111 \end{vmatrix}$$

$$A_5 = \begin{vmatrix} 111 & 000 & 000 \\ 000 & 111 & 000 \\ 000 & 000 & 111 \\ 111 & 000 & 000 \\ 000 & 110 & 001 \\ 000 & 001 & 110 \\ 011 & 100 & 000 \\ 100 & 010 & 001 \\ 000 & 001 & 110 \end{vmatrix} \qquad A_6 = \begin{vmatrix} 111 & 000 & 000 \\ 000 & 111 & 000 \\ 000 & 000 & 111 \\ 110 & 001 & 000 \\ 001 & 110 & 000 \\ 000 & 000 & 111 \\ 101 & 000 & 010 \\ 000 & 111 & 000 \\ 010 & 000 & 101 \end{vmatrix}$$

The matrix A_1 corresponds to $G_2^{(3)}$ given in Figure 3. The graph corresponding to A_2 is given in Figure 4. Similar graphs can be drawn corresponding to A_3, A_4, A_5, A_6. Note that the rows and columns of these matrices are labelled 00, 01, 02, 10, 11, 12, 20, 21, 22. We also note that the graphs corresponding to $A_1, A_2, \ldots, A_6$ are all planar.

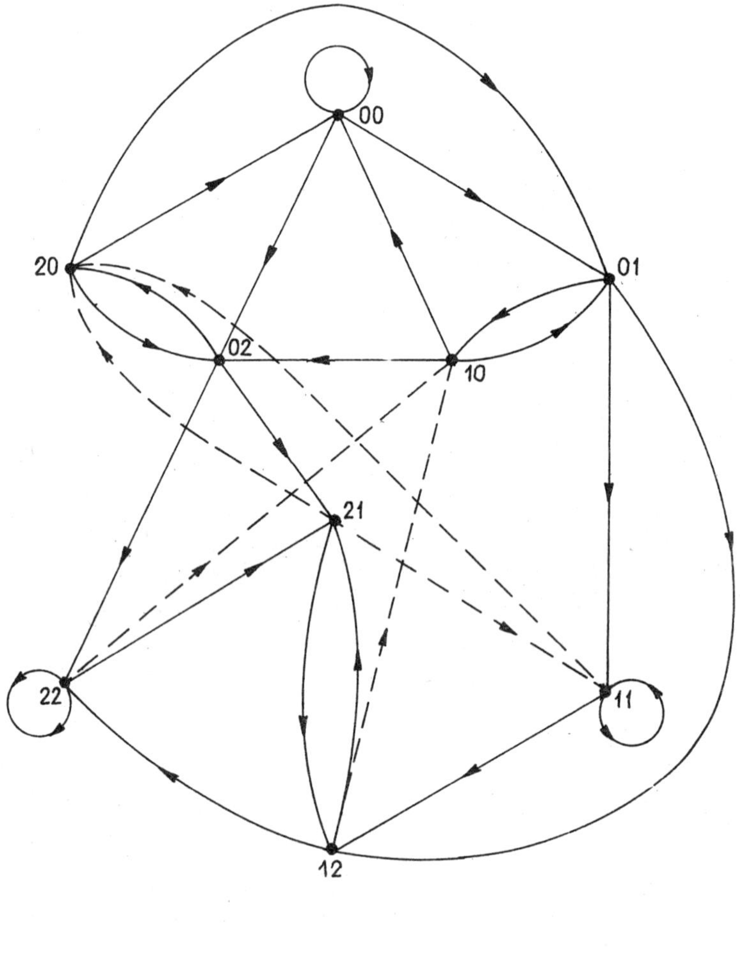

UPP graph corresponding to A_2

Figure 4.

It will be noticed from figure 4 all of the graphs have 3 2-cycles. This is a particular instance of the following fact.

Corresponding to the graph $G_t^{(r)}$ all the UPP graphs with the same r and t have the same cycle counts. For long cycles (and paths) the result is obvious since it has been obtained only from the relation $A^t = J$ without particular reference to the matrix A which was used. For short cycles we note the following. If A is a $0,1$ matrix such that $A^t = J$ then A^u is a $0,1$ matrix for $u \le t$, since if A^u had an entry ≥ 2 there would be two vertices with at least two paths of length u joining them. Also if there is a 1 in the (i,i) position, this implies that vertex i in the corresponding graph is on a cycle of length u. There can only be one such cycle, for if not, the element in position (i,i) would have a value greater than 1. It is now shown that a point on a cycle of length u is on an elementary circuit of length u or on an elementary circuit of length d where $d \mid u$.

Suppose A^r and A^s with $s > r$, $r+s = u \le t$, both have a 1 in the (i,i) position. We show that A^{s-r} has a 1 in the (i,i) position. If this were not so, then, since $A^{s-r} \cdot A^r = A^r \cdot A^{s-r} = A^s$ we have that there exist k and ℓ such that $k \neq i$, $\ell \neq i$, A^{s-r} has a 1 in the (i,k) position and A^r has a 1 in the (k,i) position. Also A^r has a 1 in the (i,ℓ) position and A^{s-r} has a 1 in the (ℓ,i) position. Also $k \neq \ell$, since if $k = \ell$, A^{2r} is a $(0,1)$ matrix $(2r \le t)$ and in the product $A^r \cdot A^r$ the number in the (i,i) position would be ≥ 2 (from the product of terms in (i,i) and (i,i) positions as well as from product terms in the (i,ℓ) and (ℓ,i) positions). Now consider $A^{r+s} = A^r \cdot A^{s-r} \cdot A^r$. The (i,i) term has a value ≥ 2 from the products in terms in respective positions $(i,\ell)(\ell,i)(i,i)$ and also in positions $(i,i)(i,k)(k,i)$. But $r+s \le t$ so that A^{r+s} is a $(0,1)$ matrix, a contradiction. It easily follows that if A^r and A^s each have a 1 in the (i,i) position, then so does A^{s-r}.

This implies A^q has a 1 in the (i,i) position where q is the g.c.d. of r and s. Hence both the r-cycle and the s-cycle are multiples of the q-cycle by the uniqueness property. Now let A be any $0,1$ matrix with $A^t = J$. Then as mentioned before A is an r^t by r^t matrix with exactly r 1's in each row and column. Hence $A^{t+1} = AJ = rJ = rA^t$. Hence, A satisfies the polynomial equation $x^{t+1} - rx^t = 0$. Hence A has characteristic roots r and 0, where the root r has multiplicity 1 since A is a primitive matrix. The root 0 appears with multiplicity $r^t - 1$. The characteristic roots of A^u $(u \le t)$ are r^u and 0. Hence $\operatorname{tr} A^u = r^u$ and as A^u is a $0,1$ matrix there are exactly r^u 1's on the main diagonal of A^u. Hence there are exactly r^u vertices on the corresponding graph which lie on cycles of length u and each vertex of the graph is on at most one such cycle. The same count now applies to a UPP graph as to a Good-de Bruijn graph and we obtain $f(n) = \frac{1}{n}\sum_{d|n} \mu(\frac{n}{d}) r^d$ for the number of elementary circuits of length n for any $n \le t$.

It is still an open question as to whether the number of elementary circuits is the same for all the UPP graphs corresponding to a Good-de Bruijn graph for $t+1 \le n \le r^t$.

CONNECTION WITH UNIVERSAL ALGEBRA

D. E. KNUTH (private communication) has shown that solutions the matrix equation $A^2 = J$ in $(0,1)$ matrices is equivalent to the existence of models of groupoids G with the single identity $(xy)(yz) = y$. We outline Knuth's argument here. In the first place we think of A as the adjacency matrix of a directed graph with the property that between any two vertices, there is a unique path of length 2. Now let x, y, z be any three vertices of such a graph.

Since there are unique vertices t and u such that $x \to t \to y$ and $y \to u \to z$, t is uniquely determined by x and y and we define $t = xy$. Hence $u = yz$. But from the path $t \to y \to u$ we have $y = tu = (xy)(yz)$. Conversely, suppose that G is a groupoid with a single identity $(xy)(yz) = y$. From this relation by reducing the expression $((zx)(xy))((xy)u)$ in two different ways we obtain $x[(xy)u] = xy$. Similarly, we obtain the identity $[x(yz)]z = yz$.

Now define a directed graph as follows: let the vertices of the graph be the elements of G and draw an arrow from x to y if and only if $y = xk$ for some k in G. Now if x and y are two elements of G, we have the path $x \to xy \to y$ since $(xy)(yy) = y$. Now let t be any vertex such that $x \to t \to y$. Then $t = xk$ and $y = tm$. Hence, $y = (xk)m$ from which $xy = x[(xk)m] = xk = t$. Hence the path of length 2 from x to y exists and is uniquely determined.

In [5] the present author has generalized these ideas to the matrix equation $A^n = J$. Here, instead of a groupoid we start with a universal algebra $G = \langle S, f_1, f_2, \ldots, f_{n-1}\rangle$ where S is a set and $f_1, f_2, \ldots, f_{n-1}$ are binary operators on S (we use left prefix bracket free notation e.g. $f_i\, xy$) satisfying the following $(n-1)^2$ identities

$$f_i f_j xy f_j yz = \lambda_{i+j}, \quad i, j = 1, 2, \ldots, n-1$$

where

$$\lambda_{i+j} = \begin{cases} f_{i+j}xy & \text{if } i+j \le n-1 \\ y & \text{if } i+j = n \\ f_{i+j-n}yz & \text{if } i+j > n. \end{cases}$$

It can be shown that there is a one-to-one correspondence between models of the algebra and directed graphs having the property that there is exactly one path of length n between any two of its vertices. The system could be reduced to a groupoid by noting that these equations allow us to express recursively $f_2 xy, f_3 xy, \ldots, f_n xy$ in terms of the operator f_1.

Such algebras enable one to construct large classes of non-isomorphic UPP graphs and to obtain some invariants of such graphs.

UNSOLVED PROBLEMS

There are a large number of open problems in connection with Good graphs which are worthy of solution. We list a few.

(1) Find the genus of $G_t^{(r)}$.

(2) Find all non-isomorphic UPP graphs corresponding to $G_t^{(r)}$.

(3) Find a complete system of invariants for the class of UPP graphs corresponding to $G_t^{(r)}$.

(4) Find the number of elementary circuits of length k, $t+2 < k < r^t$ in $G_t^{(r)}$.

(5) Is the number of elementary circuits of length k, $t < k \leq r^t$, the same for the Good-de Bruijn graph $G_t^{(r)}$ and any corresponding UPP graph?

REFERENCES

[1] N.G. DE BRUIJN, A Combinatorial Problem, Proc. Koninkl. Ned. Akad. Wetenschap (49) 758-764, 1946. (Also Indagationes Mathematicae, (8), 1946, 461.

[2] N.G. DE BRUIJN and T. VAN AARDENNE-EHRENFEST, Circuits and Trees in Oriented Linear Graphs, Simon Stevin, Vol. 28 (1951), 203-217.

[3] I.J. GOOD, Normal Recurring Decimals, J. London, Math. Soc., (21), 1946, 167-169.

[4] A. LEMPEL, M-ary Closed Sequences, to appear in J. Comb. Theory.

[5] N.S. MENDELSOHN, An Application of Matrix Theory to a problem in Universal Algebra, Linear Algebra and its applications (1), 1968. 471-478.